# APPROXIMATION THEORY AND APPLICATIONS

*Academic Press Rapid Manuscript Reproduction*

Proceedings of a workshop on Approximation Theory and Applications
Technion, Haifa, Israel
May 5-June 25, 1980

# APPROXIMATION THEORY AND APPLICATIONS

Edited by

**Zvi Ziegler**

Department of Mathematics
Technion—Israel Institute of Technology
Haifa, Israel

ACADEMIC PRESS 1981

A Subsidiary of Harcourt Brace Jovanovich, Publishers
New York London Toronto Sydney San Francisco

ACADEMIC PRESS, INC.
111 Fifth Avenue, New York, New York 10003

*United Kingdom Edition published by*
ACADEMIC PRESS, INC. (LONDON) LTD.
24/28 Oval Road, London NW1 7DX

Library of Congress Cataloging in Publication Data
Main entry under title:

Approximation theory and applications.

"Proceedings of a workshop held at the Technion, Haifa, Israel, in May-June, 1980"--Pref.
1. Approximation theory--Congresses. I. Ziegler, Zvi.
QA297.5.A68 511'.4 81-1969
ISBN 0-12-780650-4

PRINTED IN THE UNITED STATES OF AMERICA

81 82 83 84 9 8 7 6 5 4 3 2 1

# CONTENTS*



*In the Contents the name of the person who delivered the talk precedes that of his collaborator.

# CONTRIBUTORS

*Numbers in parentheses indicate the pages on which the authors' contributions begin.*

D. Amir (**1**), Department of Mathematical Sciences, Tel-Aviv University, Tel-Aviv, Israel

R. B. Barrar (**201**), Department of Mathematics, University of Oregon, Eugene, Oregon 97403

Dietrich Braess (**23, 39**), Institut fur Mathematik, Ruhr-Universitat Bochum, 4630 Bochum 1, Federal Republic of Germany

E. W. Cheney (**65**), Department of Mathematics, University of Texas, Austin, Texas 78712

Chr. Coatmelec (**89**), Laboratoire d'Analyse Numerique, Institut National des Sciences Appliquées, 35031 Rennes, Cedex, France

Carl de Boor (**13**), Mathematics Research Center, University of Wisconsin, Madison, Wisconsin 53706

Nira Dyn (**113**), Department of Mathematical Sciences, Tel-Aviv University, Tel-Aviv, Israel

S. D. Fisher (**131**), Department of Mathematics, Northwestern University, Evanston, Illinois 60201

Carlo Franchetti (**65**), Department of Mathematics, University of Genoa, Italy

B. Granovsky (**135**), Department of Mathematics, Technion-Israel Institute of Technology, Haifa 32000, Israel

Maurice Hasson (**311**), Department of Mathematics, University of Alabama, Alabama 35486

David Hill (**283**), Department of Mathematics, Temple University, Philadelphia Pennsylvania 19122

Joseph W. Jerome (**147**), Department of Mathematics, Northwestern University, Evanston, Illinois 60201

V. Knoh (**135**), ul. Vavilova 4-1-48, Leningrad, USSR

Alain Le Mehaute (**159, 171**), Laboratoire d'Analyse Numerique, Institute National des Science Appliquées, 35031 Rennes, Cedex, France

David Levin (**113**), Department of Mathematical Sciences, Tel-Aviv University, Tel-Aviv, Israel

William A. Light (**187**), Department of Mathematics, University of Lancaster, Lancaster LA1 4YL, United Kingdom

H. L. Loeb (**201**), Department of Mathematics, University of Oregon, Eugene, Oregon 97403

J. C. Mason (**207**), Department of Mathematics and Ballistics, Royal Military College of Science, Shrivenham, Swindon, SN6 8LA, United Kingdom

Jean Meinguet (**225**), Institut de Mathematique, Université Catholique de Louvain, B-1348 Louvain-La-Neuve, Belgium

Charles A. Micchelli (**131, 329**), IBM, T. J. Watson Research Center, Yorktown Heights, New York 10598

Beny Neta (**249**), Department of Mathematical Science, Northern Illinois University, DeKalb, Illinois 60115

D. J. Newman (**265**), Department of Mathematics, Temple University, Philadelphia, Pennsylvania 19122

Eli Passow (**283**), Department of Mathematics, Temple University, Philadelphia, Pennsylvania 19122

J. Prasad (**319**), Department of Mathematics, California State University, Los Angeles, California 90032

T. J. Rivlin (**291**), I.B.M., T. J. Watson Research Center, Yorktown Heights, New York 10598

Walter Schempp (**303**), Lehrstuhl fur Mathematik 1, Universitat of Siegen, D59000 Siegen 21, Federal Republic of Germany

Oved Shisha (**311**), Department of Mathematics, University of Rhode Island, Kingston, Rhode Island 02881

A. K. Varma (**319**), Department of Mathematics, University of Florida, Gainesville, Florida 32611

Grace Wahba (**329**), Department of Statistics, University of Wisconsin, Madison, Wisconsin 53706

Z. Ziegler (**1**), Department of Mathematics, Technion-Israel Institute of Technology, Haifa 32000, Israel

# PREFACE

This volume contains the proceedings of a workshop held at the Technion, Haifa, Israel, in May–June, 1980. The extended workshop lasted seven weeks, from May 5 through June 25. There were speakers from 9 countries (Belgium, France, West Germany, Hungary, Netherlands, Norway, United Kingdom, United States, and Israel). The workshop brought together specialists in Approximation Theory and its applications, and practitioners working in the industrial sectors in Israel. (A list of participants appears in the Appendix.)

There were several sessions devoted to open problems; problems of practical interest were raised by the practitioners—some of these were partially resolved on the spot and others will hopefully be resolved through further research.

The time element played a crucial role in the editing process. Thus, in endeavoring to assemble all the papers for the volume within a reasonable period of time—some errors may have escaped my attention. It is my hope that not too many have done so.

I wish to express the participants' gratitude to the various persons and organizations who made the whole endeavor possible: the Technion's president, Gen. (Res.) A. Horev, whose help was crucial in the planning stages and whose responsiveness during the whole operation is greatly appreciated; the Morton Bank Fund, whose support enabled us to launch the workshop; the Samuel Neaman Institute for Advanced Studies in Science and Technology, whose insistence on "applicability" encouraged us to increase the practitioners' participation, the European Research Office of the U.S. Army, the Israeli Academy of Sciences, and the Israeli Society for the Applications of Mathematics, for their support and sponsorship.

Finally, thanks are due to the secretarial staff, Ruth Markevich and Leoni Mahlab for their help in making the workshop run so smoothly, and to Marion Mark and Ana Burkat for the typing.

# CONSTRUCTION OF ELEMENTS OF THE RELATIVE CHEBYSHEV CENTER

D. Amir
Department of Mathematics
Tel Aviv University
Tel Aviv, Israel

Z. Ziegler*
Department of Mathematics
Technion — Israel Institute of Technology
Haifa, Israel

The characterization of Chebyshev centers in C[a,b], as well as necessary and sufficient conditions for the center to reduce to a singleton have been established by us elsewhere. Here we describe an algorithm producing an element of the center. The algorithm utilizes combinations of weighted and restricted range approximation and a limit procedure due to Rivlin and Cheney.

## I. INTRODUCTION

When E is a normed linear space and $A \subset E$ is a bounded set in E, then the Chebyshev center of A is the set in E of elements best approximating A. When G is another set in E, we may consider the set of elements in G best approximating, from amongst all elements in G, the set A. This set is called the relative Chebyshev center of A in G.

* Partially supported by the Fund for Encouragement of Research at the Technion.




ISBN 0-12-780650-4

We investigated, in [1], the structural properties of such relative centers in general spaces. In [2], we carried out a detailed analysis of the case where E is C[a,b], A is compact and G is the span of an n-unisolvent family. When G was further restricted to an extended n-unisolvent family, a full characterization of the center was established and necessary and sufficient conditions for an element to belong to the center were obtained.

We restrict ourselves here to C[a,b], a compact A and assume the approximating family F to be the span of a Chebyshev system. In this context, we obtain an algorithm, enabling us to construct an element of the center. It will yield the unique element of best simultaneous approximation when the center is a singleton. In the case of non uniqueness, the algorithm will yield a special element of the center which will be defined below.

It would have been desirable to make the "selection" of an element from the center, a continuous selection. Unfortunately, no continuous selection exists. Continuity fails at points of non uniqueness. The following example exhibits this phenomenon, that exists even in the simplest case. Consider C[0,1], and let F = [1,t] = span of {1,t}.

Consider the family

$$f_s(t) = \begin{cases} 1 + t , & 0 \leq t \leq s, \\ (1 + s)(1 - t)/(1 - s) , & s \leq t \leq 1 ; \end{cases}$$

$$f_0(t) = 1 - t , \quad 0 \leq t \leq 1.$$

We clearly have

$$\lim_{s \to \infty} \| f_s - f_0 \| = 0.$$

On the other hand, we have

$$Z(F\ ;\ f_s,\ -f_0) = \{t\}\ ,\ \text{for each } s > 0$$

$$Z(F\ ;\ f_0,\ -f_0) = \text{the set of functions } \{\alpha t\},\ \text{with } -1 \le \alpha \le 1.$$

$$Z(F\ ;\ f_0,\ -f_s) = \{-t\}\ ,\ \text{for each } s > 0.$$

Obviously, no continuous selection is possible here.

Since no continuous selection exists, we will aim at a reasonable choice of a discontinuous selection, in terms of constructibility and intuitive appeal. We will choose, in the case of non uniqueness, the nearest element to $h=(f+g)/2$ in $Z(F\ ;\ f,g)$.

We recall the simple observation to the effect that the general case of a center is reducible to the case of a center $Z(F\ ;\ f,g)$ of two elements, with $f \ge g$. We denote by $r(F\ ;\ f,g)$ the distance from F to the pair $(f,g)$, i.e.

$$r(F\ ;\ f,g) = \min_{u\ \in\ F}\ [\max(\|f-u\|,\|u-g\|)]. \tag{1}$$

We further denote $v = (f-g)/2$ , $h = (f+g)/2$, $R = \|v\| = \|(f-g)/2\|$ .

We recall, for later use in the paper, some relevant definitions (see [3]) and results (see [2] for full details):

Definition 1: The set $(t_1,\ldots,t_k)$ , $t_1 <\ldots< t_k$ is called a k-point alternance for the approximation by u to f and g (abbreviated as the (u; f,g)-approximation) if either

$$\left\{\begin{array}{l} f(t_{2i-1})-u(t_{2i-1}) = u(t_{2i})-g(t_{2i}) = r(u;\ f,g),\ \text{for all } i, \\ \quad\text{or} \\ u(t_{2i-1})-g(t_{2i-1}) = f(t_{2i})-u(t_{2i}) = r(u;\ f,g),\ \text{for all } i. \end{array}\right. \tag{2}$$

Definition 2: The point $t_0$ is called a straddle point with respect to the (u; f,g)-approximation if

$$f(t_0)-u(t_0) = u(t_0)-g(t_0) = r(u; f,g) . \tag{3}$$

Remark 3: The point $\tilde{t}$ is a straddle point of some triplet (u; f,g) if and only if

$$f(\tilde{t})-g(\tilde{t}) = 2r(F; f,g). \tag{4}$$

Thus, if $\tilde{t}$ is a straddle point for one triplet, it is a straddle point for all triplets, and $u^*(\tilde{t}) = [f(\tilde{t})+g(\tilde{t})]/2$ for all $u^* \in Z(F; f,g)$.

Remark 4: a) If (F; f,g) has no straddle points, then Z(F; f,g) is a singleton $\{u^*\}$ and $(u^*; f,g)$ has an (n+1)-point alternance.

b) If (F; f,g) has n straddle points then Z(F; f,g) is a singleton $\{u^*\}$.

c) If $(u^*; f,g)$ has an (n+1)-point alternance, then $\{u^*\} = Z(F; f,g)$.

## 2. THE ALGORITHM

Let f and g be given, $f \geq g$, $f,g \in C[a,b]$. We will proceed through discretization of the problem, describing first an algorithm for the case where the underlying space has a finite number of points. Then we will show that, through a suitable passage to a limit, an element of the center is obtained.

### I. X is a finite set of points

Let $X = \{t_1,\ldots,t_k\}$ and let F be a Chebyshev set of n functions. We start by identifying the points $s_1,\ldots,s_\ell$ where $\|f-g\|$ is attained, noting that if straddle points exist, they must be amongst the $s_i$'s. Moreover, if straddle points exist, then every element of the center must interpolate $h = (f+g)/2$ at all the $s_i$'s by Remark 3.

a) When $\ell \geq n$, we interpolate h at $s_i$, $i=1,\ldots,n$, by an element $u, u \in F$, and compute $r(u;\ f,g)$. If $r(u;\ f,g) = R$, then these are indeed straddle points, and Remark 4 implies that $Z(F;\ f,g) = \{u\}$. If, on the other hand, $r(u;\ f,g) > R$ then there are no straddle points, implying that the center reduces to a singleton $\{u^*\}$, and $(u^*;\ f,g)$ has an $(n+1)$-point alternance. The function $u^*$ can then be obtained by a slight adaptation of the Remez algorithm.

b) When $\ell < n$, we utilize the technique of weighted approximation with interpolation (see [ 4 ]). We compute the function $\bar{u}$ solving the minimum problem

$$\min\{\|(h-u)W\|_X;\ u \in F,\ u(s_i) = h(s_i),\ i = 1,\ldots,\ell\} \tag{5}$$

where the weight function W is defined on $\{t_i\}_1^n \setminus \{s_i\}_1^\ell$ by $W(t_i) = [R-V(t_i)]^{-1}$.

If $\|(h-\bar{u})W\|_X \leq 1$, then $r(\bar{u};\ f,g) = R$, i.e. the points are straddle points, and $\bar{u}$ is an element of the center. If, on the other hand, $\|(h-\bar{u})W\|_X > 1$, then every interpolant u satisfies $r(u;\ f,g) > R$, so that straddle points do not exist, and the (unique) element of $Z(F;\ f,g)$ can be obtained by a modified Remez algorithm.

We have thus completely analyzed the case where X is a finite point set, which is of independent interest as far as applications are concerned.

## II. $X = [a,b]$

The simplest case occurs where f and g are known analytically, the points $s_1,\dots,s_\ell$ where $\|f-g\|$ is attained can be readily identified, and $\ell \geq n$. Under these conditions, the search can be conducted as in subcase a) of the previous case.

When not all of the above conditions are satisfied, we discretize the problem. We choose a sequence of partitions $\{X_i\}$, with norms tending to 0. At the i-th stage, we identify the candidates $s_1^i,\dots,s_{\ell(i)}^i$ for straddle points and proceed along the lines described in I. There are two alternatives:

(i) There exists an $i_0$ such that for all partitions $X_i, i \geq i_0$, no straddle points exist. This means that at the i-th stage, for all $i \geq i_0$, either $\ell(i) \geq n$ and $r(u^i; f,g) > R$ where $u^i$ is as defined in I a), or

$$\|(h - \bar{u}^i)W\|_{X_i} \geq 1$$

where $\bar{u}^i$ is defined for $X_i$ by (5). In this case we have, for all $i \geq i_0$, a unique element of simultaneous best approximation, characterized by an (n+1)-alternance.

(ii) There exists an infinite sequence of partitions $\{X_i\}$ for which $\|(h-\bar{u}^i)W\|_{X_i} \leq 1$, where $\bar{u}^i$ is defined by (5).

Let $\{X_j\}$ be such a partition. We consider now the minimum problem

$$\min\{\|h-u\|_{X_j} \; ; \; u \in F \; , \; f-R \leq h \leq g+R\}. \tag{6}$$

This is a special case of the restricted range approximation problem, for which an algorithm has been found in [ 7 ]. Noting that $\bar{u}^j$ is within the range, we may start with $\bar{u}^j$ and, using the fact that we have a finite point set, we can obtain a simpler algorithm for our special case. The procedure needs no elaboration. Let $u_*^j$ denote the solution of (6).

We will now show that the sequence in a) converges to the unique element of simultaneous best approximation and that the sequence $u_*^j$ converges uniformly to the solution $u_*$ of the following restricted best approximation to h

$$\min\{\|h-u\|_X \; ; \; u \in F \; , \; f-R \leq u \leq g+R\}.$$

Furthermore, $(u_*; f,g)$ has straddle points, so that $u_*$ is an element of the center, and by its construction, it is indeed the element of the center closest to h.

The preceding statement is established by using a modified form of the Rivlin-Cheney theorem (cf. [6]). We will now formulate and prove the theorem suitable for our needs.

We start with a few lemmas.

Let X be a compact metric space, $Y \subset X$ a compact subset, M a finite dimensional subspace of C(X), k - a closed convex subset of M, and let $f_1, f_2 \in C(X)$. We introduce the notation

$\delta_Y = \max_x d(x,Y)$

$\omega(f,\cdot)$ - the modulus of continuity of f.

$Z^Y(K; f_1,f_2)$ - the relative center of $f_1,f_2$ in K, with respect to the Y-norm, where the Y-norm of f is the norm of $f|_Y$.

Lemma 5: Let $z^Y$ be an element of $z^Y(K; f_1,f_2)$, and let z be an element of $Z(K; f_1,f_2)$. then

$$\max\| f_i - z\| \leq \max_i \| f_i - z^Y\| \leq \tag{7}$$
$$\leq \max_i [\|f_i - z\| + \omega(f_i,\delta_Y)] + \omega(z^Y,\delta_Y).$$

Proof: The leftmost inequality is a consequence of the fact that $z \in Z(K; f_1, f_2)$. Assume $j \in \{1,2\}$ and $x_0 \in X$ satisfy

$$\max_i \|f_i - z^Y\| = |f_j(x_0) - z^Y(x_0)|$$

and choose $y \in Y$ such that $d(x_0, y) \leq \delta_Y$. Then

$$\|f_j(x_0) - z^Y(x_0)\| \leq |f_j(x_0) - f_j(y)| + |f_j(y) - z^Y(y)| + |z^Y(y) - z^Y(x_0)| \leq$$

$$\leq \omega(f_j, \delta_Y) + \|f_j - z^Y\|_Y + \omega(z^Y, \delta_Y) \leq$$

$$\leq \omega(f_j, \delta_Y) + \|f_j - z^Y\| + \omega(z^Y, \delta_Y)$$

establishing (7).

Lemma 6 (Rivlin-Cheney [6]): For each M as above there exist a number $\theta > 0$ and a mapping $\Omega : (0, \theta) \to (1, \infty)$ such that

(i) $\Omega(\delta) \to 1$, as $\delta \to 0$

(ii) for each $g \in M$ and each compact subset $Y \subset X$, $\|g\| \leq \Omega(\delta_Y) \|g\|_Y$.

Lemma 7 (Rivlin-Cheney [6]): For each M as above there exists a mapping $\overline{\omega} : [0, \infty) \to [0, \infty)$ such that

(i) $\overline{\omega}(\delta) \to 0$, as $\delta \to 0$

(ii) $\omega(g, \delta) \leq \|g\| \overline{\omega}(\delta)$, for every $g \in M$.

Lemma 8: With the above notation, assume that Y satisfies $\delta_Y < \theta$. Then

$$\max_i \|f_i - z^Y\| \leq [1 + 2\Omega(\delta_Y)]\, r(K; f_1, f_2), \tag{8}$$

$$\max_i \|f_i - z^Y\| \leq r(K; f_1, f_2) + \max_i (f_i, \delta_Y) + \Omega(\delta_Y)\overline{\omega}(\delta_Y)[\max_i \|f_i\| + r(K; f_1, f_2)]. \tag{9}$$

Proof: We have the following chain of inequalities

$$\|z^Y - f_j\| \le \|z^Y - z\| + \|z - f_j\| \le$$
$$\le \Omega(\delta_Y)\|z^Y - z\|_Y + \|z - f_j\| \le$$
$$\le \Omega(\delta_Y)[\|z^Y - f_j\|_Y + \|f_j - z\|_Y] + \|z - f_j\| \le$$
$$\le [1 + 2\Omega(\delta_Y)]\ r(K;\ f_1, f_2).$$

Using now Lemmas 5, 6, and 7, we obtain

$$\|f_j - z^Y\| \le \max_i \omega(f_i, \delta_Y) + \omega(z^Y, \delta_Y) + r(K; f_1, f_2) \le$$
$$\le \max_i \omega(f_i, \delta_Y) + \|z^Y\|\,\bar{\omega}(\delta_Y) + r(K; f_1, f_2) \le$$
$$\le \max_i \omega(f_i, \delta_Y) + \|z^Y\|_Y\,\bar{\omega}(\delta_Y)\Omega(\delta_Y) + r(K; f_1, f_2) \le$$
$$\le \max_i \omega(f_1, \delta_Y) + \bar{\omega}(\delta_Y)\Omega(\delta_Y)[\max\|f_i\| + r(K; f_1, f_2)] +$$
$$+ r(K; f_1, f_2).$$

q.e.d.

Corollary: If $\delta_Y \to 0$ then $\max_i \|f_i - z^Y\| \to r(K;\ f_1, f_2)$.

Combining these lemmas, we derive the theorem enabling us to prove convergence to an element of the center.

Theorem 9: Let X, M and K be as above. Assume that $f_1, f_2 \in C(X)$ are such that $Z(K; f_1, f_2)$ reduces to a singleton $\{z\}$. Let $\{Y_n\}_{n=1}^{\infty}$ be a sequence of closed subspaces of X, such that $\delta_{Y_n} \to 0$. Let $z^{Y_n} \in Z^{Y_n}(K; f_1, f_2)$, n=1,2,... be a sequence of elements of the corresponding relative centers. Then $\{z^{Y_n}\}$ converges uniformly to z.

Proof: Let u be an arbitrarily fixed element of K. Then $\|z^{Y_n}\| \le \|u\| + 2\max_i \|f_i - u\|$. Lemma 6 implies that $\{z^Y;\ \delta_Y \le \frac{1}{2}\theta\}$ is relatively compact in C(X). Choose therefore any convergent subsequence $\{z^{Y_n}\}$ and assume $z^{Y_n} \to g$. By the corollary of Lemma 8, $\|f_i - g\| \le r(K;\ f_1, f_2)$, i = 1,2, so that g = z.

q.e.d.

Remarks: (1) Under the above conditions on X, M, K and $\{Y_n\}$, if $Z(K; f_1,f_2)$ is not a singleton, then any limit point of a sequence $\{z^{Y_n}\}$, $z^{Y_n} \in Z^{Y_n}(K; f_1,f_2)$ will be in the relative center $Z(K; f_1,f_2)$.

(2) Restricted range approximation is obtained when $f_1 = f_2$ and K is the subset of M defined by $k_1(x) \leq P(x) \leq k_2(x)$. K is then closed and convex and the theorem can be applied to yield the desired convergence in this case.

## REFERENCES

[1] Amir, D. and Ziegler, Z., Relative Chebyshev Centers in normed linear spaces: Part I, to appear in *J.Approx.Th.*

[2] Amir, D. and Ziegler, Z., Relative Chebyshev Centers in normed linear spaces: Part II, submitted for publication in *J. Approx. Th.*

[3] Dunham, C.B., Simultaneous Chebyshev approximation of functions on an interval, *Proc. Amer. Math. Soc. 18* (1967), 472-477.

[4] Loeb, H. L., Moursund, D. G., Schumaker, L. L., and Taylor, G. D., Uniform Generalized Weight Function Polynomial Approximation with Interpolation, *SIAM J. Num. Anal. 6*(1969), 284-293.

[5] Rice, J. R., "The Approximation of Functions", vol. 1, Addison-Wesley, Palo Alto, Ca., (1964).

[6] Rivlin, T. J., and Cheney, E. W., A comparison of uniform Approximations on an interval and a finite subset thereof, *SIAM J. Numer. Anal. 3*(1966), 311-320.

[7] Schumaker, L. L., and Taylor, G. D., On Approximation by polynomials having restricted ranges, II, *SIAM J. Numer. Anal.* 6(1969), 31-36.

# THE NUMERICAL CALCULATION OF SPLINE APPROXIMATIONS ON A BIINFINITE KNOT SEQUENCE

Carl de Boor

Department of Mathematics
University of Wisconsin
Madison, Wisconsin

## I. INTRODUCTION

Standard linear approximation processes such as interpolation or $\mathbb{L}_2$-approximation require the solution of a <u>biinfinite</u> linear system

$$A\underline{\alpha} = \underline{\beta}$$

when the approximating space consists of splines (of a certain order k) on a biinfinite knot sequence. Fortunately, this linear system - which arises when the approximation problem is expressed in terms of B-splines and is to be solved by a bounded $\underline{\alpha}$ for a given bounded $\underline{\beta}$ - is not just biinfinite, but has, in addition, a <u>banded</u> and <u>totally positive</u> coefficient matrix A. It is the purpose of this talk to point out that these two properties make it possible (in principle) to approximate the solution $\alpha$ of the linear system arbitrarily well on the "interior" of any finite interval I by the solution $\underline{\alpha}_I$ of the (essentially) finite system

$$A\underline{\alpha}_I = \underline{\beta} \text{ on I-r}$$

$$\underline{\alpha}_I = \underline{0} \text{ off I}$$




ISBN 0-12-780650-4

Here, the integer $\gamma$ is the only one with this property; if the indexing in the biinfinite system has been so chosen that $(A(i,i))_{i=-\infty}^{\infty}$ is the "leftmost" nontrivial band of A, then r can be identified as the dimension of

$$N_A^+ := \{\underline{f} \in \mathbb{R}^{\mathbb{Z}} : A\underline{f} = \underline{0} \text{ and } \overline{\lim_{i \to \infty}} |\underline{f}(i)| < \infty\}.$$

## 2. LINEAR SPLINE APPROXIMATION

Let $\underline{t} := (t_i)_{-\infty}^{\infty}$ be strictly increasing and consider

$$S := m\$_{k,\underline{t}}$$

the linear space of bounded splines or <u>order</u> k with <u>knot sequence</u> $\underline{t}$. This means that S consists of bounded functions in $C^{k-2}[t_{-\infty},t_{\infty}]$ which, on each interval $[t_i, t_{i+1}]$, coincide with some polynomial of degree less than k. Equivalently, S consists of all functions of the form $\sum_{j=-\infty}^{\infty} \underline{\alpha}(j)N_j$, with $\underline{\alpha} \in \ell_{\infty}$ and $N_j = N_{j,k,\underline{t}}$ the j-th B-spline of order k for the knot sequence $\underline{t}$. This last statement identifies $N_j$ as an element of S with support $]t_i,t_{i+k}[$ and so determines it up to a scalar factor. The letter N used here customarily refers to the particular normalization for which $\sum_j N_j = 1$. For this, we have

$$D_k^{-1} \|\underline{\alpha}\|_{\infty} \leq \|\sum \underline{\alpha}(j)N_j\|_{\infty} \leq \|\underline{\alpha}\|_{\infty} .$$

For details concerning splines here and below, see, e.g., de Boor [1976].

We intend to approximate $f \in mC[t_{-\infty},t_{\infty}]$ from S by <u>linear interpolation</u>. In full generality, this means that we wish to determine $Pf \in S$ so that

$$\lambda_i Pf = \lambda_i f, \qquad \text{all } i,$$

for a given biinfinite sequence $(\lambda_i)$ of continuous linear functionals on C. Equivalently, we wish to find $\underline{\alpha} \in \ell_\infty$ for which

$$\sum_j \lambda_i N_j \, \underline{\alpha}(j) = \lambda_i f \quad , \text{ all } i$$

We consider only the specific choices

$$\lambda_{i-k+1} f := f(t_i)$$

or, more generally for n > 0,

$$\lambda_{i-k+1} f := \int \frac{n}{t_{i+n} - t_i} N_{i,n,\underline{t}} \, f .$$

The particular index relation was chosen to insure that the leftmost nontrivial band of

$$A := (\lambda_i N_j)$$

is the band $(A(i,i))_{-\infty}^{\infty}$. For these choices,

$$|\lambda_{i-k+1} f| \leq \|f\|_{\infty,[t_i, t_{i+n}]} \leq \|f\|_\infty$$

hence our problem becomes that of solving

$$A\underline{\alpha} \doteq \underline{\beta} := (\lambda_i f)$$

for some $\underline{\alpha} \in \ell_\infty$ , given that $\underline{\beta} \in \ell_\infty$. The following properties of A will be important

(i) For all i, $\sum_j A(i,j) = \sum |A(i,j)| = 1$ , hence $\|A\|_\infty = 1$.

(ii) A is <u>totally positive</u>, i.e., all its minors are non-negative.

(iii) A is <u>banded</u>. Precisely, A is m-<u>banded</u> in the sense that

$$A(i,j) \neq 0 \quad \text{implies that} \quad j \in [i, i+m] \ ,$$

with $m := k+n-1$, and is <u>strictly</u> so in the sense that

$$A(i,i)\, A(i,i+m) \neq 0.$$

The two questions concerning our approximation scheme P, viz.

(i) is P defined?

(ii) if P is defined, how can Pf be calculated?

thus have the following equivalent formulations:

Given the bounded, totally positive, strictly m-banded matrix A,

(i) is A invertible as a map in $\ell_\infty$?

(ii) if $A^{-1}$ exists, how can $A^{-1}\underline{\beta}$ be calculated?

## 3. STANDARD THEORY FOR THE TOEPLITZ CASE

It is instructive to recall the answer to these questions for the special case that A is also a Toeplitz matrix, i.e.,

$$A(i,j) = a_{j-i}, \qquad \text{all } i,j.$$

This occurs, e.g., when $\underline{t}$ is uniform or, more generally, when $\underline{t}$ is <u>geometric</u>, i.e.,

$$t_i = \rho^i, \qquad \text{all } i$$

for some positive and fixed local mesh ratio $\rho$. In this case, the theory as given in full detail in Gohberg and Feldman [8] is based on the polynomial $p_A$ given by

$$p_A(z) = \sum_i a_j z^j$$

and is as follows.

<u>Proposition 3(i)</u>. If A is a strictly m-banded Toeplitz matrix, then A is invertible (on $\ell_\infty$) iff $p_A$ does not vanish on $|z| = 1$. If A is also totally positive, then A is invertible iff $p_A(-1) \neq 0$ and, in that case, $\|A^{-1}\|_\infty = |p_A(1)/p_A(-1)|$.

Proposition 3(ii). Let A be a strictly m-banded Toeplitz matrix and let r be the number of zeros (counting multiplicity) of $p_A$ inside the unit circle. If A is invertible then, for all sufficiently large finite intervals I, the (essentially) finite system

$$A\underline{\alpha}_I = \underline{\beta} \quad \text{on } I-r$$
$$\underline{\alpha}_I = \underline{0} \quad \text{off } I$$

has exactly one solution $\underline{\alpha}_I$ for a given $\underline{\beta} \in \ell_\infty$ and $\underline{\alpha}_I \to \underline{\alpha} := A^{-1}\underline{\beta}$ pointwise exponentially fast, i.e.,

$$\exists \gamma \in [0,1[, \quad \forall i,\underline{\beta}, \quad |\underline{\alpha}(i)-\underline{\alpha}_I(i)| \leq \text{const } \underline{\beta}\ \gamma^{\text{dist}(i,\setminus I)}$$

In effect, the approximating finite system to be solved is selected from the given biinfinite one by choosing, for the given column-index set I, the row-index set I-r := {i-r : i ∈ I}; in other words, the finite section to be solved is chosen in such a way that the piece of the diagonal A(•,•+r) in it is its main diagonal. This makes it reasonable to call A(•,•+r) the main diagonal of A.

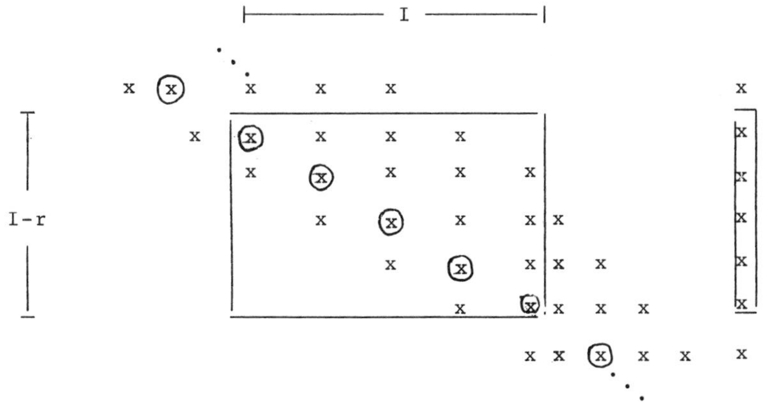


Figure 1. Selection of the approximating finite system, m = 4, r = 1.

## 4. NEW THEORY

When A is not Toeplitz, one has to decide what is to take the place of $p_A$ in the analysis. There does not seem to be a consensus on this since there does not seem to be much literature of a general nature on this. Earlier work in de Boor [1], [3] on spline interpolation, or a study of Schoenberg's [10] beautiful handling of cardinal splines (i.e., of the Toeplitz case mentioned in the preceding section), have convinced me that the kernel of A

$$N := N_A := \{\underline{f} \in \mathbb{R}^{\mathbb{Z}} : A\underline{f} = \underline{0}\}$$

as a map on the space $\mathbb{R}^{\mathbb{Z}}$ of all real biinfinite sequences is a convenient substitute for the missing $p_A$. For the case of a biinfinite strictly m-banded A considered here, dim $N$ = m. We need the following subspaces of $N$ :

$$N^* := \{\underline{f} \in N : \overline{\lim_{i \to *\infty}} |\underline{f}(i)| < \infty\} \ , \text{ for } * = + \text{ or } -$$

Then, as an example of how $N$ might be used in the analysis of $A^{-1}$, I quote from de Boor [4], [5]:

<u>Proposition 4.1</u>. There exists an $\ell_\infty$-columned matrix $A^{(-)}$ so that $AA^{(-)} = 1 = A^{(-)}A$ iff $N = N^+ \oplus N^-$.

These references also contain statements as to when $A^{-1}$ exists and when and how it can be approximated by inverses of finite sections. For the case under discussion here, though, we have the additional property that A is totally positive. For this case, the following material taken from de Boor [6] is relevant.

We use the superscript tilde to indicate multiplication of any other entry by -1 :

$\underline{\xi}^{\sim}(i) := (-1)^i \underline{\xi}(i)$ , all i.

In particular, with $\underline{1}(i) := 1$, all i, we have

$\underline{1}^{\sim}(i) = (-1)^i$ , all i.

We also use $S^+(\underline{\xi})$ to denote the number of weak sign changes in the sequence $\underline{\xi}$.

In Micchelli [9], you'll find the following striking

Conjecture. Let $\underline{\tau}$ be a strictly increasing sequence, and

$B := (N_{j,k}(\tau_i))$.

Then B is invertible (on $\ell_\infty$) iff $\exists\ \underline{\xi} \in \ell_\infty$ s.t. $B\underline{\xi} = \underline{1}^{\sim}$.

As it turns out, one has to demand that $B\underline{\xi} = \underline{1}^{\sim}$ have exactly one bounded solution, but with this modification, the conjecture is true. In fact, one can prove:

Proposition 4.(i). Let A be a bounded, totally positive, strictly m-banded matrix. Then A is invertible (on $\ell_\infty$) if the system $A\underline{\xi} = \underline{1}^{\sim}$ has exactly one bounded solution.

The proof uses the following lemmas.

Lemma 1. If $\underline{f} \in N \setminus \{0\}$, then $S^+(\underline{f}^{\sim}) < m$; i.e., $N^{\sim}$ is Haar.

Lemma 2. If $\underline{\xi} \in \ell_\infty$ and $A\underline{\xi} = \underline{1}^{\sim}$, then $S^+(\underline{\xi}^{\sim}) \leq \dim N^+ \cap N^-$.

Corollary. If $\underline{\xi}$ uniquely solves $A\underline{\xi} = \underline{1}^{\sim}$ in $\ell_\infty$, then $S^+(\underline{\xi}^{\sim}) = 0$.

Lemma 3. Let $r := \dim N^+$ and assume that $\underline{\xi}$ uniquely solves

$A\underline{\xi} = \underline{1}^{\sim}$ in $\ell_\infty$. Let

$$-\infty =: j_0 < j_1 < \ldots < j_{m+1} := \infty$$

and let $\underline{\eta}$ be the unique element of $N$ with $\underline{\eta}(j_s) = \underline{\xi}(j_s)$ , $s = 1,\ldots,m$. Then, on $J := ]j_r, j_{r+1}[$, $\underline{\eta}^{\sim}$ lies strictly between $\underline{0}$ and $\underline{\xi}$. Consequently, $\underline{\zeta} := \underline{\xi} - \underline{\eta}$ satisfies

$$|\underline{\zeta}| < |\underline{\xi}| \text{ in } J, \quad S^+(\zeta^\sim\big|_J) = 0.$$

This last lemma is the crucial one. To prove Proposition 4(i) (in the nontrivial direction), proceed as follows. Given the integer interval $I =: [\alpha,\omega]$, choose

$$j_r = \alpha-1,\ j_{r-1} = \alpha-2, \ldots, \quad j_1 = \alpha-r$$

$$j_{r+1} = \omega+1,\ j_{r+2} = \omega+2, \ldots,\ j_m = \omega + (m-r)$$

and form $\underline{\zeta} := \underline{\xi} - \underline{\eta}$ as in the lemma. Then also $A\underline{\zeta} = \underline{1}^\sim$ but, since $\underline{\zeta}$ vanishes in $j_1, \ldots, j_m$, we actually have

$$A_I(\underline{\zeta}\big|_I) = \underline{1}^\sim\big|_I \quad .$$

Here, $A_I = A_{I-r,I}$ is the matrix obtained from A by retaining only columns i with $i \in I$ and rows i with $i \in I-r$. A typical such $A_I$ is outlined in Figure 1. Since $A_I$ is totally positive by assumption, it now follows that

$$\| A_I^{-1} \|_\infty = \| \underline{\zeta}\big|_I \|_\infty < \| \underline{\xi} \|_\infty \quad .$$

This shows that the matrices $A_I$ are bounded below <u>uniformly</u> in I and this is enough to conclude that A itself is invertible.

Now, the standard argument for this conclusion (see, e.g., Gohberg and Fel'dman [8] is not directly applicable since, on $\ell_\infty$ , the truncating projection $P_I$ given by

$$(P_I\underline{f})(i) := \begin{cases} \underline{f}(i), & i \in I \\ 0 & , \text{ otherwise} \end{cases}$$

does <u>not</u> converge pointwise to 1 as $I \to \mathbb{Z}$ . But, A being <u>banded</u>, we have available the following

<u>Fact</u> (Demko [7]). Let B be a finite invertible banded matrix. Then there exists $\lambda \in [0,1[$ and const which depend only on $\| B \|$ , $\| B^{-1} \|$ and the band width such that

$$|B^{-1}(i,j)| \leq \text{const } \lambda^{|i-j|} .$$

This allows the conclusion that, e.g., $\sup_I \| A_I^{-1} \|_2 < \infty$, hence the standard argument gives invertibility of A on $\ell_2$, while Demko's Fact then gives the exponential decay away from the main diagonal of $A^{-1}$, hence the boundedness of $A^{-1}$ on $\ell_\infty$.

This settles existence of solutions, i.e., question (i), and assures the approximability of such solutions by solutions of the particular finite sections used. The pointwise exponentially fast rate of approximation attained by these approximate solutions follows from the exponential decay of $A^{-1}$. This gives

Proposition 4(ii). Let A be a strictly m-bounded totally positive matrix and let r = dim $N_A^+$. If A is invertible, then the conclusions of Proposition 3(ii) hold for every integer interval I.

REFERENCES

1. de Boor, C., On cubic spline functions that vanish at all knots, *Adv. Math. 20,* (1976) 1-17.
2. de Boor, C., Splines as linear combinations of B-splines. A survey, in "Approximation Theory II", G.G. Lorentz et al. eds., Academic Press, New York, 1976, 1-47.
3. de Boor, C., Odd-degree spline interpolation at a bi-infinite knot sequence, in "Approximation Theory, Bonn 1976", R. Schabade and K. Scherer, eds., Lecture Notes Math, 556, Springer, Heidelberg, 1976, 30-53.

4. de Boor, C., What is the main diagonal of a bi-infinite band matrix?, in "Approximation Theory, Bonn, 1979", R. DeVore and K. Scherer, eds., Academic Press, to appear.

5. de Boor, C., Dichotomies for band matrices, *SIAM J. Numer. Anal.,* to appear.

6. de Boor, C., The inverse of a totally positive bi-infinite band matrix, in preparation.

7. Demko, S., Inverses of band matrices and local convergence of splines projectors, *SIAM J. Numer. Anal., 14,* (1977), 616-619.

8. Gohberg, I.C. and Fel'dman, I.A., *Convolution equations and projection methods for their solution,* Transl. of the 1971 "Nauka" Moskva original, Amer. Math. Soc., Providence, R.I., 1974.

9. Micchelli, C.A., Infinite spline interpolation, in: Proc. Conf. Approximation Theory, Siegen, Germany, 1979, G. Meinardus, ed., 1979.

10. Schoenberg, I.J., *Cardinal Spline Interpolation,* CBMS Reg. Conf. Series in Appl. Math. Vol. 12, SIAM, Philadelphia, Pa., 1972.

# GLOBAL ANALYSIS IN NONLINEAR APPROXIMATION AND ITS APPLICATION TO EXPONENTIAL APPROXIMATION

## 1. THE UNIQUENESS THEOREM FOR HAAR-EMBEDDED MAINFOLDS

Dietrich Braess

Ruhr-Universität,
Bochum, W.Germany

## I. INTRODUCTION

In 1967 it turned out that the solution is not always unique when the best uniform approximation for a continuous function by sums of exponentials is to be determined. The classical methods in particular those of the theory of varisolvent families could not give a complete theory. Ten years later it was possible to establish a uniform bound for the number of solutions with methods from global analysis. In the theory of varisolvent families, there is a condition on the number of zeros of the difference of two functions. This condition which refers to global properties could be replaced by weaker assumptions such as connectedness etc.

It is the aim of the first part of this lecture to give an introduction to critical point theory in nonlinear approximation theory. The second part contains its application to exponential approximation. The emphasis is put on the ideas and the motivation. The reader interested in the details is referred to [2,3].




ISBN 0-12-780650-4

Throughout the paper, let I be a compact interval in $\mathbb{R}$ and C(I) be equipped with the uniform norm $\|f\|=\sup\{|f(x)| ; x \in I\}$. Some considerations are correct for an arbitrarily normed linear space E. But from Wolfe's investigations in [8] it is clear that our final result on the bound of the number of solutions cannot be extended to $L_p$-spaces.

## II. CRITICAL POINTS

When looking for a method with which to estimate the number of solutions it is natural to think of critical point theory. The following example, however, shows that this requires an extension of the classical theory.

Example 2.1. Let I = [-1,+1]. Consider the approximation of f(x) = x from the set of constant functions:

$$V = \{u_a \in C(I) ; u_a(x) \equiv a , a \in \mathbb{R}\}.$$

The distance function

$$\rho(u_a) = \|f-u_a\| = \max_{-1\leq x\leq 1} |x-a| = 1 + |a|$$

is just not differentiable at the minimum $u = u_0$.

As a consequence, we will perform the analysis in such a way that we need differentiability only in connection with the sets of the approximating functions, but not with the distance function.

For the study of the local structure of a set it is natural to introduce the concept of tangent cones [3].

Definition 2.2. Let V be a non-empty set in a normed linear space E. Then the tangent cone $C_uV$ at u to V consists of the elements h $\in$ E with the following property: there is a continuous mapping of the unit interval [0,1], $\lambda \to u_\lambda \in V$ such that $u_0 = h$ and $\| u_\lambda - u - \lambda h\| = o(\lambda)$ as $\lambda \to 0$.

If the tangent cone is a linear space, it is said to be a tangent space.

The following lemma (which has been rediscovered again and again during the last two decades) is immediate.

Definition and Lemma 3.3. (Lemma of first variation).
Let V be a non empty set in a normed linear space E. If u is a local best approximation to f from V, then u is a critical point to f in V, i.e., a critical point of the map $V \to \mathbb{R} : v \to \| f-v\|$. In other words, zero is then the best approximation to (f-u_ in $C_u V$.

Proof. Assume that u is not a critical point. Then by definition

$$\| f-u-h\| < \| f-u\|$$

for some $h \in C_u V$. Let $(u_\lambda)$ be a corresponding path. Hence,

$$\begin{aligned} \| f-u_\lambda\| &\le \| f-u-\lambda h\| + \| u_\lambda - u - \lambda h\| \\ &\le \| (1-\lambda)(f-u)+\lambda(f-u-h)\| + o(\lambda) \\ &\le \| f-u\| - \lambda[\|f-u\| - \| f-u-h\|] + o(\lambda) \\ &\le \| f-u\| - \text{const}\cdot\lambda + o(\lambda) < \| f-u\| , \end{aligned}$$

for sufficiently small positive $\lambda$. Consequently, u is not a local best approximation. □

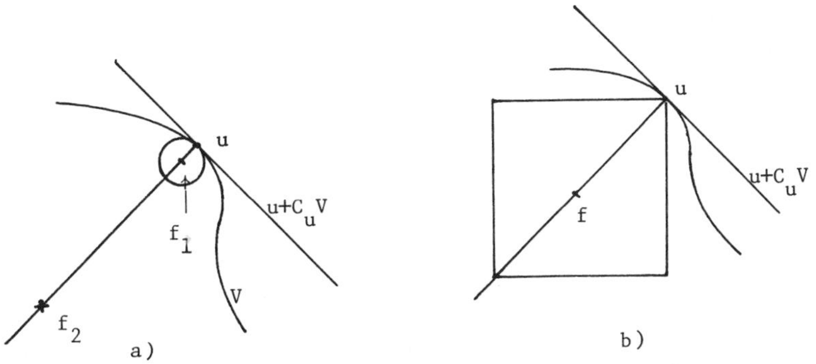


Fig. 1. Critical points in 2-Space. a) Euclidean norm b) Sup-norm

This lemma is useful since the tangent cones are often easier to handle than the original set. Often the tangent cones are linear spaces or convex cones and the result from linear and convex theory are applicable.

To get some more insight into the lemma we consider some examples in 2-space. Let V be a smooth curve in Euclidean 2-space as shown in Fig. 1a. Then u is a local best approximation only if (u-f) is orthogonal to the tangent line. Then u is the closest point on the tangent line. The latter is found also in Fig. 1 b) where the situation for the supremum-norm is shown.

From Fig. 1 we also get a first answer to the question of whether the converse of the lemma also holds. In the Euclidean case the critical point u is a local best approximation to $f_1$ but not to $f_2$. On the other hand, in the sup-norm case u is a local best approximation whenever it is a critical point. We will see that this is not restricted to 2-space but is true more generally whenever the tangent cones satisfy the Haar condition.

## III. HAAR EMBEDDED MANIFOLDS

A set with a countable basis for the topology is called an n-dimensional manifold (without boundary), if to each point there is an open neighborhood which is homeomorphic to an open set in n-space. The mappings from sets in $\mathbb{R}^n$ into the manifold are called parametrizations and their domains are called charts. The differentiable structure is then defined by conditions on the change of coordinate systems.

In our framework, we do not consider abstract manifolds but manifolds which are embedded into a normed linear space E. The manifold is given by a collection of maps {F} from certain subsets of $\mathbb{R}^n$ into E and in this case, only the differentiability of these F's needs to be checked. It is not necessary to control the differentiability of those maps which correspond to the change of coordinate systems (parametrizations). This makes things much easier, particularly in the case of boundaries.

Before we give the definition of a manifold (with boundary) we will motivate one of the postulates. Let $A \subset \mathbb{R}^n$ be an open set and $F : A \to V \subset E$ be a Fréchet-differentiable parametrization for a neighborhood of $u = F(a) \in V$. Moreover assume that $D_aF$ is not singular [3]. Given $u_1 \in F(A)$, we may write $u_1 = F(a+b)$. From $\|F(a+b)-F(a)-D_aFb\| = o(\|b\|)$ we obtain

$$\| u_1 - u - h \| = o(\| h \|) \tag{3.1}$$

if we put $h = D_aFb$. From (3.1) we can conclude that h is a tangent vector and $C_u v \supset D_aF(\mathbb{R}^n)$. More important is the following geometric interpretation of (3.1): all elements of the manifold in a neighborhood are very close to the tangent plane. When boundaries are admitted, this property can no longer be deduced from simpler assumptions. Therefore it will be explicitly postulated.

Definition 3.1. A subset V in a normed linear space E is an n-dimensional $C^1$-submanifold (with boundary) of E if to each $u \in V$ there is an open neighborhood $U \subset V$ with the following properties:

(i) There is a closed convex subset $C \subset \mathbb{R}^n$, an open set $W \subset \mathbb{R}^n$ and a homeomorphism $F : C \cap W \to U$.

(ii) F is Fréchel-differentiable and the derivative $D_aF$ is continuous in a.

(iii) Without loss of generality assume that $u = F(0)$. There is a continuous mapping

$$x : U \to D_0F\ (\overline{\bigcup_{\lambda>0} \lambda C}) \tag{3.2}$$

such that

$$x(u) = 0,$$

$$\| u_1-u-x(u_1) \| = o\| x(u_1) \|), \quad u_1 \in U. \tag{3.3}$$

In contrast to the definition of $C^1$-manifolds without boundary, the property (3.3) is now postulated explicitly, cf. the discussion above. The tangent cone is given by the tangent set in (3.2).

In 1977, Wulbert [9] introduced the concept of Haar embedded manifolds. A $C^1$-manifold without boundary embedded into C(I) was called Haar embedded if each tangent space is a Haar subspace of C(I). The reason for introducing them was the fact that best approximations in Haar spaces have the strong unicity property [9].

Definition 3.2. An element $u \in V$ is a strongly unique best approximation to f in V if there is a number $c > o$ such that

$$\| f-u_1 \| \geq \|f-u\| + c\|u_1 -u\| \tag{3.4}$$

for all $u_1 \in V$. An element $u \in V$ is a strongly unique local best approximation to $f \in V$ if u is a strongly unique best approximation to f in some open neighborhood of u in V.

Fortunately, strong unicity does not only hold in linear Haar spaces but also in some special cones.

Definition 3.3. (1) Let $v_1, v_2, \ldots, v_n \in C(I)$ and $m \le n$. The convex cone

$$\{h \; ; \; h(x) = \sum_{i=1}^{n} a_i v_i(x) \; , \; a_i \ge 0 \text{ for } i = m+1, \ldots, n\}$$

has the Haar property if the functions $\{v_i, \; i \in J\}$ span a Haar subspace whenever $\{1,2,\ldots,m\} \subset J \subset \{1,2,\ldots,n\}$.

(2) A $C^1$-submanifold of $C(I)$ is Haar embedded if each tangent cone has the Haar property.

Examples 3.4. The following sets are cones with the Haar property:

(1) The subset of polynomials

$$\{h \; ; \; h(x) = \sum_{i=0}^{n} a_i x^i \; , \; a_i \in \mathbb{R} \; , \; a_n \ge 0\}.$$

(2) The subset of exponentials

$$\{h \; ; \; h(x) = \sum_{i=0}^{n} a_i x^i e^x + \sum_{i=0}^{m} b_i x^i, \; a_i, b_i \in \mathbb{R}, a_n \ge 0, \; b_m \ge 0\}$$

(3) The subset of positive $\gamma$-polynomials with fixed characteristic numbers $t_i$

$$\{h \; ; \; h(x) = \sum_{i=1}^{n} a_i \gamma(t_i, x) \; , \; a_i \ge 0, \; i = 1,2,\ldots,n\}.$$

Observe that the k-dimensional boundaries of Haar cones are subsets of k-dimensional Haar subspaces.

The example 3.4(2) contains cones of a form which we will meet again when we apply the theory to the approximation by exponential sums.

Concerning uniform approximation, the classical results for Haar subspaces can be extended to cones with the Haar property [3].

Lemma 3.5. To each $f \in C(I)$ there is a strongly unique best approximation in a Haar cone with the Haar property.

Outline of proof: If $U = \Sigma_i a_i u_i$ is a best approximation in the cone, then u is also optimal in the linear space spanned by $\{v_i \; ; \; i \leq m \text{ or } a_i > 0\}$. Apply this observation to $u = (u_1+u_2)/2$ if $u_1$ and $u_2$ are best approximations in the cone. Now uniqueness is immediate. Moreover, by combining the kind of argument above with the well known technique for the proof of the strong unicity in Haar subspaces, we obtain the same property for the cone. □

Now we are ready to prove the main result for Haar embedded manifolds.

Theorem 3.6. For an element u in a Haar embedded manifold V the following are equivalent:

1° u is a critical point of f in V.

2° u is a local best approximation to f in V.

3° u is a strongly unique best approximation to f in V.

Proof. Since 3° ⇒ 2° is obvious and 2° ⇒ 1° is a consequence of the lemma of first variation, it remains to prove the implication 1° ⇒ 3°. Let u be a critical point. Then by Lemma 3.5

$$\| f-u-h \| \geq \| f-u \| + c\| h \|$$

holds for all $h \in C_u V$ and some $c > 0$. By 3.1 (iii) there is a neighborhood U of u in V, such that to each $u_1 \in U$ there exists an $h \in C_u V$ satisfying

$$\| u_1-u-h \| \leq \frac{c}{3} \| h \| . \tag{3.5}$$

Since $c \leq 1$ we conclude from (3.5) that $\| u_1 - u \| \leq \frac{4}{3} \| h \|$. By combining all inequalities we obtain

$$\begin{aligned} \| f-u_1 \| &\geq \| f-u-h \| - \| u_1-u-h \| \\ &\geq \| f-u \| + c\| h \| - \frac{c}{3} \| h \| \\ &\geq \| f-u \| + \frac{c}{2} \| u_1-u \| . \end{aligned}$$

This proves strong local uniqueness. □

## IV. THE UNIQUENESS THEOREM FOR HAAR EMBEDDED MANIFOLDS

Let V be a smooth curve in 2-space with no tangent being parallel to one of the coordinate axes. The latter

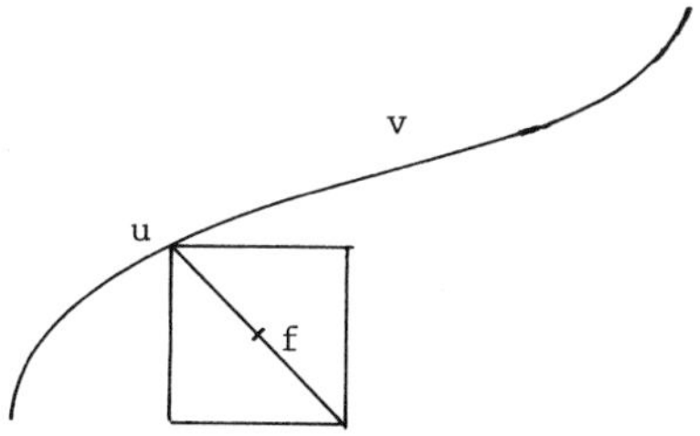


Fig. 2. Approximation to a curve in 2-space when each tangent satisfies the Haar condition.

condition is just the Haar condition. From the shape of the curve (see Fig. 2) it follows that V is a Chebyshev set with respect to the uniform norm. We get enough information on the global shape of the curve from the local property mentioned above, since V is a connected set.

The uniqueness result which was illustrated for the one-dimensional case, will be proved for Haar embedded manifolds of arbitrary finite dimension.

To become familiar with the tools needed in the proof, we will discuss another example.

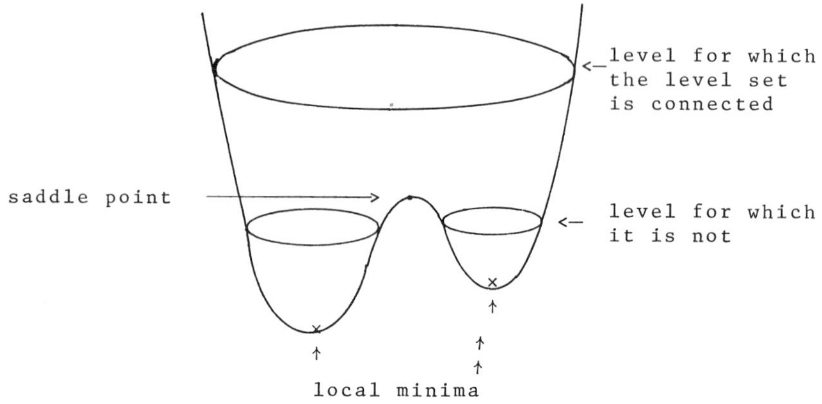


Fig. 3. Function with two local minima

Let $\rho:\mathbb{R}^2 \to \mathbb{R}$ be a function with two isolated local minima and assume that $\rho$ tends to $+\infty$ for sufficiently large arguments (see Fig. 3). Then $\rho$ must have a saddle point. The existence of a saddle point follows from the fact that some level sets are connected and some are not.

We will restrict ourselves to the approximation problem when giving the precise definition of level sets. When f is to be approximated in a set V, the level set

$$V^{\alpha} = \{u \in V \; ; \; \| f-u \| \le \alpha\},$$

$\alpha \in \mathbb{R}$, contains the points for which the distance from f does not exceed $\alpha$. The comparison of different level sets and their topological structure will be performed by considering maps which are called flows.

<u>Definition 4.1</u>. A continuous mapping $\phi : [0,1] \times V \to V$ is a (descending) flow if

$$\| f-\phi(t,u) \| \le \| f-\phi(s,u)\| \text{ for } 0 \le s < t \le 1 \; , \; u \in V$$

and if $\phi_0$ is the identity. Here $\phi_t = \phi(t,.) : V \to V$ for $t \in [0,1]$.

A flow $\phi$ generates for each $u \in V$ a streamline $(u_t)_{t \in [0,1]}$, $u_t = \phi_t u$, along which the distance function $\| f-.\|$ is a non-increasing function.

The lemma of first variation was proved by constructing such a streamline, starting at a non-critical point. In a $C^1$-manifold, this streamline can be extended to a flow.

Lemma 4.2. Let $u_0$ be a non-critical point in a $C^1$-submanifold V of a normed linear space E. Then there is a flow $\phi$ such that

$$\| f-\phi(u_0)\| < \| f-u_0\|$$

Proof. First we will establish a homotopy map

$$\psi : [0,1] \times U \to V$$

such that $\psi_0$ is the identity map and

$$\|f-\psi_t u\| < \|f-\psi_s u\| \tag{4.1}$$

for all $u \in U$ and $0 \leq s < t \leq 1$.

To this end, choose a parametrization F such that $F(0) = u_0$. Since $u_0$ is not a critical point, there is a tangent ray $h = D_0Fb$ such that

$$\lambda := \| f-u_0\| - \| f-u_0 - h\| > 0. \tag{4.2}$$

After multiplying b by a small scale factor (if necessary) we may assume that b is contained in the parameter set. Continuity implies that

$$\| f-F(a) \| - \| f-F(a) - D_0F(b-a)\| > \frac{\lambda}{2} \tag{4.3}$$

for a in a sufficiently small neighborhood $A_1$ of 0 in the chart A. With the splitting technique which we have already used in the proof of the lemma of first variation we obtain from (4.3)

$$\| f-F(a)\| - \| f-F(a)-tD_0 f(b-a)\| \geq t \frac{\lambda}{2} \tag{4.4}$$

for $0 \leq t \leq 1$. After reducing $A_1$ if necessary, we may assume that

$$\| D_a F - D_0 F\| \cdot \| b-a \| < \frac{\lambda}{4}$$

holds for $a \in A_1$. If we choose a sufficiently small number $\delta > o$ and a sufficiently small neighborhood $V_2 \subset V_1$, we have

$$a + t(b-a) \in V_1$$

if $0 \leq t \leq \delta$ and $a \in V_2$. Here the convexity of the chart has been used. Consequently

$$\begin{aligned} &\| F(a+t(b-a))-F(a+s(b-a)) - (t-s)D_0 F(b-a)\| \\ &= \| \int_s^t \{D_{a+\tau(b-a)} F(b-a) - D_0 F(b-a)\} d\tau\| \leq (t-s) \frac{\lambda}{4} \end{aligned} \tag{4.5}$$

for $0 \leq s < t \leq \delta$. Combining (4.4) and (4.5), it follows that

$$\begin{aligned} \| f-F(a+t(b-a))\| &\leq \| f-F(a+s(b-a)-(t-s)D_0 F(b-a)\| \\ &\quad + \| F(a+t(b-a)-F(a-s(b-a))-(t-s)D_0 F(b-a)\| \\ &\leq \| f-F(a+s(b-a))\| - \frac{\lambda}{2}(t-s) + \frac{\lambda}{4} (t-s) \\ &< \| f-F(a+s(b-a))\| - \frac{\lambda}{4} (t-s). \end{aligned}$$

Hence, by going back into the manifold via

$$u \xrightarrow{F^{-1}} a \longrightarrow a+t\cdot\delta(b-a) \xrightarrow{F} u_t =: \psi(t,u)$$

a homotopy map satisfying (4.1) is established on $U=F(A_2)$.

In order to construct from this $\psi$ a flow on the whole manifold, we must modify $\psi$ near the boundary of U. Choose $v > o$ such that $u \in U$ whenever $u \in V$ and $\| u-u_0\| < r$. By

$$\chi(u) = \begin{cases} 1 - \| u-u_0\| / r, & \text{if } \| u-u_0\| < r \\ 0 & \text{otherwise} \end{cases}$$

a continuous cut-off function $\chi : V \to \mathbb{R}$ is defined.

Now

$$\phi(t,u) = \begin{cases} \psi(t\cdot\chi(u),u) & \text{if } \| u-u_0 \| < r, \\ u & \text{otherwise} \end{cases}$$

yields a continuous extension of the cut homotopy map. This proves the lemma. □

Roughly speaking, in the preceding discussion, flows with a local support were constructed. From these one can obtain global flows, i.e., flows which act in a nontrivial way on large domains. The main tool is the following

Theorem 4.3: Deformation Theorem for the Non-Critical Case.

Let $0 < \alpha < \beta$ and let V be a $C^1$-submanifold of a normed linear space. Assume that the set

$$V_1 = \{u \in V \; ; \; \alpha \leq \| f-u \| \leq \beta\}$$

is compact and contains no critical point. Then $V^\alpha$ is a strong deformation retract of $V^\beta$, i.e., there is a flow $\phi$ such that $V^\alpha = \Phi\ (V^\beta)$.

Proof. First observe how flows may be glued together. Let $\phi$ and $\psi$ be flows. Then by

$$\tilde{\phi}_t = \begin{cases} \phi_{2t}, & 0 \leq t \leq 1/2, \\ \psi_{2t-1}\circ\phi_1, & 1/2 \leq t \leq 1, \end{cases}$$

a composite flow is established.

Now by the preceding lemma for any $u \in V_1$ there is a flow $\phi = \phi^u$ and an open neighborhood $U = U^{(u)}$ of u such that

$$\| f-\phi_1(u_1) \| < \| f-u_1 \| \text{ for each } u_1 \in U.$$

Since $V_1$ is compact, a finite number of points $u^{(1)}, u^{(2)}, \ldots, u^{(m)}$ can be chosen such that the corresponding open sets defined above cover $V_1$. Let $\psi$ be the flow which is obtained when we glue $\phi^{u^{(1)}}, \phi^{u^{(2)}}, \ldots, \phi^{u^{(m)}}$ together. Since $V_1$ is compact we have

$$c = \inf_{u \in V_1} \{ \| f-u \| - \| f-\psi_1(u) \| \} > 0 .$$

Hence, $kc \geq \beta - \alpha$ for some natural number k. If we apply $\psi$ just k times, then each element in $V^\beta$ is sent to $V^\alpha$.

At this point, we have established a flow that associates to each $u \in V^\beta$ an orbit $(u_t)$ which ends in $V^\alpha$. Moreover, $u_t$ is constant on each part of the orbit on which the distance to f is not strictly decreasing. The flow is modified by the following rule. As soon as the orbit $u_t$ enters $V^\alpha$, it shall be stopped and considered fixed. The modified flow is also continuous and yields the deformation described in the theorem. □

We recall that a set in a normed linear space is boundedly compact if all its closed bounded subsets are compact.

Now we may state and prove the

Theorem 4.4: Uniqueness Theorem for Haar Embedded Manifolds. Let $V \subset C(I)$ be a connected, boundedly compact, Haar embedded manifold. Then there is a unique best approximation to each $f \in C(I)$ in V and no other local best approximation.

Remark: A first version of the uniqueness theorem was established in 1971 by Wulbert [9]. But the conditions on the compactness were so restrictive that it actually could only be applied to Tornheim families.

Proof of Theorem 4.4. Let $u_1$ and $u_2$ be two local best approximations and let $\| f-u_1 \| \leq \| f-u_2 \|$. Since V is connected, $u_1$ and $u_2$ belong to the same connected component $C^\beta$ of some level set $V^\beta$. The set of critical points in $C^\beta$ is closed. Hence, the distance function assumes its maximum on the critical set at some $u_3$. We may assume $u_1 \neq u_3$. From local strong unicity we conclude that

$$\| f-u \| \geq \| f-u_3 \| + c \| u-u_3 \|$$

for all $u \in V$, provided $\| u-u_3 \| \leq r$. Here $c, r > 0$. This implies $r < \| u_1-u_3 \|$. Put $\alpha = \| f-u_3 \| + \frac{1}{2} cr$. By the deformation theorem, $C^\alpha$ is a deformation retract of $C^\beta$ and is connected. It contains $u_1$ and $u_3$, but no element u with $\| u-u_3 \| = r$. This is a contradiction. □

References: See Part II.

# GLOBAL ANALYSIS AND NONLINEAR APPROXIMATION AND ITS APPLICATION TO EXPONENTAIL APPROXIMATION

## 2. APPLICATIONS TO EXPONENTIAL APPROXIMATION

Dietrich Braess

Ruhr-Universität,
Bochum, W.Germany

## INTRODUCTION

In the first part of the paper we presented a uniqueness theorem for uniform approximation in Haar embedded manifolds. Now we will apply that uniqueness result to uniform approximation by exponential sums, i.e., to a case in which we do not always have uniqueness. A bound for the number of solutions will be derived which is independent of the given function. We have to restrict ourselves to uniform approximation (on an interval I), because the $L_p$-case is quite different [8].

In 1967 it was shown by the author that a best approximation is not always unique. The question arose whether the number of solutions is always finite and whether there is a finite bound:

$$c_n = \sup_{f \in C(I)} \{\text{number of local best approximations to f in } V_n\} \qquad (0,1)$$

(For a definition of the families $V_n$, n=1,2,... see below.) Here we have already taken into account that it is mathematically more elegant and gives more insight to consider




ISBN 0-12-780650-4

not only the global solutions but also the local ones. We note that the local solutions coincide with the critical points in the families of interest.

How can one construct all local solutions and count them? Though the rigorous treatment of the problem must be done in a very abstract setting, the basic idea may be better described from the numerical viewpoint. When one intends to compute a best approximation and uses Newton's method, a sequence is generated which in general will converge only to a local best approximation. Which particular one of the (possibly many) local solutions is found greatly depends on the starting point of the iteration. Therefore the following question is the key to the solution: Is it possible to characterize a set of starting points from which all solutions are reached when applying Newton-like algorithms? - Note that the descending flows (introduced in part I) just define a continuous analog of the Newton method.

It will turn out that such a complete set of starting points for $V_n$ is obtained from the solutions in $V_{n-1}$. Thus $c_n$ can be estimated by an induction. It is typical for the application of global methods that the main result is achieved after the elimination of some nasty exceptional cases by considering the problem from the *generic* viewpoint.

## V. γ-POLYNOMIALS

When sums of exponentials are considered in physics, biology, and in other sciences, usually the set of *proper exponential sums* of order $\leq n$ is considered:

$$E_n^o = \{u(x) = \sum_{\nu=1}^{n} a_\nu e^{t_\nu x}; a_\nu, t_\nu \in \mathbb{R}, \nu=1,2,\dots,n\}. \tag{5.1}$$

For a mathematical treatment of the approximation problem it is necessary to regard the closure of $E_n^o$ and to introduce (*generalized* or) *extended exponential sums* [4]:

$$E_n = \{u(x) = \sum_{\nu=1}^{\ell} p_\nu(x) e^{t_\nu x}; t_\nu \in \mathbb{R}, p_\nu \in \Pi_n, \quad k := \sum_{\nu=1}^{\ell} (1+\partial p_\nu) \leq n\}. \tag{5.2}$$

Here, $\partial p$ denotes the degree of the polynomial p. One should think of $E_n \setminus E_n^o$ as the set of the sums with coalescing characteristic numbers. Specifically the value of $(1+\partial p_\nu)$ is the multiplicity of the characteristic number $t_\nu$.

Exponential sums are a special case of the $\gamma$-polynomials introduced by Hobby and Rice in 1967 [4]. Let $T \subset \mathbb{R}$ be an open interval and assume that the *kernel* $\gamma \in C(T \times I)$ has continuous derivatives $\gamma^{(\mu)}(t,x) := (\partial/\partial t)^\mu \gamma(t,x)$ up to $\mu=n$. Then the $\gamma$-polynomials of order $\leq n$ form the set

$$V_n = \{u(x) = \sum_{\nu=1}^{\ell} \sum_{\mu=1}^{m_\nu} a_{\nu\mu} \gamma^{(\mu-1)}(t_\nu, x); a_{\nu\mu} \in \mathbb{R}, \quad k := \sum_{\nu=1}^{\ell} m_\nu \leq n\}. \tag{5.3}$$

Obviously, $V_n = E_n$ if $\gamma(t,x) = e^{tx}$ and $T = \mathbb{R}$. The set $V_n^o$ is defined similarly to $E_n^o$.

To specify the assumptions on the kernel which are usually postulated for $\gamma$-polynomials we need some notation: In (5.2) and (5.3) two integers are associated to each function (here it is understood that dummy terms are eliminated from the sums):

$\ell$ = the number of distinct characteristic numbers,

k = the order = the number of characteristic numbers counting multiplicities.

The spectrum of u, denoted by spect(u), is the set of characteristic numbers.

In the rest of the paper we assume that the following two conditions hold.

Assumptions 5.1.

1. *Sign regularity:* Each $\gamma$-polynomial of order $k \leq 2n$ has at least k-1 zeros or vanishes identically. This means that $\gamma$ is extended sign regular of order 2n in t in the sense of Karlin [5] for short: $\gamma$ is $ESR_{2n}(t)$.
2. *Normality:* To each $u \in V_n \setminus V_{n-1}$ there is a neighborhood U and a compact set $T_o \subset T$ such that

$$\text{spect}\ (u_1) \subset T_o \quad \text{for all } u_1 \in U.$$

This means that nothing serious can happen when the characteristic numbers approach the boundary of T.

The exponential kernel $\gamma(t,x)=e^{tx}$ is extended sign regular (even extended totally positive) of any order [5]. It follows from E.Schmidt's compactness results [6] that this kernel is normal in the sense of Assumption 5.1.

Before Wulbert had introduced the concept of Haar embedded manifolds, exponential sums were treated in the framework of varisolvent families or by using the local Haar condition of Meinardus (cf.[4] and [2, part I]). With those classical theories the following results were obtained.

Theorem 5.2. Let $\gamma$ be extended sign regular of order 2n.

(1) $u \in V_n$ is a best approximation to f in $V_n$ if there is an alternant of length n+k+1, and it is a (local) best approximation only if there is an alternant of length $n+\ell+1$.

(2) For each $f \in C(I)$ there is at most one best approximation in $V_n^o$.

Obviously there is a gap in the characterization of best approximations when there are characteristic numbers of multiplicity greater than one. The gap would not be serious if this rarely happened. But it will turn out in Section 7 that coalescence in the spectrum is not an exceptional case.

## VI. The manifold $V_n \setminus V_{n-1}$

Since there may be more than one best approximation in $V_n$, $n \geq 2$, we cannot expect $V_n$ to be a Haar embedded manifold. But we will see soon that $V_n \setminus V_{n-1}$ is such a manifold. Therefore the main result on approximation in $V_n$ will not be derived directly, but by an induction argument.

First we will give a motivation for the restriction to non degenerate $\gamma$-polynomials. Here the $\gamma$-polynomials from $V_{n-1}$ are considered degenerate in $V_n$, with degeneracy being understood as in rational approximation. Consider the canonical representation for the elements of $E_1$:

$$\begin{aligned} F : \mathbb{R}^2 &\longrightarrow E_1 \\ (a,t) &\longmapsto a\, e^{tx} \end{aligned} \tag{6.1}$$

Obviously, F is bijective only if the line in 2-space with the elements $(0,t), t\in\mathbb{R}$, is contracted to a single point. But then this point has no open neighborhood with a compact closure. Hence, the resulting set is not locally compact and cannot be a manifold. Similarly, $u\in V_{n-1}$ is degenerate in $V_n$, because it can be indentified with $u+a\gamma(t,.)\in V_n$ where $a=0$ and t is an arbitrary point in T.

In order to see that $V_n\setminus V_{n-1}$ is a Haar embedded manifold we have to compute tangent cones. The simplest case is when $u\in V_n^o$. Consider the parametrization

$$F : \mathbb{R}^2 \longrightarrow V_n^o,$$

$$(a_1,a_2,\ldots,a_n,t_1,\ldots,t_n) \longmapsto \sum_{\nu=1}^{n} a_\nu \gamma(t_\nu,.). \tag{6.2}$$

This is a bijection on the subset of the points sent to nondegenerate $\gamma$-polynomials if we allow only parameters for which

$$t_1<t_2<\ldots<t_n. \tag{6.3}$$

It follows from $\frac{\partial F}{\partial a_\nu} = \gamma(t_\nu,.)$, $\frac{\partial F}{\partial t_\nu} = a_\nu \gamma^{(1)}(t_\nu,.)$, $\nu=1,2,\ldots,n$, and the sign regularity of the kernel that the partial derivatives span a Haar subspace of dimension 2n. Hence, the span is the tangent space at u.

Next we regard the simplest case of coalescing characteristic numbers, i.e. an exponential sum $E_2$ with $\ell=1$. Recalling (6.3) it is natural to require

$$t_1<t_2<\ldots<t_n \tag{6.4}$$

when the characteristic numbers of *extended* $\gamma$-polynomials are repeated corresponding to their multiplicities. Consider the set of those points in n-space whose coordinates satisfy (6.4). Its boundary points are characterized by the fact that equality holds for at least one relation. Therefore parametrizations for extended $\gamma$-polynomials have domains with boundaries. An appropriate parametrization for $E_2$ is given by

$$F : \mathbb{R}^3\ \mathbb{R}_+ \longrightarrow E_2,$$
$$(a,b,t,0) \longmapsto e^{tx}(a \cosh \sqrt{\theta}x + b \frac{\sinh\sqrt{\theta}x}{\sqrt{\theta}}). \qquad (6.5)$$

or more generally for $\gamma$-polynomials

$$(a,b,t,\theta) \longmapsto a\ \frac{1}{2}[\gamma(t+\sqrt{\theta},.)+\gamma(t-\sqrt{\theta},.)]$$
$$+\ 6\frac{1}{2}\ \theta^{-1/2}\ [\gamma(t+\sqrt{\theta},.)-\gamma(t-\sqrt{\theta},.)].$$

The quotient in 6.5 is interpreted in an obvious way for $\theta=0$. Then F is a differentiable function of the parameters and the partial derivatives are linearly independent. If $\theta=0$,

$$\frac{\partial F}{\partial a} = e^{tx},\ \frac{\partial F}{\partial b} = xe^{tx},\ \frac{\partial F}{\partial t} = e^{tx}(ax+bx^2),$$
$$\frac{\partial F}{\partial \theta} = e^{tx}(\frac{a}{2}\ x^2 + \frac{b}{3}\ x^3).$$

The tantent cone consists of the functions

$$\frac{\partial F}{\partial a}\ \Delta a + \frac{\partial F}{\partial b}\ \Delta b + \frac{\partial F}{\partial t}\ \Delta t + \frac{\partial F}{\partial \theta}\ \Delta\theta,$$

where $\Delta a, \Delta b, \Delta t \in \mathbb{R}$ and $\Delta\theta \geq 0$. The restriction on $\Delta\theta$ is due to the fact that $\theta \in \mathbb{R}_+$. Simple calculations yield the cone for $u=F(a,b,t,o)$:

$$C_u E_2 = \{h;h(x) = e^{tx} \sum_{\mu=o}^{3} \delta_\mu x^\mu,\ \delta_\mu \in \mathbb{R},\quad \delta_3 \geq 0\} \qquad (6.6)$$

From Example 3.4(1) we know that $C_u E_2$ is a cone with the Haar property.

The consideration above is not restricted to $E_2$ and $V_2$. Given a $\gamma$-polynomial $u \in V_n$, $n>2$, each term with multiplicity 2 gives rise to a portion in the tangent cone comparable to (6.6). The terms with multiplicity $m_\nu > 2$ are treated with more sophisticated parametrizations [2, Section 15]. If all terms are put together, one obtains for $u \in V_n \setminus V_{n-1}$:

$$C_u V_n = \{h = \sum_{\nu=1}^{\ell} \sum_{\mu=1}^{\tilde{m}_\nu} \delta_{\nu\mu} \gamma^{(\mu-1)}(t_\nu, .),\ \delta_{\nu\mu} \in \mathbb{R},$$
$$\delta_{\nu\tilde{m}_\nu} \cdot a_{\nu m_\nu} \geq 0 \quad \text{if } m_\nu \geq 2\}, \tag{6.7}$$

where

$$\tilde{m}_\nu = \begin{cases} m_\nu + 2 & \text{if } m_\nu \geq 2, \\ 2 & \text{if } m_\nu = 1. \end{cases} \tag{6.8}$$

The tangent cones satisfy the Haar condition (cf. Examples 3.4). Finally the completeness of the tangent cones in the sense of Definition 3.1(iii) is a consequence of Lemma 10.2 in [2]. Hence, $V_n \setminus V_{n-1}$ is a Haar embedded manifold.

Fortunately, best approximations in Haar cones can be characterized in terms of alternants. One has to evaluate the maximal number of zeros of the functions in the cone (6.7). This may be done by a *generalized Descartes' rule*, which is of independent interest. To this end we will define recursively for every $\gamma$-polynomial of order k a sign vector $\{s_1, s_2, \ldots, s_k\}$ with k components.

<u>Definition 6.1.</u> (a) Let u be a $\gamma$-polynomial having one characteristic number of multiplicity m

$$u = \sum_{\mu=1}^{m} a_\mu \gamma^{(\mu-1)}(t, .), \quad a_m \neq 0, \ \text{sign } a_m = \sigma.$$

Then

$$\text{sign } u = ([-1]^{m-1}\sigma, [-1]^{m-2}\sigma, \ldots, \sigma, -\sigma, \sigma).$$

(b) If all characteristic numbers of $u_1$ are smaller than those of $u_2$, the following composition rule applies:

$$\text{sign}(u_1+u_2) = (\text{sign } u_1, \text{ sign } u_2)$$

Example 6.2. If $u_o(x)=(3-x)e^{-x}+4e^{2x}+(x+2x^2)e^{5x}$ then

$$\text{sign } u_o = (+1,-1,+1,+1,-1,+1).$$

The origin of the components becomes clearer if we write them in a clustered form: sign $u_o = ((+1,-1),(+1),(+1,-1,+1))$.

For mnemotechnic reasons it is useful to draw the signs (as it is done for spectra in physics) on a t-axis. Multiple lines are resolved and drawn here with alternating signs.

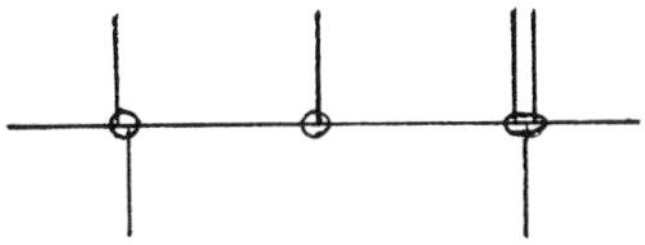

Fig.4. Generalized signs of the γ-polynomial in Example 6.2

Definition 6.1 makes sense because each extended γ-polynomial may be approximated arbitrarily closely by proper ones with the same sign vector [2,Section 3]. Because of this approximation property the well known Descartes' rule [5,Chapter 5,Theorem 1.2] may be extended.

<u>Lemma 6.2</u> (Descartes'Rule for Extended γ-Polynomials). Assume that γ is extended sign regular of order n. If a γ-polynomial u of order k with sign $u=(s_1,s_2,...,s_k)$ has m zeros $x_1<x_2<...<x_m$, then there are at least m sign changes in the sequence $s_1,s_2,...,s_k$. Moreover, if the number of sign changes equals m, then

$$\text{sign } u(x) = \varepsilon_m \cdot s_k \quad \text{for } x>x_m,$$

where whether $\varepsilon_m=+1$ or $-1$ depends only on m.

At this point we note that a sign regular kernel is called totally positive, if all $\varepsilon_m$'s are +1. We will restrict ourselves to these kernels in the remainder of this section in order not to have to consider these extra sign factors.

Now we are in a position to calculate the maximal number of zeros for the elements of the tangent cone (6.7). To do this we write the $\gamma$-polynomials in a form where terms with multiplicity one and greater than one are separated:

$$u = \sum_{\nu=1}^{\ell_1} \sum_{\mu=1}^{m_\nu} \alpha_{\nu\mu} \gamma^{(\mu-1)}(t_\nu,.) + \sum_{\nu=\ell_1+1}^{\ell} \alpha_{\nu 1} \gamma(t_\nu,x),$$

$$t_1 < t_2 < \ldots < t_{\ell_1}.$$

The characteristic numbers with multiplicity greater than one are ordered. Put

$$\begin{aligned} \sigma_\nu &= \text{sign } a_{\nu m_\nu}, \quad \nu = 1,2,\ldots,\ell_1, \\ r_\nu &= 1 \quad \text{if } \sigma_\nu \sigma_{\nu+1}(-1)^{m_{\nu+1}} > 0, \\ &= 0 \quad \text{otherwise}, \\ L &= \ell + \sum_{\nu<\ell_1} r_\nu . \end{aligned} \tag{6.9}$$

By induction on $\ell_1$ one sees that each $h \in C_u V_n$ has at most $n+L-1$ zeros. Note that

$$\ell \leqslant L \leqslant k(=n).$$

Therefore it is reasonable to extend the definition of L and put $L(u)=\ell(u)=k(u)$ for $u \in V_n^o$. With the integer L we may fill the gap which was left in the characterization of best approximations given in Theorem 5.2.

Theorem 6.3. Let $f \in C(I)$ and let $\gamma$ be extended totally positive of order 2n. Then $u \in V_n \setminus (V_n^o \cup V_{n-1})$ is a local best approximation to f if f-u has an alternant with n+L+1 points $x_o < x_1 < \ldots < x_{n+L}$ and if

$$\text{sign}(f-u)x_{n+L}) = -\sigma_{\ell_1}.$$

At first the characterization theorem and the introduction of the integer L seem to be technical details. But we shall need (6.9) for the consideration of generic properties in the next section.

Finally we recall a property of Haar embedded manifolds which is very useful when perturbation arguments are applied [2,Theorem 17.1].

Lemma 6.4. Let $u_o$ be a local best approximation to $f_o$ in a Haar embedded manifold $V \subset C(I)$. Then there is a neighborhood W of $f_o$ in C(I) and a neighborhood U of $u_o$ in V, such that U contains exactly one local best approximation u to each $f \in W$ in U. Moreover, the map that sends f to u is continuous in W.

## VII. GENERIC PROPERTIES

In the last section an integer L was associated to each $\gamma$-polynomial u. The parameter L is not a continuous function of u on $V_n \setminus V_{n-1}$. The non-continuity leads to some exceptional cases. It seems that the possible pathologies are so nasty that their study is not very promising at present.

In this section we will see that for *most* local best approximations $L=\ell$ holds. Therefore the analysis is simplified considerably if we disregard the exceptional cases on the way to the main result. Fortunately the final result applies to the exceptional cases as well.

The term "most" means that the corresponding statement is *generic* in the following sense.

Definition 7.1. (1) A subset of a topological space is said to be residual if it can be expressed as a countable intersection of dense, open sets. The points of the complement of a residual set are said to be exceptional points.
(2) A property is called generic if the set of points with this property contains a residual set.

A subset of a normed linear space is obviously residual if its complement is a set of first category in the sense of Baire; e.g., the set of irrational numbers is a residual subset of $\mathbb{R}$.

To acquaint the reader with the concept of generic properties we give a simple example from approximation theory.

Example 7.2. The set of functions in C(I) for which the best approximation in $\Pi_n$ is characterized by an alternant of exact length n+2 is residual.

We will prove this statement because we need a similar property and a similar argument when discussing $\gamma$-polynomials: Assume that $f_o - u_o$ has an alternant of length m>n+2, with $u_o \in \Pi_n$. Given $\varepsilon > 0$, there is a $g \in C(I)$, $\|g\| < \varepsilon$ such that $f_o - u_o + g$ has an alternant of exact length n+2. Then the best approximation fo $f = f_o + g$ is also $u_o$ and is characterized by an alternant of minimal length. Consequently, the set under consideration is dense.

Assume that $f_o - u_o$ has an alternant of exact length n+2. Then for all functions in some neighborhood of $f_o - u_o$ in C(I) the length of the alternant is at most n+2. Since the metric projection of C(X) onto $\Pi_n$ is continuous, the set under consideration is also open. □

Now we return to approximation by γ-polynomials. We ask: what can be said about a local solution $u_o \in V_n \setminus V_{n-1}$ having an alternant of exact length $n+\ell(u_o)+1$? (Note that this condition implies $L(u_o)=\ell(u_o)$).

1. Assume that the alternant of $f_o - u_o$ has exactly the mentioned length. We have $\ell(u) \geq \ell(u_o)$ for all γ-polynomials in some neighborhood uf $u_o$. From Lemma 6.4 and the arguments in the discussion of Example 7.2 we conclude that the alternant of f-u is not longer than $n+\ell(u_o)+1$ if $\|f-f_o\|$ is sufficiently small and u is the best local approximation u to f in a neighborhood of $u_o$. Consequently, $\ell(u)=\ell(u_o)$ and the multiplicities $m_\nu$ of the characteristic numbers do not change if small perturbations are added to f. We may say that the multiplicities are here *structurally stable*.

If on the other hand $u_o$ is a local solution to f such that $L(u_o) > \ell(u_o)$ than we do not have structural stability. It is possible to construct a γ-polynomial $u_1$ satisfying $\ell(u_1)=L(u_1)=L(u_o)>\ell(u_o)$ with $\|u_1 - u_o\|$ arbitrarily small. To this end let $t_\nu$ be any characteristic number of $u_o$ with multiplicity $m_\nu$ for which the parameter $r_\nu$ of (6.9) equals 1. Let $u_1$ have the characteristic numbers $t_\nu$ and $t_\nu+\delta$, ($\delta>o$ but small) with multiplicity $m_\nu - 1$ and one, resp. By Theorem 6.3

$u_1$ is a local solution to $f_1=f_0+(u_1-u)$. By modifying $f_1$ once more if necessary, we get a function such that the length of the alternant is only $n+\ell(u_1)+1$.

The considerations above apply to all local solutions as long as their number is bounded. We can assume that this is true because we will apply the following generic result later in an inductive proof.

Lemma 7.3. Let $\gamma$ be extended sign-regular of order 2n. Moreover assume that $c_k<\infty$, $k=1,2,\ldots,n$. Then the functions f for which all local best approximations in $V_k$, $k=1,2,\ldots,n$, have the following properties form a residual set in C(I).

(i) For each local best approximation u to f in $V_k$ there is an alternant of exact length $k+\ell(u)+1$.

(ii) Each local best approximation to f in $V_k$ has the maximal order k.

(iii) There is a neighborhood U of f in C(I) such that the number of local best approximations in $V_k$ is the same for all $g\in U$.

Proof. It follows from the preceding discussion that (i) is a generic peroperty. If there is an alternant of length only $k+\ell(u)+1$ for some $u\in V_k$, it follows from Theorem 5.2 that u cannot be a local best approximation in $V_{k+1}$. Therefore (ii) is a consequence of (i). Finally we conclude from Lemma 6.4 that the number of local solutions in $V_k\setminus V_{k-1}$ is a lower semicontinuous function on C(I). From this and the boundedness of $c_k$ we see that (iii) is a generic property too. □

Lemma 7.3(i) says how the gap in Theorem 5.2(i) is filled in *most* cases. Furthermore, from the classical theory we know that each local solution which is not a global one, has at least one characteristic number with multiplicity greater than one. According to Lemma 7.3, in *most* cases the coalescence of characteristic numbers happens in such a way that L is not greater than $\ell$.

## VIII. MAXIMAL COMPONENTS

It was already pointed out that the uniqueness theorem (Theorem 4.4) cannot be applied to the whole family $V_n$. But we can apply it to certain subsets which are level sets. Those of these sets of maximal level will play a central role in the following.

We recall that

$$V^\alpha = \{u \in V; \|f-u\| \leq \alpha\}$$

denotes the level set with level $\alpha > 0$, when the approximation of a given $f \in C(I)$ in a given family V is studied.

Remark 8.1. Let $Cp^\alpha$ a (non-void) component of the level set $V_n^\alpha$. If $Cp^\alpha$ is disjoint from $V_{n-1}$ then $Cp^\alpha$ contains exactly one local best approximation.

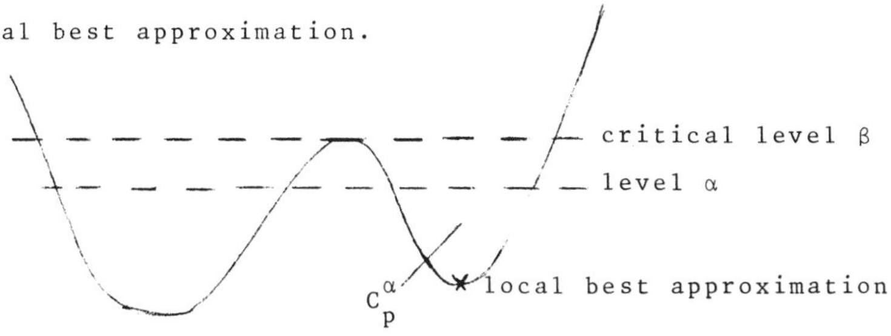


Fig.5. Component of a level set.

This remark is not complete unless we specify the topology of $V_n$ carefully. Typical is the situation with exponential kernel. According to E. Schmidt [6] a subset of $E_n \subset C(I)$ is compact with respect to the norm topology if it is disjoint from $E_{n-1}$ and if it is closed with respect to the topology of compact convergence on the open interval $\mathring{I}$. Complications arise in sets of exponential sums with unbounded spectrum.

Once more we will simplify the analysis by restricting the problem. We will see at once that the restriction does not spoil the result. Consider those $\gamma$-polynomials whose spectrum is contained in a given compact interval $T_1 \subset T$.

$$V_n(T_1) = \{u \in V_n;\ \operatorname{spect}(u) \subset T_1\}.$$

The restriction of the spectrum induces two additional boundary pieces, but the tangent cones still have the Haar property. Hence, $V_n(T_1) \setminus V_{n-1}(T_1)$ is also a Haar embedded manifold. If u is a local best approximation to f in $V_n$ and if $u \in V_n(T_1)$ then u is also a local solution in $V_n(T_1)$. Consequently a bound on the number of local solutions in $V_n(T_1)$ which is independent of $T_1$ is also valid for the whole family $V_n$.

Because we do not want to overload the notation, we will write $V_n$ instead of $V_n(T_1)$ whenever no confusion is possible.

Proof of Remark 8.1. Since $Cp^\alpha$ is closed and bounded, it is a compact subset of $V_n$. Hence, it is a compact level set in some subset of $V_n \setminus V_{n-1}$ which again is a Haar embedded manifold. Therefore, the uniqueness theorem applies. □

Given a local best approximation $u^*\in V_n\setminus V_{n-1}$ we consider the collection $Cp^\alpha$, $\alpha\in\mathbb{R}$, of the components of the level sets containing $u^*$. From Remark 8.1 we know that $u^*$ is the unique local solution in $Cp^\alpha$ if $\alpha>\|f-u^*\|$ is sufficiently small. We will make the level as high as possible. Put

$$\beta=\sup\{\alpha; Cp^\alpha\cap V_{n-1}=\emptyset\}$$

and

$$Cp=\bigcup_{\alpha<\beta} Cp^\alpha.$$

Cp is called the maximal component associated to $u^*$. Instead of searching for all local best approximations we will search for all maximal components.

<u>Assertion 8.2.</u> The closure $\overline{Cp}$ of a maximal component intersects $V_{n-1}$.

Proof. Assume that $\overline{Cp}\cap V_{n-1}=\emptyset$. We have $\|f-u\|=\beta$ for all u on the boundary $\partial Cp$. Since all critical points in $V_n\setminus V_{n-1}$ are strong local solutions, $\partial Cp$ contains no critical point. By Lemma 4.2 for each $u_o\in\partial Cp$ there is an open neighborhood U and a flow

$$\psi:[o,1]\times\overline{U}\longrightarrow V_n$$

such that $\|f-\psi_1(u_o)\|<\|f-u_o\|=\beta$. After making U smaller if necessary the following holds:

(i) $\overline{U}\cap V_{n-1}=\emptyset$,

(ii) $\overline{U}$ is compact and connected,

(iii) $\|f-\psi_1(u)\|<\beta$ for each $u\in\overline{U}$.

Note that $U\cap Cp$ is not empty and is sent to Cp by $\psi_1$. Since $\psi_1(\overline{U})$ is connected, from (iii) it follows that $\psi_1(\overline{U})\subset Cp$.

Consequently we have $\|f-u\|>\beta$ for all $u\in\bar{U}\backslash\overline{Cp}$. Otherwise the orbit $\psi_1(t,u)$, $0<t<1$ would establish a connecting are between u and $\psi_1(u)$ which runs below the level $\beta$.

Since $\partial Cp$ is compact, a finite number of such open sets, say $U_1,U_2,\ldots,U_m$, cover $\partial Cp$.

$$U = \bigcup_{j=1}^{m} U_j \supset \partial Cp.$$

(U is a tubular neighborhood of $\partial Cp$.) The set $M=\bar{U}\backslash(U\cup Cp)$ is compact and the distance function $\|f-.\|$ achieves its minimum at some $u_1\in M$. By the construction we have $\beta_1:=\|f-u_1\|>\beta$. Since $Cp\cup U$ is connected it contains $Cp^{\beta+(\beta_1-\beta)/2}$, which contradicts the maximality of $\beta$. Hence, $\overline{Cp}$ intersects $V_{n-1}$. ☐

The proof of the assertion is typical, it shows the application of flows as efficient tools. This holds also for the next proof.

Assertion 8.3. If $\hat{u}\in\overline{Cp}$ has order $k<n$, then $\hat{u}$ is a critical point in $V_k$.

Outline of proof. If $\hat{u}$ is not a critical point, then there is a flow $\phi$ which pushes $\hat{u}$ down to some $u_1\in V_k$. This flow can be lifted to a neighborhood of $\hat{u}$ in $V_n$. The flow sends elements of Cp which are close to $\hat{u}$ to elements close to $u_1\in V_k$. Since $\|f-u_1\|<\|f-\hat{u}\|=\beta$ this contradicts $Cp^{\alpha}\cap V_{n-1}=\emptyset$ for $\alpha<\beta$. ☐

Assertion 8.3 can be improved on in the generic case. Then $\hat{u}\in\overline{Cp}\cap V_{n-1}$ must be a critical point in $V_{n-1}$ and not only in $V_k$. Consequently, $\overline{Cp}\cap V_{n-2}$ is empty.

## IX. THE STANDARD CONSTRUCTION

In the last section we associated to each local best approximation in $V_n$ a maximal component and a critical point in $V_{n-1}$ on its boundary. Now we will argue in the other direction. Given any critical point in $V_{n-1}$ can we find all maximal components which branch at it? In order to answer this question we will study the following construction. For convenience, we use the abbreviation

$$\begin{aligned} &V_n(t_1,t_2,\ldots,t_m) \\ &= \text{closure}\,\{u\in V_n;\ \{t_1,t_2,\ldots,t_m\}\subset \text{spect}(u)\}. \end{aligned} \tag{9.1}$$

<u>Standard Construction.</u> Assume that f is not exceptional with respect to $V_{n-1}$. Take a local best approximation $\hat{u}$ to f in $V_{n-1}$ and a $\tau\notin\text{spect}(\hat{u})$. Consider

$$u+0\cdot\gamma(\tau,.)\in V_{n-1}+\text{span}\{\gamma(\tau,.)\}\subset V_n(\tau).$$

Then there is a tangent vector

$$h=h_1+\delta\gamma(\tau,.)\in C_uV_n(\tau). \tag{9.2}$$

$$h_1\in C_uV_{n-1},\qquad \delta\in\mathbb{R},$$

such that $\|f-u-h\|<\|f-u\|$. Let $(u_\lambda)$ be the curve in $V_n(\tau)$ associated to the tangent vector h. Then $u_\lambda$ is a better approximation than u for sufficiently small $\lambda$ and the curve

a) either represents a maximal component

b) or belongs to a level set which intersects $V_{n-1}$.

In Case b) we will discard it.

In this way all local best approximations $\hat{u}_1,\hat{u}_2,\ldots$ are scanned and for each $\hat{u}_i$ the additional characteristic number $\tau$ scans a set such that each interval of $\mathbb{R}\setminus\text{spect}(\hat{u}_i)$ is met.

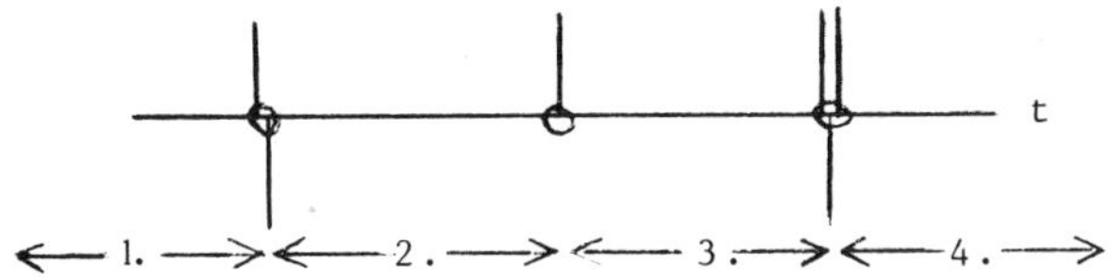


Fig.6. Intervals of $\mathbb{R}\setminus\text{spect}\,(\hat{u})$

It is the aim of the paper to show that this procedure yields all maximal components. This result contains the statement that no more than n components branch from each critical point in $V_{n-1}$. First we show that this estimate cannot be improved.

Example 9.1. Let $r \in C(I)$ have an alternant $x_1 < x_2 < \ldots < x_{2n-1}$ of length 2n-1. Moreover, assume that $r(x_{2n-1}) = -\|r\|$. Choose $t_1 < t_2 < \ldots < t_{n-1}$ and put

$$\hat{u}(x) = \sum_{\nu=1}^{n-1} e^{t_\nu x}.$$

Obviously, $\operatorname{sign}(\hat{u}) = (+1, +1, \ldots, +1)$. By Theorem 5.2 $\hat{u}$ is the unique (local and global) best approximation to $f = \hat{u} + r$ in $E_{n-1}$. On the other hand there are n local solutions $u_1, u_2, \ldots, u_n$ to f in $V_n$, which may be distinguished by their sign vectors:

$$\operatorname{sign} u_m = (+1, +1, \ldots, +1, \underset{\substack{\uparrow \\ m\text{-th position}}}{-1}, +1, \ldots, +1). \tag{9.3}$$

The maximal component containing $u_m$ is constructed by choosing $\tau$ according to

$$t_{m-1} < \tau < t_m, \quad m = 1, 2, \ldots, n.$$

Here, the inequalities referring to $t_0$ and $t_n$ are ignored. By Descartes' rule the constructed exponential sums have the sign vectors (9.3).

Assertion 9.2. The constructed component depends only on $\hat{u}$ and $\tau$ but not on the choice of the tangent vector h or the associated curve.

Proof. Consider an approximation of f by a $\gamma$-polynomials on the constructed curve. The error function has the form $f-\delta\gamma(\tau,.)-u$, where $|\delta|$ is small and $u\in V_{n-1}$ is close to $\hat{u}$. We may compare this with the approximation of

$$f_{\delta,\tau}=f-\delta\gamma(\tau,.)$$

by elements in the neighborhood of $\hat{u}\in V_{n-1}$. From Lemma 6.4 it follows that there is a unique critical point $u_{\delta,\tau}$ for $f_{\delta,\tau}$, $|\delta|$ small, in a neighborhood U of $\hat{u}$. Thus the level sets for this approximation problem have only one connected component which intersects U. Hence, there is an arc from u to $u_{\delta,\tau}$ such that the distance to $f_{\delta,\tau}$ decreases along it. Moreover, $u_{\delta,\tau}$ is a continuous function of $\delta$. Therefore it is always possible to establish a connecting arc below the level $\|f-\hat{u}\|$. □

<u>Assertion 9.3.</u> Let $\tau_1$ and $\tau_2$ belong to the same interval of $\mathbb{R}\setminus\mathrm{spect}(\hat{u})$. Then the standard construction with $\hat{u}$ and $\tau_1$ leads to the same component as the construction with $\hat{u}$ and $\tau_2$.

Proof. We may assume $\tau_1<\tau_2$. For any $\tau\notin\mathrm{spect}(\hat{u})$ there is a small $\delta=\delta(\tau)$, such that

$$\|f_{\delta,\tau}-u_{\delta,\tau}\|<\|f-\hat{u}\|. \tag{9.4}$$

Since $[\tau_1,\tau_2]$ is compact we may choose a $\delta\neq 0$, such that (9.4) holds uniformly for all $\tau\in[\tau_1,\tau_2]$. Since the map $\tau\longmapsto\delta\gamma(\tau,.)+u_{\delta,\tau}$ is continuous, the assertion follows with the same arguments as for the preceding one. □

Assertion 9.3 helps us to understand once more why we have to restrict ourselves to the generic case in the sense of Section 7. The assertion says that the constructed component depends only on ($\hat{u}$ and on) the interval of $\mathbb{R}\setminus\mathrm{spect}(\hat{u})$ in which the new characteristic number is chosen. Note that the number of those intervals depends on the multiplicities of the characteristic numbers. Their (structural) stability is necessary if we want to exclude uncontrollable changes.

Now we are ready to sketch the proof that each maximal component is obtained by the standard construction in the generic case.

Let $u^*$ be any local best approximation in $V_n \setminus V_{n-1}$ with maximal component Cp and let $\hat{u} \in \partial Cp \cap V_{n-1}$. Then there is a sequence $(u_r) \subset Cp$ which converges to $\hat{u}$. We distinguish two cases.

*Case 1.* The sequence of the spectra $(\mathrm{spect}(u_r))$ has an accumulation point $\tau \notin \mathrm{spect}(\hat{u})$.

After passing to a subsequence if necessary we may split the $\gamma$-polynomials of the sequence

$$u_r = v_r + w_r,$$
$$v_r \in V_{n-1}, \qquad \lim v_r = \hat{u},$$
$$w_r = a^{(r)} \gamma(t^{(r)}, .), \qquad \lim a^{(r)} = 0, \qquad \lim t^{(r)} = \tau.$$

By Lemma 6.4 there is a unique local best approximation to $(f - w_r)$ in a neighborhood of $\hat{u}$ in $V_{n-1}$, provided that $\|w_r\|$ is sufficiently small. Consequently, Cp is obtained by the standard construction with $t^{(r)}$, r sufficiently large, being the additional characteristic number. Thus the proof for Case 1 is complete.

*Case 2.* The sequence $(\mathrm{spect}(u_r))$ has an accumulation point $\tau$ with multiplicity $m > 1$ and $\tau$ is an $(m-1)$-fold characteristic number uf $\hat{u}$.

In this case it is necessary to find a way to consider m characteristic numbers fixed close to $\tau$. (We note that in the standard construction only one is considered fixed). Recall (9.3). Since $C_{\hat{u}}V_n(\underbrace{\tau,\dots,\tau}_{m \text{ times}})\subset C_{\hat{u}}V_{n-1}$, the $\gamma$-polynomial $\hat{u}$ is also a critical point in $V_n(\tau,\dots,\tau)$. Moreover a neighborhood of $\hat{u}$ in $V_n(\tau,\dots,\tau)$ is a Haar embedded manifold and $\hat{u}$ is a strongly local best approximation in this set. It follows by perturbation arguments that there is a unique local best approximation $\tilde{u}$ to f in

$$V_n(t_1,t_2,\dots,t_n)$$

close to $\hat{u}$, provided that

$$|t_\nu-\tau|<\delta_o,\delta_o \text{ small}, \qquad \nu=1,2,\dots,m.$$

Moreover we may assume that $f-\tilde{u}$ like $f-\hat{u}$ has an alternant of length $(n-1)+\ell(\hat{u})+1$ only. Specifically, let $\tilde{u}$ be a local best approximation close to $\hat{u}$ in the set

$$W=\cup V_n(t_1,t_2,\dots,t_m)$$

where the union is taken over

$$\begin{aligned} &|t_i-\tau|\leqslant\delta, \qquad i=1,2,\dots,m-1,\\ &|t_m-\tau|\leqslant\delta_o \end{aligned} \tag{9.5}$$

and $0<\delta<\delta_o$. Then $\tilde{u}$ is also locally optimal in some $V_n(\tilde{t}_1,\tilde{t}_2,\dots,\tilde{t}_m)$. Because of the length of the alternant, $\tilde{u}$ cannot be a local solution of a problem where $\tilde{t}_m$ is considered a free parameter. Hence, we have $\tilde{t}_m=\tau+\delta_o$ or $\tilde{t}_m=-\delta_o$. We conclude that we can connect all elements of W close to $\hat{u}$ with some $\tilde{u}\in V_n(\tau+\delta_o)\cup V_n(\tau-\delta_o)$ by an arc running below the level $\|f-\hat{u}\|$. If $\delta$ is sufficiently small, $\tilde{u}$ is unique by

Lemma 6.4. Because of $u_r \to u$ one has $u_r \in W$ for sufficiently large r. Therefore we obtain the maximal component which contains $u_r$, by applying the standard construction with the additional characteristic number $\tau+\delta$ or $\tau-\delta$. This proves the completeness for Case 2.

After these preparations we are ready to prove the main theorem.

Theorem 9.3. If the kernel $\gamma$ is extended sign regular of order 2n, then

$$c_n < n! \tag{9.6}$$

Proof. Since $V_1 = V_1^o$ is a uniqueness set, $c_1 = 1$ is clear. Assume that $c_{n-1} \leq (n-1)!$. By Lemma 7.3 there is a function $f \in C(I)$ which is not exceptional for $V_{n-1}$ and which has $c_n$ or (at least) $n!+1$ local solutions in $V_n$. Let $T_1$ be a compact interval which contains all the spectra. The Assumption 5.1(2) holds for the restriction of $\gamma$ to $T_1 \times I$. Now it follows from the completeness of the standard construction that f has at most $n \cdot c_{n-1}$ local solutions in $V_n$. Hence, $c_n \leq n \cdot c_{n-1}$. □

The bound (9.6) seems to be too pessimistic. But recently, R. Verfürth [7] established a lower bound

$$c_n \geq \frac{1}{4} \cdot n^{-1/2} \cdot 2^{\frac{2}{3}n} .$$

Therefore $c_n$ increases faster than any polynomial when $n \to \infty$.

REFERENCES

[1] R. B. Barrar and H. L. Loeb, On the continuity of the nonlinear Tschebyscheff operator. *Pacific J. Math. 32*, 593-601 (1970).

[2] D. Braess, Chebyshev approximation by γ-polynomials. *J. Approx. Theory*. Part 1: 9,20-43(1973), part II: 11,16-37 (1974), part III: 24,119-145(1978).

[3] D. Braess, Kritische Punkte bei der nichtlinearen Tschebyscheff-Approximation, *Math. Z. 132*, 327-341 (1973).

[4] C.R.Hobby and J.R.Rice, Approximation from a curve of functions. Arch. Rational *Mech.Anal.27*,91-106 (1967).

[5] S. Karlin, *Total Positivity*. Stanford Univ. Press, 1968.

[6] E. Schmidt, Zur Kompaktheit bei Exponential summen. *J. Approx. Theory 3*, 445-454 (1970).

[7] R. Verfürth, to be published.

[8] J. M. Wolfe, On the unicity of nonlinear approximation in smooth spaces, *J. Approx, Theory 12*,165-181 (1974).

[9] D. Wulbert, Uniqueness and differential characterization of approximation from manifolds of functions, *Amer. J. Math. 93*, 350-366 (1971).

# SIMULTANEOUS APPROXIMATION AND RESTRICTED CHEBYSHEV CENTERS IN FUNCTION SPACES

Carlo Franchetti and E. W. Cheney*

Department of Mathematics
University of Texas
Austin, Texas

## 1. INTRODUCTION

Let X be a Banach space, and Y a subset of X (often a closed linear subspace). In the classical study of best approximation, one defines, for each $x \in X$, the *distance* from x to Y:

$$d(x,Y) = \inf_{y \in Y} \|x-y\|$$

The first important question is whether this infimum is *attained* for each x. If $y \in Y$ and $\|x-y\| = d(x,Y)$, we say that y is a *best approximation* to x in Y. If at least one best approximation exists for every $x \in X$, the subset Y is said to be *proximinal.*

If a *set* of elements S is given in X, it is natural to approximate all of the elements of S simultaneously by a single element in Y. This type of problem arises when a function being approximated is not known precisely, but is only known to belong to a set. Also, the approximation of a function of two variables by a function of one variable leads to the approximation of a *set* of functions. Thus, for example,

$$\sup_{s,t} |x(s,t) - y(t)| = \sup_{s} \|x_s - y\|_T$$

---

*The first author was supported by a NATO Senior Fellowship administered through the CNR (National Research Council of Italy). The second author was supported in part by the U.S. Army Research Office under Grant DAAG29-77-G-0215.




ISBN 0-12-780650-4

where $x_s$ denotes a cross-section of $x$ defined by $x_s(t) = x(s,t)$, and the norm with subscript $T$ is the supremum norm over $T$.

Returning to the general problem, one is led to define the *radius* (or *Chebyshev radius*) of a set $S$ with respect to $Y$ by the equation

$$r_Y(s) = \inf_{y \in Y} \sup_{x \in S} \|x-y\|$$

In approximating simultaneously all elements of $S$ by a single element of $Y$, we try to identify elements of the set

$$E_Y(S) = \{y \in Y: \sup_{x \in S} \|x-y\| = r_Y(S)\}$$

The elements of $E_Y(S)$ are solutions to the simultaneous approximation problem. The set $E_Y(S)$ is called the *restricted center* of $S$ (relative to $Y$). These concepts were introduced by Garkavi [3] and have been investigated by a number of authors. See, for example, [1-4, 6, 8, 10-12, 15, 16, 18-20]. In the case $Y = X$, we write $E(S)$ in place of $E_X(S)$. This is called the *Chebyshev center*.

The elements of $E_Y(S)$ are also called "simultaneous approximants of $S$," or "global approximants of $S$," or "restricted centers of $S$." A basic problem, as in classical approximation, is to determine whether $E_Y(S)$ is nonempty. If $E_Y(S)$ is nonempty for every bounded set $S$ in $X$, we shall say that "$Y$ is proximinal for the global approximation of bounded sets." A similar terminology is used when attention is restricted to the *compact* subsets of $X$.

This article is partly expository. In proving the new results which we wanted, it was necessary to develop some techniques which then turned out to be useful in giving shorter proofs of known results and in some cases improving them.

Some of our results, while technically new, are easy extensions of known results. In many cases the extension weakens the assumptions made on the space $T$ in $C(T)$.

Among the new results are

(1) a formula for the restricted Chebyshev radius (Theorem 3.11);

(2) a useful lemma concerning proximal subspaces of finite-codimension (Lemma 2.3);

(3) a theorem asserting that Chebyshev centers in $C(T)$ are non-empty, for all topological spaces $T$ (Theorem 3.1);

(4) a theorem on proximinality of extended order intervals in $C(T)$, when $T$ is normal (Theorem 3.3);

(5) a theorem on proximinality of order intervals in any Banach lattice (Lemma 3.6);

(6) a measure-theoretic lemma having applications in other situations (Lemma 4.3);

(7) a new proof of an important theorem of Garkavi on proximinality in the global approximation problem (Theorem 4.1);

(8) a duality theorem for global approximation (Theorem 2.2);

(9) a theorem on the preservation of proximinality under isomorphism (Theorem 4.8).

## 2. SIMULTANEOUS APPROXIMATION IN GENERAL SPACES

In this section we collect a few results that pertain to arbitrary Banach spaces. First we show that the problem of simultaneous approximation can always be reduced to a problem of approximating a single element by an appropriate subset in an appropriate space.

Suppose that $S$ is a bounded set in a Banach space $X$, and that $Y$ is a subset of $X$. The space $C(S,X)$ is defined as the set of all bounded continuous maps $f$ from $S$ into $X$, normed by writing

$$\|f\| = \sup_{s\in S} \|f(s)\|_X \qquad f \in C(S,X)$$

For each $y \in Y$ we define $\bar{y} \in C(S,X)$ by the rule $\bar{y}(s) = y$. The set $\bar{Y}$ of all such functions $\bar{y}$ is a subset of $C(S,X)$ and an isometric copy of $Y$. Finally, the identity function, $e$, is defined in $C(S,X)$ by the rule $e(s) = s$. Then $\|e-\bar{y}\| = \sup_{s\in S} \|s-y\|$. Therefore, if we find the best approximation of $e$ in the subset $\bar{Y}$ we will have obtained the best *global* approximation of the set $S$ in the subset $Y$. To summarize:

2.1 Theorem. *The problem of globally approximating a set* $S$ *in a Banach space* $X$ *by the elements of a subset* $Y$ *is equivalent to the ordinary approximation of the identity function by functions* $\bar{y}(s) = y$, $y \in Y$, *within the Banach space* $C(S,X)$.

2.1a Theorem. *Let* $S$ *be a compact set and* $Y$ *a subspace in a Banach space* $X$. *In order that an element* $y_0$ *of* $Y$ *belong to the restricted center* $E_Y(S)$ *it is necessary and sufficient that for each* $y \in Y$ *there exist a pair* $(\phi,a)$ *with* $a \in S$, $\phi$ *an extreme point of the unit cell in* $X^*$, $\phi(y_0-y) \geq 0$ *and* $\phi(a-y_0) = \sup_{s\in S} \|s-y_0\|$.

Proof. This is a consequence of Theorem 2.1 and two results of Ivan Singer, namely, Lemma 1.7, page 197, and Theorem 1.13, page 62 in [14].∎

2.2 Duality Theorem. *Let* $A$ *be a totally bounded subset, and* $Y$ *a subspace in a normed space* $X$. *Then*

$$\inf_{y\in Y} \sup_{a\in A} \|a-y\| = \max_{f\in Y^\perp} \inf_{y\in f^{-1}(0)} \sup_{a\in A} \|a-y\|$$

Proof. Let $\varepsilon$ be an arbitrary positive number. Since $A$ is totally bounded, there is a finite subset $\{x_1,\ldots,x_n\} \subset A$ such that the $\varepsilon$-spheres centered at $x_1,\ldots,x_n$ cover $A$. Then

$$\sup_{a\in A} \|a-y\| \le \varepsilon + \max_{1\le i\le n} \|x_i - y\| \le \varepsilon + \sup_{a\in A} \|a-y\| \tag{1}$$

Let $X^n$ be the n-fold direct sum of $X$, normed by putting

$$\|(u_1,u_2,\ldots,u_n)\| = \max_{1\le i\le n} \|u_i\|$$

In $X^n$, define the subspace $Y_0 = \{(y,y,\ldots,y) : y \in Y\}$. Then $\inf_{y\in Y} \max_{1\le i\le n} \|x_i - y\| = d[(x_1,\ldots,x_n),Y_0]$. By the Hahn-Banach Theorem

$$d[(x_1,\ldots,x_n),Y_0] = \max\{\Phi(x_1,\ldots,x_n) : \Phi \in Y_0^{\perp}, \|\Phi\| = 1\}.$$

It is easily seen that $\Phi \in (X^n)^*$ if and only if there exist $\phi_1,\ldots,\phi_n \in X^*$ such that $\Phi(x_1,\ldots,x_n) = \sum \phi_i(x_i)$ and $\|\Phi\| = \sum\|\phi_i\|$. Hence

$$\inf_{y\in Y} \max_{1\le i\le n} \|x_i - y\| = \max \left\{ \sum_{i=1}^n \phi_i(x_i) : \phi_i \in X^*, \sum_{i=1}^n \phi_i \in Y^{\perp}, \sum_{i=1}^n \|\phi_i\| \right\}. \tag{2}$$

A particular case of this is when $Y = f^{-1}(0)$ for some $f \in X^*$:

$$\inf_{y\in f^{-1}(0)} \max_{1\le i\le n} \|x_i - y_i\|$$
$$= \max \left\{ \sum \phi_i(x_i) : \phi_i \in X^*, \sum \phi_i \in [f^{-1}(0)]^{\perp}, \sum \|\phi_i\| = 1 \right\}. \tag{3}$$

We can write (2) in the alternative form

$$\inf_{y\in Y} \max_{i} \|x_i - y\|$$
$$= \max_{f\in Y^{\perp}} \sup \left\{ \sum \phi_i(x_i) : \phi_i \in X^*, \sum \phi_i = f, \sum \|\phi_i\| = 1 \right\} \tag{4}$$

For any $g \in X^*$, the conditions

(a) $g \in Y^{\perp}$

(b) $\exists f \in Y^{\perp}$ such that $g \in [f^{-1}(0)]^{\perp}$

are equivalent. Hence, from (3) and (4) we obtain the equation

$$\inf_{y\in Y} \max_{i} \|x_i - y\| = \max_{x\in Y^{\perp}} \inf_{y\in f^{-1}(0)} \max_{i} \|x_i - y\| .$$

The inequality (1) shows that $\max\|x_i - y\|$ can be replaced by $\sup\|a-y\|$ with an error of at most $\varepsilon$. Since $\varepsilon$ was arbitrary, this completes the proof. ■

2.3 Lemma. *Let* $V$ *be a bounded set in a Banach space* $X$. *Let* $Y$ *be a proximinal subspace of finite codimension in* $X$. *If* $d(V,Y) = \delta$ *then there is an* $x \in X$ *such that* $\|x\| = \delta$ *and* $d(V-x,Y) = 0$.

Proof. By the Hahn-Banach Theorem

$$\delta = \inf_{v\in V} \inf_{y\in Y} \|v-y\| = \inf_{v\in V} \sup_{\phi\in\Sigma} \langle\phi,v\rangle$$

where $\Sigma$ denotes the unit cell in $Y^{\perp}$. Select $v_n \in V$ so that $d(v_n,Y) \to \delta$. For each $n$ let $\phi_n$ be an element of $\Sigma$ for which $\langle\phi_n,v_n\rangle = d(v_n,Y)$. Define $f_n \in (Y^{\perp})^*$ by the equation $\langle f_n,\phi\rangle = \langle\phi,v_n\rangle$ for all $\phi \in Y^{\perp}$. Since $|\langle f_n,\phi\rangle| \le \|v_n\| \, \|\phi\|$ and since $(Y^{\perp})^*$ is finite-dimensional, we can assume (by passing to a subsequence if necessary) that $f_n \to f \in (Y^{\perp})^*$. At the same time, we can assume that $\phi_n \to \phi_0 \in \Sigma$. From the inequality $\langle\phi_n,v_n\rangle \ge \langle\phi_0,v_n\rangle = \langle f_n,\phi_0\rangle$ we conclude that $\delta \ge \langle f,\phi_0\rangle$. Since $Y$ is proximinal and factor-reflexive, a theorem of Garkavi [14, p.292] implies that there exists an $x \in X$ for which $\|x\| = \|f\|$ and $\langle\phi,x\rangle = \langle f,\phi\rangle$ for all $\phi \in Y^{\perp}$.

The inequality
$$\|f\| \ge \langle\phi_n,f\rangle = \langle\phi_n,f_n\rangle - \langle\phi_n,f_n-f\rangle$$
$$\ge \langle\phi_n,v_n\rangle - \|\phi_n\| \, \|f_n-f\|$$

shows that $\|f\| \ge \langle\phi_0,f\rangle \ge \delta$. On the other hand, for $\phi \in \Sigma$ we have $\langle\phi,f_n\rangle = \langle\phi,v_n\rangle \le \langle\phi_n,v_n\rangle$. Hence $\langle\phi,f\rangle \le \delta$, and $\|f\| \le \delta$. Thus we have $\|x\| = \|f\| = \delta$.

To complete the proof, we show that $d(V-x,Y) = 0$. For any $\phi \in \Sigma$ we have $\langle\phi,v_n-x\rangle = \langle f_n,\phi\rangle - \langle\phi,x\rangle = \langle f,\phi\rangle + \langle f_n-f,\phi\rangle - \langle\phi,x\rangle = \langle f_n-f,\phi\rangle \le \|f_n-f\|$. It follows that $d(V-x,Y) \le d(v_n-x,Y) \le \|f_n-f\| \to 0$. ■

2.4 Lemma. *Let* $K$ *be a convex set and* $Y$ *a (closed) subspace in a normed space* $X$. *Then*

$$d(K,Y) = \max\{d(K,\ker(f)) : f \in Y^{\perp}, \|f\| = 1\} .$$

Proof. If $f \in Y$ and $\|f\| = 1$, then $Y \subset \ker(f)$. Hence for any $x$, $d(x,Y) \geq d(x,\ker(f))$. Taking infima, we get $d(K,Y) \geq d(K,\ker(f))$. Taking a supremum, we get $d(K,Y) \geq \sup_f d(K,\ker(f))$. If $d(K,Y) = 0$, this proves the theorem. To prove the reverse inequality we can assume $r \equiv d(K,Y) > 0$. We will enhibit an $f \in Y^{\perp}$ such that $\|f\| = 1$ and $d(K,\ker(f)) \geq d(K,Y)$. Put $r = d(K,Y)$. Let $S_r$ be the open ball of radius $r$ at $0$. Then $K + S_r$ is a convex set with nonempty interior which is disjoint from $Y$. By the Tukey Separation Theorem [9, p.118], there exists a nonzero $f \in X^*$ such that $f(y) \leq f(x+z)$ whenever $y \in Y$, $x \in K$, and $\|z\| < r$. We can assume $\|f\| = 1$. Taking $y = 0$ we get $f(x+z) \geq 0$ whence $f(x) \geq \sup_{\|z\|<r} f(z) = r$. If $v \in \ker(f)$ then $\|x-v\| \geq f(x-v) = f(x) \geq r$. Hence $d(K,\ker(f)) \geq r$. Finally, for a fixed $x_0 \in K$ and any $y \in Y$ we have $f(y) \leq f(x_0)$. This implies that $f \in Y^{\perp}$.■

## 3. SIMULTANEOUS APPROXIMATION OF BOUNDED SETS IN C(T) BY ARBITRARY SUBSPACES

We begin by considering unrestricted centers. Here $T$ will denote an arbitrary topological space, and $C(T)$ will be the Banach space of all bounded, real-valued, continuous functions defined on $T$. The norm in $C(T)$ is the usual supremum norm.

3.1 Theorem. *Let* $T$ *be any topological space, and let* $A$ *be any bounded subset of* $C(T)$. *Then the Chebyshev center of* $A$ *is nonempty.*

Proof. The space $C(T)$ is an abstract M-space with unit. By Kakutani's Theorem [9][13] there exists a *compact* Hausdorff space $T'$ and a (linear) isometry $i$ of $C(T)$ onto $C(T')$. It is easy to verify that for any subset $Y$ in $C(T)$, the center $E_Y(A)$ obeys the equation

$$iE_Y(A) = E_{iY}(iA) .$$

This equation reduces the proof to the special case of a *compact* topological space $T'$. The theorem has been proved in this case by Zamjatin [20]. See also [6]. ∎

Remarks. As the proof shows, the Chebyshev center of a bounded set in an abstract M-space with unit is always nonempty. If $T$ is completely regular, $T'$ is the Stone-Cech compactification of $T$.

Now we recall the description of Chebyshev centers in $C(T)$ that has been given by Holmes [6] and others. Let $A$ be any bounded subset of $C(T)$. The *upper* and *lower envelopes* of $A$ are the functions

$$A^+(t) = \inf_{\mathcal{N}} \sup_{s\in\mathcal{N}} \sup_{a\in A} a(s)$$

$$A^-(t) = \sup_{\mathcal{N}} \inf_{s\in\mathcal{N}} \inf_{a\in A} a(s)$$

where $\mathcal{N}$ runs over all the neighborhoods of $t$ in $T$. It is known that $A^- \le A$, $A^-$ is lower semicontinuous, and $A^+$ is upper semicontinuous. Define also

$$w = \tfrac{1}{2}(A^+ - A^-), \quad b = \|w\| - w, \quad c = \tfrac{1}{2}(A^+ + A^-) .$$

Then $w$, $w+c$, $w-c$, and $c-b$ are upper semicontinuous, while $c+b$ is lower semicontinuous. All these functions are bounded, inasmuch as $A$ is a bounded set. By the results of Laurent and Tuan [10] and Diaz and McLaughlin [2],

$$r_Y(A) = \inf_{y\in Y} \sup_{a\in A} \|a-y\| = \inf_{y\in Y} \|w+|c-y|\| = \inf_{y\in Y} \max\{\|A^+-y\|, \|A^--y\|\} .$$

3.2 Lemma [6,p.186]. *The Chebyshev center of a bounded set* $A$ *in* $C(T)$ *is an "extended" order interval:*

$$E(A) = \{x \in C(T) : |x-c| \le b\} .$$

*Furthermore,* $\|w\| = r(A)$.

We refer to the set as an *extended* order interval because the endpoints $c-b$ and $c+b$ do not necessarily belong to $C(T)$.

3.3 Theorem. *Let* $u$ *be a bounded and upper semicontinuous function on a normal space* $T$. *Let* $v$ *be bounded and lower semicontinuous. Assume that* $u \le v$. *Denoting the space of bounded functions on* $T$ *by* $B(T)$, *let*

$$F = \{x \in B(T) : u \le x \le v\}, \quad F_0 = F \cap C(T).$$

*Then* $F_0$ *is nonempty and proximinal in* $C(T)$, *and for each* $y \in C(T)$, $d(y,F_0) = d(y,F)$.

Proof. Given $y \in C(T)$ we define $z = (y \vee u) \wedge v$. Then $z$ is bounded, and the number $\delta = \|y-z\|$ is well-defined. If $x \in F$ then $|y-x| \ge |y-z|$. Indeed, at points $t$ where $y(t) < u(t)$, we have $x(t) - y(t) \ge u(t) - y(t) = z(t) - y(t) > 0$. At points where $u(t) \le y(t) \le v(t)$, we have $y(t) = z(t)$. At points where $y(t) > v(t)$ we have $y(t) - x(t) \ge y(t) - v(t) = y(t) - z(t) > 0$. It follows that $\|y-x\| \ge \|y-z\| = \delta$ and that

$$d(y,F_0) \ge d(y,F) \ge \delta$$

In order to complete the proof, we use the Hahn-Tong Theorem [13,p.100 and 17] to establish the existence of an $x \in F_0$ such that $\|y-x\| = \delta$. Such an $x$ must be a continuous function satisfying $|x-y| \le \delta$ and $u \le x \le v$. Equivalently,

$$u^* = u \vee (y-\delta) \le x \le v \wedge (y+\delta) = v^* .$$

Since $u^*$ is upper semicontinuous and $v^*$ is lower semicontinuous, the Hahn-Tong Theorem will apply if $u^* \le v^*$. For this, it suffices to prove that $u \le y + \delta$ and that $y - \delta \le v$. This follows from the inequalities $u-y \le z-y \le \delta$ and $y-v \le y-z \le \delta$. ∎

3.4 Corollary. *If* $T$ *is any topological space, then the Chebyshev center of any bounded subset of* $C(T)$ *is proximinal. The same is true if* $C(T)$ *is replaced by any abstract* M*-space with unit.*

Proof. Since the first assertion is a special case of the second, we prove only the latter. Let $A$ be a bounded set in an abstract M-space $X$ with unit. By Kakutani's Theorem there is an isometry $i : X \to C(S)$, for some compact space $S$. Then $iA$ is bounded in $C(S)$. By Lemma 3.2, $E(iA)$ is an order interval in $C(S)$. (Its endpoints are continuous.) By Theorem 3.3, $E(iA)$ is proximinal in $C(S)$. Hence $i^{-1}E(iA)$ is proximinal in $X$. As in the proof of Theorem 3.1, $i^{-1}E(iA) = E(A)$. ■

In a lattice, $L$, an *order interval* is any set of the form $[a,b] = \{x : a \le x \le b\}$, where $x,a,b \in L$.

3.5 Lemma. *In a Banach lattice, every order interval is proximinal. One best approximation to* $c$ *from* $[a,b]$ *is* $(b \wedge c) \vee a$, *and the minimum distance is* $\|0 \wedge (a-c) \vee (c-b)\|$.

Proof. If $a \le x \le b$, then $|x-c| \ge 0$, $|x-c| \ge x-c \ge a-c$, and $|x-c| \ge c-x \ge c-b$. Thus $|x-c| \ge 0 \vee (a-c) \vee (c-b) \equiv m \ge 0$ and $\|x-c\| \ge \|m\|$. Now put $u = (b \wedge c) \vee a$. Then $u - c = [(b-c) \wedge 0] \vee (a-c)$. Since $a-c \le m$ and $(b-c) \wedge 0 \le 0 \le m$, we have $u - c \le m$. Also, $c - u = -(u-c) = [(c-b) \vee 0] \wedge (c-a) \le (c-b) \vee 0 \le m$. This establishes that $|c-u| \le m$ and $\|c-u\| \le \|m\|$. Since $a \le u \le b$, this shows that $u$ is a best approximation to $c$ and that $\mathrm{dist}(c,[a,b]) = \|m\|$. ■

3.6 Lemma. *In the space* $C(T)$, *or in any abstract M-space with unit, if* $k \ge b \ge 0$ *and* $k$ *is a constant, then*

$$\inf_{|x| \le b} \|z-x\| = \max\{0, \|k+|z|-b\| - k\}$$

Proof. By Lemma 3.5,

$$\rho = \inf_{|x| \le b} \|z-x\| = \|0 \vee (-b-z) \vee (z-b)\| = \|0 \vee (-b+|z|)\|$$

Since $k \geq 0$ and $k + |z| - b \geq 0$, we have

$$\rho = -k + \| k + [0 \vee (-b+|z|)] \| = -k + \| k \vee (k + |z| - b) \|$$
$$= -k + (\|k\| \vee \|k+|z|-b\|) = \max\{0, \|k+|z|-b\| - k\} .$$

Note that Kakutani's Theorem is useful here to conclude that

$\|k+x\| = k + \|x\|$ when $k \geq x \geq 0$.■

The following theorem has been stated for paracompact $T$ by Ward and Smith [15]. An extension to arbitrary $T$ is therefore possible by means of the Kakutani Theorem. We prefer to give an complete proof, however, emphasizing elementary lattice techniques.

3.7 Theorem. *Let* $T$ *be a topological space,* $Y$ *a subset of* $C(T)$, *and* $A$ *a bounded subset of* $C(T)$. *Then*

$$r_Y(A) = r(A) + d(Y,E(A))$$

*If* $Y$ *is proximinal, then in order that the restricted center* $E_Y(A)$ *be non-empty it is necessary and sufficient that the function* $z \to d(z,Y)$ *attain its infimum on* $E(A)$.

Proof. Since the statement of the theorem is in Banach-space terms, it suffices to prove it for compact $T$ and then use the Kakutani Theorem. By the Diaz-McLaughlin result cited earlier,

$$r_Y(A) = \inf_{y \in Y} \| w + |c-y| \|$$

In Lemma 3.6, take $z = c-y$, $k = \|w\|$, and $b = \|w\| - w$ to get

$$\inf_{|w| \leq b} \|c-y-x\| = \max\{0, \|w+|c-y|\| - \|w\|\} \qquad (1)$$

Since $w \geq 0$

$$\|w\| + \inf_{y \in Y} \inf_{|x| \leq b} \|c-y-x\|$$
$$= \max\{\|w\|, \inf_{y \in Y} \|w+|c-y|\|\} = \inf_{y \in Y} \|w+|c-y|\| .$$

By Lemma 3.2, $\|w\| = r(A)$, and $E(A) = \{c-x : |x| \leq b\}$. Hence the formula in the theorem is established. Observe that no use of extension or selection theorem is required.

In order to prove the second assertion of the theorem, suppose first that $x_0$ is a point of $E(A)$ for which $d(X,Y)$ is a minimum. Let $y_0$ be a best approximation to $x_0$ in $Y$. Then $y_0 \in E_Y(A)$ because for all $a \in A$ and $x \in E(A)$,

$$\|y_0-a\| \le \|y_0-x_0\| + \|x_0-a\| \le d(x_0,Y) + r(A)$$

$$\le d(E(A),Y) + r(A) = r_Y(A)$$

Now suppose that $y_0 \in E_Y(A)$. By equation (1),

$$\|w\| + \inf_{|x|\le b} \|c-y_0-x\| = \|w+|c-y_0|\,\| = r_Y(A)$$

By Theorem 3.3, the infimum in this equation is attained at some point $x_0$. Then, since $\|w\| = r(A)$ by Lemma 3.2,

$$r(A) + \|c-y_0-x_0\| = r_Y(A)$$

By the equation in the statement of this theorem, we have

$$\|c-y_0-x_0\| = d(E(A),Y).$$

Hence $c-x_0$ minimizes the expression $d(z,Y)$. ■

3.8 Lemma. *If* $f \in C(T)^*$, $c \in C(T)$, $b \in C(T)$, *and* $b \ge 0$, *then*

$$\inf_{|x|\le b} |f(c-x)| = \max\{0, |f(c)| - |f|(b)\}.$$

Proof. By the Bishop-Phelps Theorem [7,p.169], every functional is the limit of functionals which attain their supremum on the unit cell. A continuity argument shows that it suffices to prove the lemma for functionals having this additional property. If $f$ attains its supremum on the unit cell, then the supports of $f^+$ and $f^-$ are (closed) disjoint sets. Using the Tietze Theorem, we find an element $u \in C(T)$ such that $u(t) = b(t)$ on $\mathrm{supp}(f^+)$ and $u(t) = -b(t)$ on $\mathrm{supp}(f^-)$. Put $z = (u \wedge b) \vee (-b)$. Then $z(t) = u(t)$ on $\mathrm{supp}(f)$ and $|z| \le b$. Hence $f(z) = |f|(b)$. There is no loss of generality in assuming $f(c) \ge 0$.

If $|x| \le b$, then $|f(c)| - |f|(b) \le f(c) - f(x) \le |f(c-x)|$. Hence $|f(c-x)| \ge \max\{0, |f(c)| - |f|(c)\} = r$. If $f(c) > |f|(b)$ then $f(c-z) = |f(c)| - |f|(b) = r$. If $f(c) \le |f|(b)$ there is a $\lambda \in [0,1]$ such that $|f(c-\lambda z)| = 0 = r$.∎

3.9 Theorem. *Let* T *be a compact space,* Y *a subspace of* C(T), *and* $I = \{x : |x-u| \le v\}$, *with* $u,v \in C(T)$ *and* $v \ge 0$. *Then*

$$d(Y,I) = \max_{\substack{f\in Y^\perp \\ \|f\|=1}} \max\{0, |f(u)| - |f|(v)\}$$

Proof. Use Lemmas 2.4 and 3.8.

3.10 Theorem. *Let* T *be a compact space,* Y *a subspace of* C(T) *and* A *a bounded subset of* C(T) *for which the upper and lower envelopes* $A^+$ *and* $A^-$ *are continuous. Then*

$$r_Y(A) = \inf_{y\in Y} \sup_{a\in A} \|a-y\| = \max_{\substack{f\in Y^\perp \\ \|f\|=1}} \max\{\|w\|, |f(c)| + |f|(w)\}$$

*where* $w = \frac{1}{2}(A^+ - A^-)$ *and* $c = \frac{1}{2}(A^+ + A^-)$.

Proof. By Theorem 3.7, Lemma 3.2, and Theorem 3.9,

$$r_Y(A) = r(A) + d(Y,E(A)) = \|w\| + \max_{\substack{f\in Y^\perp \\ \|f\|=1}} \max\{0, |f(c)| - |f|(b)\} =$$

$$\max_{\substack{f\in Y^\perp \\ \|f\|=1}} \max\{\|w\|, |f(c)| + |f|(w)\}$$

Here we wrote $\|w\| - |f|(b) = |f|(\|w\|-b) = |f|(w)$.∎

3.11 Characterization Theorem. *Let* A *be a compact set and* Y *a subspace in* C(T), T *arbitrary. A point* $y_0$ *of* Y *belongs to the restricted center* $E_Y(A)$ *if and only if either*

(1) $c + w - \|w\| \leq y_0 \leq c - w + \|w\|$, *or*

(2) *there is an* $f \in Y^{\perp}$ *such that* $\|f\| = 1$ *and*

$$|f(c)| + |f|(w) = \| w + |c-y_0| \|$$

Proof. Assume that $y_0 \in E_Y(A)$. If $r_Y(A) = r(A)$, then by the definition of $E_Y(A)$, $y_0 \in E(A)$. By Lemma 3.2, Condition (1) holds. If $r_Y(A) > r(A)$, then by Theorem 3.10 there is an $f \in Y^{\perp}$ such that $\|f\| = 1$ and

$$r_Y(A) = \max\{\|w\|, |f(c)| + |f|(w)\} = \| w + |c-y_0| \|.$$

Since $r_Y(A) > r(A) = \|w\|$ (by hypothesis and by Lemma 3.2), condition (2) holds.

For the converse, suppose first that condition (1) is true. Then $y_0 \in E(A)$ by Lemma 3.2. Hence $r_Y(A) = r(A)$ and $y_0 \in E_Y(A)$.

If condition (2) is true, then by Theorem 3.10,

$$r_Y(A) \geq \max_{\substack{g \in Y^{\perp} \\ \|g\|=1}} \{|g(c)| + |g|(w)\} \geq |f(c)| + |f|(w) = \| w + |c-y_0| \| \geq r_Y(A).$$

Hence $y_0 \in E_Y(A)$. ■

## 4. SIMULTANEOUS APPROXIMATION OF A COMPACTUM IN C(T) BY SUBSPACES OF FINITE CODIMENSION

In this section we give a new proof of the following theorem, due essentially to Garkavi [5].

4.1 Theorem. *Let* T *be an arbitrary topological space,* C(T) *the space of bounded continuous real functions on* T, *and* Y *a subspace of finite codimension in* C(T). *In order that* Y *be proximinal for the global approximation of compact subsets of* C(T) *it is necessary and sufficient that it be proximinal in the ordinary sense.*

The theorem is true for the class of sets A such that $A^+$ and $A^-$ are continuous. Examples are balls, order intervals, and compact sets.

Theorem 4.1 stands in contrast to the following result of Zamjatin concerning global approximation of bounded sets.

4.2 Theorem. *Let* Y *be a subspace of finite codimension in* C(T), *where* T *is an arbitrary topological space. In order that* Y *be proximinal for the global approximation of bounded subsets of* C(T) *it is necessary and sufficient that each* $f \in Y^{\perp}$ *be a (finite) linear combination of extreme points of the unit cell in* $C(T)^*$.

Proof. In the case that T is compact, this theorem has been proved by Zamjatin [19]. Since the theorem is stated in purely Banach space terms, the general case follows from Kakutani's Theorem. ■

Remark. Zamjatin's original formulation of this theorem is in terms of point-evaluation functionals, and does *not* generalize to all non-compact T. This was pointed out by Smith and Ward [15].

Corollary. *Let* T *be a locally compact space, and* $C_0(T)$ *the space of continuous functions on* T *which vanish at infinity. Then every bounded subset of* $C_0(T)$ *has a nonempty Chebyshev center.*

Proof. Let S be the one-point compactification of T, $S = T \cup \{\infty\}$. Then $C_0(T) = \{x \in C(S) : x(\infty) = 0\}$. This is a hyperplace. By Theorem 4.2, the infimum of $\sup_{a \in A} \|a-y\|$ as y ranges over $C_0(T)$ is attained whenever A is a bounded subset of C(S). A fortiori, the infimum is attained when A is a bounded subset of $C_0(T)$. ■

In the remainder of this section, T will denote a compact Hausdorff space.

4.3. Lemma. *Let* f *and* g *be positive functionals on* C(T), *each absolutely continuous with respect to the other. Then in the norm topology,*

$$\lim_{\varepsilon \to 0^+} \varepsilon^{-1}\{|f-\varepsilon g| - f\} = -g.$$

Proof. Let the measures associated with f and g be $\mu$ and $\nu$, respectively. If $x \in C(T)$ and $x \geq 0$, then [9,p.239]

$$|f-\varepsilon g|(x) = \sup_{0 \leq y \leq x} (f-\varepsilon g)(y) = \sup_{0 \leq y \leq x} \left[\int y\,d\mu - \int \varepsilon y\,d\nu\right].$$

Since $\mu$ is absolutely continuous with respect to $\nu$, the Radon-Nikodym Theorem implies the existence of a positive function $z \in L_1(\nu)$ such that $\int y\,d\mu = \int yz\,d\nu$ for all y. Then

$$|f-\varepsilon g|(x) = \sup_{0 \leq y \leq x} \int y(z-\varepsilon)\,d\nu.$$

Now let $A = \{t \in T : z(t) > \varepsilon\}$ and $B = T\backslash A$. Since $z - \varepsilon \leq 0$ on B,

$$f(x) - \varepsilon g(x) \leq |f-\varepsilon g|(x) = \sup_{0 \leq y \leq x} \left[\int_A y(z-\varepsilon)\,d\nu + \int_B y(z-\varepsilon)\,d\nu\right] \leq$$

$$\sup_{0 \leq y \leq x} \int_A y(z-\varepsilon)\,d\nu = \int_A x(z-\varepsilon)\,d\nu =$$

$$\int x(z-\varepsilon)\,d\nu - \int_B x(z-\varepsilon)\,d\nu \leq f(x) - \varepsilon g(x) + \varepsilon \int_B x\,d\nu$$

It follows that

$$0 \leq \varepsilon^{-1}\{|f-\varepsilon g|(x) - f(x)\} + g(x) \leq \int_B x\,d\nu.$$

Consequently, the functional $\varepsilon^{-1}(|f-\varepsilon g|-f) + g$ is positive, and its norm is attained at the function $x = 1$. Thus

$$\|\varepsilon^{-1}(|f-\varepsilon g|-f) + g\| \leq \nu(B).$$

Finally, we note that

$$\lim_{\varepsilon \to 0^+} \nu(B) = \nu(z^{-1}(0)) = 0.$$

Indeed, the equation

$$\int x\,d\mu = \int xz\,d\nu = \int_{T\setminus z^{-1}(0)} xz\,d\nu$$

is valid for $x \in L_1(\mu)$, and shows that $\mu(z^{-1}(0)) = 0$. By the absolute continuity of $\nu$ with respect to $\mu$, we conclude that $\nu(z^{-1}(0)) = 0$. ∎

4.4 <u>Lemma</u>. *Let* f *and* g *be elements of* $C(T)^*$, *each absolutely continuous with respect to the other on the support of the other. Then in the norm topology*

$$\lim_{\varepsilon\to 0} \varepsilon^{-1}\{|f+\varepsilon g| - |f|\} = |g| - 2g_2^+ - 2g_1^-$$

*where* $g_2^+(x) = \int_{\text{supp}(f^-)} x(t)\,d\nu^+$, $\nu^+$ *is the measure associated with* $g^+$, $g_1^-(x) = \int_{\text{supp}(f^+)} x(t)\,d\nu^-$, *and* $\nu^-$ *is the measure associated with* $g^-$.

<u>Proof</u>. Denote by $F, F^+, F^-, G, G^+, G^-$ the supports of $f, f^+, f^-$, $g, g^+$, and $g^-$, respectively. We decompose g and f according to the chart

| | | | |
|---|---|---|---|
| $g^+ = g_0^+ + g_1^+ + g_2^+$ | $G^+\setminus F$ | $G^+ \cap F^+$ | $G^+ \cap F^-$ |
| $g^- = g_0^- + g_1^- + g_2^-$ | $G^-\setminus F$ | $G^- \cap F^+$ | $G^- \cap F^-$ |
| $f^+ = f_0^+ + f_1^+ + g_2^+$ | $F^+\setminus G$ | $F^+ \cap G^+$ | $F^+ \cap G^-$ |
| $f^- = f_0^- + f_1^- + f_2^-$ | $F^-\setminus G$ | $F^- \cap G^+$ | $F^- \cap G^-$ |

The meaning of the table is illustrated by

$$g_0^+(x) = \int_{G^+\setminus F} x(t)\,d\nu(t), \qquad g_1^-(x) = -\int_{G\cap F^+} x(t)\,d\nu(t)$$

where $\nu$ is the measure associated with the functional g. Now we observe that if $\varepsilon > 0$,

$$|f+\varepsilon g| = f_0^+ + f_0^- + f_1^+ + f_2^- + \varepsilon g_0^+ + \varepsilon g_0^- + \varepsilon g_1^+ + \varepsilon g_2^- + |-f_1^- + \varepsilon g_2^+| + |f_2^+ - \varepsilon g_1^-|.$$

It follows that

$$|f+\varepsilon g| - |f| = |f_1^- - \varepsilon g_2^+| - f_1^- + |f_2^+ - \varepsilon g_1^-| - f_2^+ +$$
$$\varepsilon(g_0^+ + g_0^- + g_1^+ + g_2^-)$$

Applying Lemma 4.3, we have

$$\lim_{\varepsilon\to 0^+} \varepsilon^{-1}\{|f+\varepsilon g| - |f|\} = -g_2^+ - g_1^- + (g_0^+ + g_0^- + g_1^+ + g_2^-) =$$
$$|g| - 2(g_2^+ + g_1^-)$$

The hypotheses of Lemma 4.3 require that $f_1^-$ and $g_2^+$ be absolutely continuous with respect to each other, and similarly for $f_2^+$ and $g_1^-$. This is true because $\mathrm{supp}(f_1^-) \subset G$, $\mathrm{supp}(f_2^+) \subset G$, $\mathrm{supp}(g_1^-) \subset F$, $\mathrm{supp}(g_2^+) \subset F$, $f$ is absolutely continuous with respect to $g$ on $G$, and $g$ is absolutely continuous with respect to $f$ on $F$. $\blacksquare$

The following theorem of Garkavi [14,p.302] will be needed in the sequel.

4.5. <u>Theorem</u>. *Let* $Y$ *be a subspace of finite codimension in* $C(T)$, $T$ *being compact. In order that* $Y$ *be proximinal the following condition (in three parts) is necessary and sufficient.*

(1) *For each* $f \in Y^\perp$, $\mathrm{supp}(f^+)$ *and* $\mathrm{supp}(f^-)$ *are disjoint from each other.*

(2) *For each pair* $f,g \in Y^\perp$, $\mathrm{supp}(f)\setminus \mathrm{supp}(g)$ *is closed.*

(3) *For each pair* $f,g \in Y^\perp$, $f$ *is absolutely continuous with respect to* $g$ *on* $\mathrm{supp}(g)$.

4.6 <u>Lemma</u>. *Let* $Y$ *be a proximinal subspace of finite codimension in* $C(T)$, $T$ *being an arbitrary topological space. In order that* $Y$ *be proximinal for the global approximation of compact sets it is sufficient that* $E_Y(A) \neq \emptyset$ *for each compact set* $A$ *satisfying* $d(Y,E(A)) = 0$.

Proof. Let $A$ be a compact set. By Lemma 2.3 there is an $x \in X$ such that $\|x\| = d(Y,E(A))$ and $d(E(A)-x,Y) = 0$. Since $E(A-x) = E(A) - x$ we have $d(E(A-x),Y) = 0$. If we assume the stated condition, there exists a point $y \in Y \cap E(A-x)$. Hence $y + x \in E(A)$ and $\sup_a \|y+x-a\| = r(A)$. Now $y \in E_Y(A)$ because $\|y-a\| \leq \|y+x-a\| + \|x\| \leq r(A) + d(Y,E(A)) = r_Y(A)$ for all $a \in A$, by Theorem 3.5.■

4.7 Lemma. *Let* $f$ *and* $g$ *be two elements of* $C(T)^*$ *such that the linear span of* $f$ *and* $g$ *satisfies Garkavi's conditions in Theorem 4.5. If* $c,b \in C(T)$, $b \geq 0$, $f(c) = |f|(b)$ *and* $(f+\lambda g)(c) \leq |f+\lambda g|(b)$ *for all real* $\lambda$, *then*

$$|g(c) + g_2(b) - g_1(b)| \leq |g_0|(b)$$

*Here the subscripts* 0,1,2 *indicate restriction of the integrals to* $\mathrm{supp}(g)\setminus \mathrm{supp}(f)$, $\mathrm{supp}(f^+)$, *and* $\mathrm{supp}(f^-)$, *respectively.*

Proof. For $\varepsilon > 0$ we have, by Lemma 4.4,

$$|f|(b) + \varepsilon g(c) = (f+\varepsilon g)(c) \leq |f+\varepsilon g|(b)$$

$$g(c) \leq \varepsilon^{-1}\{|f+\varepsilon g| - |f|\}(b)$$

$$g(c) \leq (|g| - 2g_2^+ - 2g_1^-)(b)$$

A similar argument using $-g$ produces the inequality

$$g(c) \geq (-|g| + 2g_2^- + 2g_1^+)(b)$$

The two inequalities thus obtained for $g(c)$ can now be combined as in the statement of the lemma.■

Proof of Theorem 4.1. Because of Kakutani's Theorem, it suffices to give a proof in the case that $T$ is compact. The proof is by induction on the codimension of the subspace, and only in the inductive step is the proof new.

Suppose that the theorem has been proved for subspaces of codimension $n$. Let $Y$ be a subspace of codimension $n+1$. We will use Lemma 4.6. Let $A$ be a compact set in $C(T)$ with $d(Y,E(A)) = 0$. By Lemma 3.2

$$E(A) = \{c-x : |x| \le b\}.$$

In order to prove that $Y \cap E(A)$ is nonempty we must find an $x$ such that $|x| \le b$ and $c-x \in Y$; i.e., $h(c-x) = 0$ for all $h \in Y^{\perp}$. Since $d(Y,E(A)) = 0$, we have $d(h^{-1}(0), E(A)) = 0$ for all $h \in H^{\perp}$. By Theorems 3.7 and 3.10,

$$r_{h^{-1}(0)}(A) = r(A) + d(h^{-1}(0), E(A)) = r(A) = \|w\| =$$

$$\max\{\|w\|, |h(c)| + |h|(w)\}, \quad (\|g\| = 1)$$

Consequently, $|h(c)| + |h|(w) \le \|w\| = |h|(w)$, $(\|h\| = 1)$, and $|h(c)| \le |h|(b)$.

If $c \in Y$ then $c \in Y \cap E(A)$ and we are finished. Otherwise, $c \notin Y$, and we put

$$B = \{\lambda \in \mathbb{R} : h(c) \le \lambda |h|(b) \quad \text{for all} \quad h \in Y^{\perp}\}$$

Let $\theta = \inf B$. An elementary argument shows that $0 \le \theta \le 1$ and that $h(c) \le |h|(\theta b)$ for all $h \in Y^{\perp}$. In the remainder of the proof we use $b$ in place of $\theta b$. Since $\theta \le 1$ the new order interval is contained in the original one. Select $f \in Y^{\perp}$ so that $\|f\| = 1$ and $f(c) = |f|(b)$. Let

$$S = \bigcup \{\text{supp}(h) : h \in Y^{\perp}\} = \bigcup_{i=1}^{n+1} \text{supp}(h_i), \ \{h_1,\ldots,h_{n+1}\} \text{ a basis for } Y^{\perp}.$$

Then $S$ is closed, and by the 3rd condition in Theorem 4.5, so is $S\backslash\text{supp}(f)$. Define

$$x(t) = \begin{cases} b(t) & \text{on} \quad \text{supp}(f^+) \\ -b(t) & \text{on} \quad \text{supp}(f^-) \\ 0 & \text{on} \quad S\backslash\text{supp}(f) \end{cases}$$

Then $|x| \le b$ and $f(x) = |f|(b)$. By the Tietze theorem and simple lattice operations $x$ can be extended to all of $T$, keeping the inequality $|x| \le b$.

Write $Y^{\perp} = [f] \oplus G$ where $G$ is of dimension $n$. Let $Y_0 = \{z \in C(T) : g(z) = 0$ for all $g \in G\}$. Then $Y \subset Y_0$. By Lemma 4.7, each $g \in G$ satisfies

$$|g(c) + g_2(b) - g_1(b)| \le |g_0|(b) \tag{1}$$

where the subscripts 0,1,2 denote the restrictions to $S\backslash supp(f)$, $supp(f^+)$, and $supp(f^-)$, respectively. Now define

$$c'(t) = \begin{cases} c(t) \\ c(t) - b(t) \\ c(t) + b(t) \end{cases} \qquad b'(t) = \begin{cases} b(t) & \text{on } S\backslash supp(f) \\ 0 & \text{on } supp(f^+) \\ 0 & \text{on } supp(f^-). \end{cases}$$

Then inequality (1) can be written simply as

$$g(c') \le |g|(b') \qquad (g \in G) \tag{2}$$

By Theorem 3.9, inequality (2) implies that the distance from $Y_0$ to the order interval $\{c' - x' : |x'| \le b'\}$ is 0. This order interval is $E(A')$ for an appropriate compact set $A'$. Now $Y_0$ is proximinal, and so by the induction hypothesis, it is proximinal for global approximation of all compact sets. By Lemma 4.6, there exists a point $c' - x'$ in $Y_0 \cap E(A')$, with $|x'| \le b'$.

We assert that $c - (x+x') \in Y \cap E(A)$. First note that $x'(t) = 0$ on $supp(f)$ because $b'(t) = 0$ there. Hence $f(x') = 0$ and $f(x+x') = f(x) = |f|(b) = f(c)$. Next, using the definitions of $x$, $c'$ and the fact that $c' - x' \in Y_0$, we have

$$g(x+x') = g(x) + g(x') = g_1(b) - g_2(b) + g(c') =$$

$$g_1(b) - g_2(b) + g(c) + g_2(b) - g_1(b) = g(c).$$

Thus $c - (x+x') \in Y$. Finally, we note that $|x+x'| \le b$, so that $c - (x+x') \in E(A)$. ■

A general question concerning proximinality in C(T) is this: If Y is a proximinal subspace and L is an isomorphism of C(T) onto C(T), is L[Y] proximinal? There are two obvious cases in which the answer is "yes". Namely, when L is an isometry or when Y is finite dimensional. Another affirmative case is given in the next theorem.

Theorem 4.8. *Let* Y *be a proximinal subspace of finite codimension in* C(T). *Let* L *be an isomorphism of the form* $Lx = x \circ h\beta$ *where* h *is a homeomorphism of* T *and* β *is a positive element of* C(T). *Then* L[Y] *is proximinal.*

Proof. Since Y is proximinal, $Y^{\perp}$ satisfies Garkavi's condition in Theorem 4.5. We want to show that $(LY)^{\perp}$ also satisfies these three conditions. The first step is to note that $L^{-1}x = (x/\beta) \circ h^{-1}$ and that

$$(LY)^{\perp} = \{\phi \circ L^{-1} : \phi \in Y^{\perp}\}. \tag{1}$$

A short calculation shows that

$$\text{supp}(\phi \circ L^{-1}) = h^{-1}[\text{supp}(\phi)]. \tag{2}$$

The second Garkavi condition on $(LY)^{\perp}$ follows from this equation.

Next, one proves that $(\phi \circ L^{-1})^{+} = \phi^{+} \circ L^{-1}$. This is easily done with the definition

$$\phi^{+}(x) = \sup\{\phi(z) : 0 \le z \le x\} \qquad (x \ge 0) \tag{3}$$

The same type of equation is true for $\phi^{-}$. From this the first Garkavi condition on $(LY)^{\perp}$ follows.

Last, we prove that if $\mu_i$ is the measure associated with the functional $\phi_i$, then the measure $\nu_i$ associated with $\phi_i \circ L^{-1}$ is given by

$$\nu_i(A) = \int_{h^{-1}[A]} [1/\beta(t)]d\mu_i \qquad (A \text{ is a Borel set}) \tag{4}$$

If $\phi_1, \phi_2 \in Y^{\perp}$, then by the 3rd Garkavi condition, $|\mu_1|(A) = 0$ when $A \subset \text{supp}(\phi_2)$ and $|\mu_2|(A) = 0$. Using equations (2) and (4), it follows readily that $\nu_1$ and $\nu_2$ also have this property. ■

REFERENCES

(1) P. Belobrov, "On the problem of the Chebyshev center of a set", *Izv. Vys. Ucheb. Zaved.* (1964), 3-9. (Russian)

(2) J. B. Diaz and H. W. McLaughlin, "Simultaneous approximation of a set of bounded real functions", *Math. Comp.* 23(1969), 538-584.

(3) A. L. Garkavi, "The best possible net and the best possible cross-section of a set in a normed space", *Izv. Akad. Nauk SSSR Ser. Mat.* 26(1962), 87-106. English transl., *Amer. Math. Soc. Transl.* (2), 39(1964), 111-132. MR25#429.

(4) A. L. Garkavi, "The Helly problem and best approximation in a space of continuous functions", *Izv. Akad. Nauk SSSR Ser. Math.* 31(1967), 641-656 = *Math. USSR Izv.* 1(1967), 623-638. MR36#590.

(5) A. L. Garkavi, "The conditional Chebyshev center of a compact set of continuous functions", *Mat. Zam.* 14(1973), 469-478. (Russian) = *Math. Notes of the USSR* (*1973*), 827-831.

(6) R. B. Holmes, "A Course on Optimization and Best Approximation", *Lecture Notes in Mathematics*, vol. 257, 1972. Springer-Verlag, New York.

(7) R. B. Holmes, *Geometric Functional Analysis and Its Applications*, Springer-Verlag, New York, 1975.

(8) M. I. Kadec and V. Zamjatin, "Chebyshev centers in the space C[a,b]", Teor. Funkcii Funkcional. Anal. Prilozen. Vyp. 7(1968), 20-26. (Russian) MR42#3480.

(9) J. L. Kelley, I. Namioka, *et al.*, *Linear Topological Spaces*, van Nostrand, Princeton, New Jersey, 1963. Reprinted, Springer-Verlag, New York.

(10) P. J. Laurent and P. D. Tuan, "Global approximation of a compact set by elements of a convex set in a normed space", *Numer. Math.* 15(1970), 137-150.

(11) J. M. Lambert and P. D. Milman, "Restricted Chebyshev centers of bounded subsets in arbitrary Banach spaces", *J. Approximation Theory* 26(1979), 71-78.

(12) J. Mach, "On the existence of best simultaneous approximation", *J. Approximation Theory* 25(1979), 258-265.

(13) Z. Semadeni, *Banach Spaces of Continuous Functions*, Polish Scientific Publishers, Warsaw, 1971.

(14) I. Singer, *Best Approximation in Normed Linear Spaces by Elements of Linear Subspaces*, Springer-Verlag, New York, 1970. MR38#3677 and 42#4937.

(15) Philip W. Smith and J. D. Ward, "Restricted Centers in $C(\Omega)$", *Proc. Amer. Math. Soc.* 48(1975), 165-172.

(16) Philip W. Smith and J. D. Ward, "Restricted centers in subalgebras of C(X)", *J. Approx. Theory* 15(1975), 54-59.

(17) H. Tong, "Some characterizations of normal and perfectly normal spaces", *Duke Math. J.* 19(1952), 289-292. MR14#304.

(18) Joseph D. Ward, "Chebyshev centers in spaces of continuous functions", *Pacific J. Math.* 52(1974), 283-287.

(19) V. M. Zamjatin, "Relative Cebysev centers in a space of continuous functions", *Dokl. Akad. Nauk SSSR* 209(1973), 1267-1270 = *Soviet Math. Dokl.* 14(1973), 610-614.

(20) V. M. Zamjatin, "Cebysev centers in the space C(S)", *First Scientific Conference of Young Scholars of the Akygei, 1971*, pp. 28-35. (Russian)

# QUELQUES PROPRIETES D'UNE FAMILLE D'OPERATEURS POSITIFS SUR DES ESPACES DE FUNCTIONS REELLES DEFINIES PRESQUE PARTOUT SUR [0, + ∞[

Chr. Coatmelec

Institut National des Sciences Appliquées
Rennes, Cedex, France

Dans cet article, on introduit une suite d'opérateurs linéaires positifs définis sur certains espaces de fonctions définies presque partout sur [0,+∞[. Ces opérateurs sont obtenus à partir des sommes de Bernstein-Poisson étudiees par Szasz [5] de la même manière que Durrmeyer [3] obtenait des sommes de Bernstein modifiées qui sont etudiees dans Derriennic [2].

## I, INTRODUCTION

Soit $L^\infty$ l'espace des fonctions réelles, définies presque partout sur [0, +∞[ et qui possedent sur [0,+∞[ un supremum essentiel fini. Nous posons pour $f \in L^\infty$:

$$\|f\|_\infty = \underset{t \in [0,+\infty[}{\text{ess. sup}} |f(t)|$$

Pour $1 \leq p < +\infty$, soit $L^p$ l'espace des fonctions réeles f telles que $|f|^p$ soit sommable sur [0,+∞[. Nous posons alors




ISBN 0-12-780650-4

$$\| f \|_p = \left( \int_0^{+\infty} |f(t)|^p \, dt \right)^{1/p}$$

Pour $f \in L^1$ nous introduisons $f_n = C_n f$ définie par

$$f_n(x) = C_n f(x) = \sum_{k=0}^{+\infty} \frac{n^{2k+1}}{(k!)^2} x^k e^{-nx} \int_0^{+\infty} t^k e^{-nt} f(t) dt$$

$$= \int_0^{+\infty} K_n(x,t) \, f(t) dt$$

Notons dès maintenant que l'opérateur $f \to C_n f$ a un sens pour des fonctions f n'appartenant pas à $L^1$. En particulier si $\Pi$ désigne l'ensemble des fonctions polynomiales l'opérateur $f \to C_n f$ est défini sur $\Pi$,

Nous étudierons dans la suite diverses propriétes des opérateurs $C_n$ définis sur certains espaces $\mathcal{B}$ munis de différentes topologies.

## I.1 - Les espaces $L_K^p$

Soit $K > 0$ et $1 \leq p < +\infty$: on désigne par $L_K^p$ l'espace des fonctions réelles telles que $|f|^p$ soit localement sommable sur $[0,+\infty[$ et telles que pour tout $f \in L_K^p$:

$$\int_0^{+\infty} |f(t)|^p \, K \, e^{-Kt} dt \quad \text{soit fini.}$$

On pose pour $f \in L_K^p$: $\| f \|_{p,K} = \left( \int_0^{+\infty} |f(t)|^p \, K \, e^{-Kt} dt \right)^{1/p}$.

$L_K^\infty$ désigne l'espace des fonctions réelles, definies presque partout sur $[0,+\infty[$ et telles que, pour tout $f \in L_K^\infty$, il existe M tel que $|f(t)| e^{-Kt} \leq M$ presque partout sur $[0,+\infty[$.

On pose pour $f \in L^\infty_K$: $\| f\|_{\infty,K} = \operatorname{ess.sup}_{t \in [0,+\infty[} |f(t)|e^{-Kt}$.

Pour $f \in L^\infty$ nous avons $\| f\|_{p,K} \leq \| f\|_\infty$.

Nous poserons $K\, e^{-K|t|}dt = d\mu_t$.

Nous introduisons aussi les <u>modules de régularité</u> suivants: pour $1 \leq p < +\infty$ et pour $f \in L^p_K$:

$$\omega_{p,K}(f,\delta) = \sup_{0\leq h\leq\delta}\left(\int_{-\infty}^{+\infty} |\tilde{f}(x+h)-\tilde{f}(x)|^p d\mu_x\right)^{1/p}$$

$$= \sup_{0\leq h\leq\delta} \| \tau_h\tilde{f}-\tilde{f}\|_{p,K}$$

où $\tau_h\tilde{f}$ désigne la fonction $x \to \tau_h\tilde{f}(x) = \tilde{f}(x+h)$ et où $\tilde{f}$ désigne un prolongement de f á $\mathbb{R}$ tout entier tel que

$$\left(\int_{-\infty}^{+\infty} |\tilde{f}(x)|^p d\mu_x\right)^{1/p} = \| \tilde{f}\|^{\vee}_{p,K} < +\infty .$$

La fonction $\delta \to \omega_{p,K}(f,\delta)$ est telle que:

(I.1.1) $\omega_{p,K}(f,\cdot)$ est definie sur $\mathbb{R}^+$, croissante au sens large sur $\mathbb{R}^+$

(I.1.2) $\omega_{p,K}(f,0) = 0$

(I.1.3) $\omega_{p,K}(f,.)$ <u>est continue à droite en 0</u>. En effet si g désigne une fonction continue sur $\mathbb{R}$ à support compact dans $\mathbb{R}$ nous avons

$$\omega_{p,K}(g,\delta) \leq \left(\sup_{0\leq h\leq\delta} |g(x+h) - g(x)|\right) \cdot \left(\int_{-\infty}^{+\infty} d\mu_x\right)^{1/p}$$

et donc:

$$\omega_{p,K}(g,\delta)\leq 2^{1/p}\omega_\infty(g,\delta) .$$

Comme $\lim\limits_{\substack{\delta\to 0\\ \delta>0}} \omega_\infty(g,\delta) = 0$ pour g continue sur $\mathbb{R}$ a support compact il en résulte que $\lim\limits_{\substack{\delta\to 0\\ \delta>0}} \omega_{p,K}(g,\delta) = 0$.

Soit maintenant $f \in L^p_K$ : on peut approcher $\tilde{f}$ à $\varepsilon$ près sur $\mathbb{R}$ par g continue sur $\mathbb{R}$ a support compact sur $\mathbb{R}$, de sorte que $\| \tilde{f}-g \|_{\widetilde{p,K}} < \varepsilon$. On a donc

$$\omega_{p,K}(f,\delta) \leq \sup_{0\leq h\leq\delta} \| \tau_h\tilde{f}-\tilde{f} \|_{\widetilde{p,K}} \leq_{0\leq h\leq\delta} \| \tau_h(\tilde{f}-g)-(\tilde{f}-g) \|_{\widetilde{p,K}}$$

$$+ \sup_{0\leq h\leq\delta} \| \tau_h g-g \|_{\widetilde{p,K}}$$

et

$$\omega_{p,K}(f,\delta) \leq 2\| \tilde{f}-g \|_{\widetilde{p,K}} + 2\omega_\infty(g,\delta).$$

On a donc $\overline{\lim\limits_{\substack{\delta\to 0\\ \delta>0}}}\, \omega_{p,K}(f,\delta) \leq 2\varepsilon$ et (I.1.3) en résulte.

(I.1.4) $\omega_{p,K}(f,.)$ <u>est sous additive</u> c'est-à-dire que:

$$\forall\delta_1 \geq 0 \quad \forall\delta_2 \geq 0 : \omega_{p,K}(f,\delta_1+\delta_2) \leq \omega_{p,K}(f,\delta_1)+\omega_{p,K}(f,\delta_2)$$

Il suffit pour le démontrer d'écrire:

$$|\tilde{f}(x+h)-\tilde{f}(x)| \leq |\tilde{f}(x+\frac{h\delta_1}{\delta_1+\delta_2})-\tilde{f}(x+h)| + |\tilde{f}(x+\frac{h\delta_1}{\delta_1+\delta_2})-\tilde{f}(x)|$$

avec

$$|h-\frac{h\delta_1}{\delta_1+\delta_2}| \leq \delta_2 \quad \text{et} \quad |\frac{h\delta_1}{\delta_1+\delta_2}| \leq \delta_1 \quad \text{pour} \quad 0 \leq h \leq \delta_1+\delta_2$$

Il en résulte que:

(I.1.5) $\forall\delta \geq 0$, $\forall\lambda \geq 0$ : $\omega_{p,K}(f,\lambda\delta) \leq (\lambda+1)\, \omega_{p,K}(f,\delta)$.

Pour $f \in L^p$ nous désignerons de même par $\omega_p(f,\delta)$ l'expression

$$\omega_p(f,\delta) = \sup_{0\leq h\leq\delta} \left( \int_{-\infty}^{+\infty} |\tilde{f}(x+h) - \tilde{f}(x)|^p dx \right)^{1/p} .$$

Pour $f \in C^o([a,b])$:

$$\omega_{\infty,[a,b]}(f,\delta) = \sup_{\substack{x \in [a,b] \\ x+h \in [a,b] \\ 0 \leq h \leq \delta}} |f(x+h) - f(x)|$$

avec $0 \leq a < b < +\infty$

Si f est uniformément continue sur $[0,+\infty[$ nous poserons

$$\omega_\infty(f,\delta) = \sup_{\substack{x \in [0,+\infty[ \\ 0 \leq h \leq \delta}} |f(x+h) - f(x)| .$$

## II. QUELQUES PROPOSITIONS PRELIMINAIRES

### II.1 - Approximation des fonctions $e_j : t \to t^j$

Soit $e_j$ la fonction de $\mathbb{R}^+$ dans $\mathbb{R}^+$ : $t \to t^j$ où j est un entier $\geq 0$.

$$C_n e_j(x) = \sum_{k=0}^{+\infty} \frac{n^{2k+1}}{(k!)^2} x^k e^{-nx} \int_0^{+\infty} t^{k+j} e^{-nt} dt$$

$$C_n e_j(x) = \sum_{k=0}^{+\infty} \frac{n^k x^k}{(k!)^2} e^{-nx} \cdot \frac{1}{n^j} \int_0^{+\infty} \cdot v^{k+j} e^{-v} dv$$

$$= \sum_{k=0}^{+\infty} \frac{n^k x^k}{k!} e^{-nx} \circ \frac{(k+j)!}{k!} \cdot \frac{1}{n^j}$$

Mais $n^j x^j e^{nx} = \sum_{k=0}^{+\infty} \frac{(nx)^{k+j}}{k!}$

et donc $\frac{d^j}{dx^j}(x^j e^{nx}) = \sum_{k=0}^{+\infty} \frac{(k+j)!}{(k!)^2} (nx)^k$

d'où $C_n e_j(x) = \frac{e^{-nx}}{n^j} \left[ \frac{d^j}{dx^j}(x^j e^{nx}) \right] = \ell_{jn}(x)$ .

Proposition (II.1.1): $C_n e_j$ est un polynôme de degré j

et

$$\|e_j - C_n e_j\|_{p,K} \leq \frac{(j+1)!(j-1)!}{n} \frac{e^K}{K^{j-1}} \quad \text{pour} \quad 1 \leq p \leq +\infty$$

En effet par la formule de Leibniz:

$$\ell_{jn}(x) = \frac{e^{-nx}}{n^j}\left(\sum_{r=0}^{j} \binom{j}{r} \frac{j!}{r!} (nx)^r e^{nx}\right)$$

$$= \frac{1}{n^j} \sum_{r=0}^{j} \binom{j}{r} \frac{j!}{r!} n^r x^r = e_j(x) + \ldots + \frac{j!}{n^j}$$

$$= e_j(x) + \frac{1}{n} j^2 x^{j-1} + \ldots + \frac{j!}{n^j}$$

Il en résulte que:

$$0 < \ell_{jn}(x) - x^j < \frac{1}{n^j} \sum_{r=0}^{j-1} \frac{j!}{(j-r)!r!} \times \frac{j!}{r!} n^r x^r$$

$$0 < \ell_{jn}(x) - x^j < \frac{j(j!)}{n^j} \sum_{r=0}^{j-1} \frac{(j-1)!}{(j-1-r)!r!} n^r x^r$$

et donc

$$0 < \ell_{jn}(x) - x^j < \frac{(j+1)!}{n^j} (1 + nx)^{j-1}$$

$$0 < \ell_{jn}(x) - x^j < \frac{(j+1)!}{n} \left(x + \frac{1}{n}\right)^{j-1}$$

Il en résulte que

$$\| e_j - C_n e_j \|_{\infty,K} < \frac{(j+1)!}{n} \sup_{x \in [0,+\infty[} e^{-Kx} \left(x + \frac{1}{n}\right)^{j-1}$$

$$\| e_j - C_n e_j \|_{\infty,K} < \frac{(j+1)!(j-1)!}{n} \frac{e^{\frac{K}{n}}}{K^{j-1}}$$

en effet la fonction $v \to v^i e^{-Kv}$ est maximum lorsque $v = \frac{i}{K}$ et alors son maximum est égal à $\frac{i^i}{K^i} e^{-i} \leq \frac{i!}{K^i}$.

D'où $\sup_{v \in [0,+\infty[} e^{-Kv} v^i \leq \frac{k!}{K^i}$

et

$$\| e_j - C_n e_j \|_{\infty,K} < \frac{(j+1)!(j-1)!}{n} \times \frac{e^{\frac{K}{n}}}{K^{j-1}} < \frac{(j+1)!(j-1)!\, e^K}{K^{j-1}\, n} .$$

Il en résulte que, pour tout polynôme $\pi$, il existe $\lambda(\pi,K)$ tel que:

$$\| \pi - C_n \|_{\infty,K} < \frac{\lambda}{n} \qquad \text{pour } n \geq 1 .$$

Notons que, en particulier:

$$C_n e_0(x) = e_0(x) = 1$$

$$C_n e_1(x) = e_1(x) + \frac{1}{n}$$

$$C_n e_2(x) = e_2(x) + \frac{4x}{n} + \frac{2}{n^2}$$

II.2 - Approximation des fonctions $t \rightarrow e_{jN}(x) = x^j e^{Nx}$

$$C_n e_{jN}(x) = \sum_{k=0}^{+\infty} \frac{x^k n^k}{k!} e^{-nx} \cdot \frac{n^{k+1}}{k!} \int_0^{+\infty} t^{k+j} e^{-(n-N)t} dt$$

et pour $n > N$, on a

$$C_n e_{jN}(x) = \sum_{k=0}^{+\infty} \frac{x^k n^k}{k!} e^{-nx} \cdot \frac{n^{k+1}}{k!} \cdot \frac{(k+j)!}{(n-N)^{k+j+1}}$$

$$= \frac{e^{-nx}}{(n-N)^j} \sum_{k=0}^{+\infty} \frac{x^k n^k}{k!} \left(\frac{n}{n-N}\right)^{k+1} \cdot \frac{(k+j)!}{k!}$$

$$= \frac{ne^{-nx}}{(n-N)^{j+1}} \sum_{k=0}^{+\infty} \left(\frac{n^2x}{n-N}\right)^K \cdot \frac{1}{k!} \cdot \frac{(k+j)!}{k!}$$

Mais $e^{\frac{n^2x}{n-N}} = \sum_{k=0}^{\infty} \left(\frac{n^2x}{n-N}\right)^k \cdot \frac{1}{k!}$

et donc $\frac{d^j}{dx^j}\left[\frac{n^{2j}x^j}{(n-N)^j} e^{\frac{n^2x}{n-N}}\right] = \frac{n^{2j}}{(n-N)^j} \sum_{k=0}^{+\infty} \left(\frac{n^2x}{n-N}\right)^k \frac{(k+j)!}{(k!)^2}$

$$C_n e_{jN}(x) = \frac{ne^{-nx}}{(n-N)^{j+1}} \frac{d^j}{dx^j}\left(x^j e^{\frac{n^2x}{n-N}}\right)$$

Proposition II.2.1 : Pour tout j entier et tout N réel $P_n e_{jN}$ converge vers $e_{jN}$ sur tout compact de $[0,+\infty[$ lorsque n tend vers $+\infty$.

Par la formule de Leibniz:

$$C_n e_{jN}(x) = \frac{ne^{-nx}}{(n-N)^{j+1}} \left( \sum_{r=0}^{j} \binom{j}{r} \frac{j!}{r!} x^r \left(\frac{n^2}{n-N}\right)^r \right) e^{\frac{n^2 x}{n-N}}$$

$$C_n e_{jN}(x) = \frac{ne^{\frac{nNx}{n-N}}}{(n-N)^{j+1}} \times \frac{x^j n^{2j}}{(n-N)^j} + \rho_{jnN}(x)$$

$$0 < C_n e_{jN}(x) - \left(\frac{n}{n-N}\right)^{2j+1} x^j e^{\frac{nNx}{n-N}} = \rho_{jnN}(x)$$

et

$$\rho_{jnN}(x) < \frac{n\ e^{\frac{nNx}{n-N}}}{(n-N)^{j+1}} \cdot (j+1)! \cdot \left(1 + \frac{n^2 x}{n-N}\right)^{j-1}$$

$$< \frac{n}{n-N} e^{\frac{nNx}{n-N}} \left(x + \frac{n-N}{n^2}\right)^{j-1} (j+1)! \ \frac{n^{2j-2}}{(n-N)^{2j-1}}$$

$$< \frac{(j+1)!}{n-N} \left(\frac{n}{n-N}\right)^{2j-1} \left(x + \frac{n-N}{n^2}\right)^{j-1} e^{\frac{nNx}{n-N}}$$

$$< \frac{(j+1)!}{n-N} \left(\frac{n}{n-N}\right)^{2j-1} \left(\ell + \frac{n-N}{n^2}\right)^{j-1} e^{\frac{nN}{n-N}\ell}$$

$$< \frac{1}{n-N} K(\ell) \quad .$$

$$0 < C_n e_{jN}(x) - e_{jN}(x) < -x^j e^{Nx} + \left(\frac{n}{n-N}\right)^{2j+1} x^j e^{\frac{nNx}{n-N}} + K(\ell) \times \frac{1}{n-N}$$

et donc

$$0 \leq \overline{\lim_{n\to+\infty}} \left(C_n e_{jN}(x) - e_{jN}(x)\right) \leq 0 \quad .$$

## II.3 - Expression de $C_n f(x)$ en fonction des dérivées successives, au point d'abscisse n, de la transformée de Laplace de f

Soit $F_o(p) = \int_0^{+\infty} e^{-pt} f(t)dt = L[f](p)$ .

Nous savons que $L[t\to t^k f(t)](p) = (-1)^k \left[L[f]\right]^{(k)}(p)$

et donc

$$C_n f(x) = \sum_{k=0}^{+\infty} \frac{n^{2k+1}}{(k!)^2} x^k e^{-nx} (-1)^k F_o^{(k)}(n)$$

$$C_n f(x) = e^{-nx} \sum_{k=0}^{+\infty} (-1)^k \frac{n^{2k+1}}{(k!)^2} F_o^{(k)}(n)\, x^k$$

$$C_n f(x) = \left(\sum_{r=0}^{+\infty} (-1)^r \frac{n^r x^r}{r!}\right)\left(\sum_{k=0}^{+\infty} (-1)^k \frac{n^{2k+1}}{(k!)^2} F_o^{(k)}(n)\, x^k\right)$$

$$= \sum_{s=0}^{+\infty} (-1)^s \left(\sum_{k+r=s} \frac{n^{2k+r+1}}{r!(k!)^2} F_o^{(k)}(n)\right) x^s$$

$$= \sum_{s=0}^{+\infty} (-1)^s \left(\sum_{k+r=s} \frac{n^{k+1}}{k!} \binom{s}{k} F_o^{(k)}(n)\right) \frac{n^s x^s}{s!}$$

$$= \sum_{s=0}^{+\infty} (-1)^s \left[\sum_{k=0}^{s} \frac{n^{k+1}}{k!} \binom{s}{k} F_o^{(k)}(n)\right] \frac{n^s x^s}{s!}$$

II.4 - Une relation entre $(C_n f)'$, $C_n f'$ et $(C_n f')'$

(II.4.1) Théorème: Si f possède presque partout sur $[0,+\infty[$ une dérivée première $f' \in L^1$ on a

$$(C_n f)' - (C_n f') = \frac{(C_n f')'}{n} .$$

Démonstration

$$(C_n f)(x) = \sum_{k=0}^{+\infty} \frac{n^{2k+1}}{(k!)^2} x^k e^{-nx} \int_0^{+\infty} t^k e^{-nt} f(t)dt$$

$$(C_n f)'(x) = \sum_{k=0}^{+\infty} \frac{n^{2k+1}}{(k')^2} (k-nx)\, e^{-nx} x^{k-1} \int_0^{+\infty} t^k e^{-nt} f(t)dt$$

Mais

$$\int_0^{+\infty} t^k f(t) e^{-nt}dt = \frac{1}{n}\int_0^{+\infty} \left[k\, t^{k-1} f(t) + t^k f'(t)\right] e^{-nt}dt \text{ pour } k\geq 1$$

d'où

$(C_n f)'(x) =$

$$= \sum_{k=1}^{+\infty} \frac{n^{2k+1}}{(k!)^2} x^{k-1} \; e^{-nx} \frac{k}{n} \int_0^{+\infty} \left[ k t^{k-1} f(t) + t^k f'(t) \right] e^{-nt} - n C_n f(x)$$

$$= \frac{1}{n} \sum_{r=0}^{+\infty} \frac{n^{2r+3}}{(r!)^2} x^r \; e^{-nx} \int_0^{+\infty} \left( t^r f(t) e^{-nt} + \frac{t^{r+1}}{r+1} f'(t) e^{-nt} \right) dt - n C_n f(x)$$

$$= \frac{1}{n} \sum_{k=0}^{\infty} \frac{n^{2k+1}}{(k!)^2} \; x^{k-1} \; e^{-nx} \times k \int_0^{+\infty} t^k f'(t) \; e^{-nt} \; dt$$

$$= \frac{1}{n} \sum_{k=0}^{\infty} \frac{n^{2k+1}}{(k!)^2} \; (k-nx) \; x^{k-1} \; e^{-nx} \int_0^{+\infty} t^k f'(t) e^{-nt} dt + (C_n f')(x)$$

D'où

$$(C_n f)'(x) - (C_n f')(x) = \frac{(C_n f')'(x)}{n} \quad .$$

(II.4.2) <u>Corollaire</u>: Si $f$ possède, presque partout sur $[0,+\infty[$, une dérivée $f^{(k)}$ intégrable sur $[0,+\infty[$, on a:

$$(C_n f)'(x) =$$
$$= (C_n f')(x) + \frac{1}{n} (C_n f'')(x) + \ldots + \frac{1}{n^{k-1}} (C_n f^{(k)})(x) +$$
$$+ \frac{1}{n^k} (C_n f^{(k)})'(x)$$

Il suffit de raisonner par récurrence

$$(C_n f)'(x) =$$
$$= (C_n f')(x) + \ldots + \frac{1}{n^{k-2}} (C_n f^{(k-1)})(x) + \frac{1}{n^{k-1}} (C_n f^{(k-1)})'(x)$$

et

$$(C_n f^{(k-1)})'(x) = (C_n f^{(k)})(x) + \frac{1}{n} (C_n f^{(k)})'(x).$$

(II.4.3) <u>Proposition</u>: Si $f' \in L^\infty \cap L^1$

alors

$$|(C_n f')'(x)| < \| f' \|_\infty \sqrt{\frac{n}{x}} \quad \text{pour } x > 0 \; .$$

En effet

$$C_n f'(x) = \sum_{k=0}^{+\infty} \frac{n^{2k+1}}{(k!)^2} x^k e^{-nx} \int_0^{+\infty} t^k f'(t)\, e^{-nt} dt$$

et

$$|(C_n f')'(x)| \leq \sum_{k=0}^{+\infty} \frac{n^{2k+1}}{(k!)^2} |k-nx| x^{k-1} e^{-nx} \int_0^{+\infty} t^k |f'(t)| e^{-nt} dt$$

$$|(C_n f')'(x)| \leq \|f'\|_\infty \sum_{k=0}^{+\infty} \frac{n^{2k+1}}{(k!)^2} |k-nx| x^{k-1} e^{-nx} \frac{k!}{n^{k+1}}$$

$$\leq \|f'\|_\infty \left( \sum_{k=0}^{+\infty} \frac{n^k}{k!} |k-nx| x^{k-1} \right) e^{-nx} \quad .$$

Mais

$$\sum_{k=0}^{+\infty} \frac{n^k x^k}{k!} = e^{nx}$$

et

$$x|(C_n f')'(x)| \leq e^{-nx} \|f'\|_\infty \sum_{k=0}^{+\infty} \frac{n^k x^k}{k!} |k-nx|$$

et par l'inégalite de Cauchy-Schwarz:

$$x^2 |(C_n f')'(x)|^2 \leq e^{-2nx} \|f'\|_\infty^2 \left[ e^{nx} \sum_{k=0}^{\infty} \frac{n^k x^k}{k!} |k-nx|^2 \right]$$

$$x^2 |(C_n f')'(x)|^2 \leq e^{-nx} \|f'\|_\infty^2 \left[ n^2 x^2 e^{nx} - 2nx \sum_{k=1}^{\infty} \frac{n^k x^k}{(k-1)!} + \sum_{k=1}^{\infty} \frac{n^k x^k}{(k-1)!} k \right]$$

$$x^2 |(C_n f')'(x)|^2 \leq e^{-nx} \|f'\|_\infty \left[ n^2 x^2 e^{nx} - 2n^2 x^2 e^{nx} + nx \sum_{r=0}^{+\infty} \frac{n^r x^r}{r!} (r+1) \right]$$

$$x^2 |(C_n f')'(x)|^2 \leq nx \|f'\|_\infty^2$$

et donc

$$|(C_n f')'(x)| \leq \|f'\| \sqrt{\frac{n}{x}} \quad .$$

(II.4.4) <u>Proposition</u>: Si $f' \in L^p \cap L^1$ on a

$$\frac{|(C_n f')'(x)|}{n} \leq \frac{1}{n^{(p-1)/2p}} \frac{\|f'\|_p}{\sqrt{x}}$$

En effet

$$x|(C_nf')'(x)| \leq e^{-nx}\sum_{k=0}^{+\infty}\frac{n^{k+1}}{(k!)}|k-nx|x^k\left(\int_0^{+\infty}\frac{n^kt^ke^{-nt}}{k!}\right)^{1/q}\|f'\|_p$$

Car $\frac{n^kt^k}{k!}e^{-nt} \leq 1$ et donc on peut écrire

$$x|(C_nf')'(x)| \leq e^{-nx}\|f'\|_p\, n^{1/p}\left(\sum_{k=0}^{+\infty}\frac{n^kx^k}{k!}|k-nx|^2\right)^{1/2}$$

car

$$\int_0^{+\infty}\frac{n^kt^k}{k!}e^{-nt}dt = \frac{1}{n}$$

$$x|(C_nf')'(x)| \leq e^{-nx}\|f'\|_p\, n^{1/p}[\sqrt{nx}\,]\,e^{nx}$$

et donc:

$$\frac{1}{n}|(C_nf')'(x)| \leq \|f'\|_p\frac{1}{\sqrt{x}}\times\frac{1}{n^{1/2-1/p}} = \|f'\|_p\frac{1}{\sqrt{x}}\cdot\frac{1}{n^{(p-1)/2p}}$$

Corollaires:

(1) Il en résulte que, pour $1 \leq p < 2$

$$\frac{1}{n}\|(C_nf')'\|_{p,K} \leq H\frac{\|f'\|_p}{n^{(p-1)/2p}}$$

(2) Les inegalités obtenues en II.4.4 et II.4.3 ainsi que le théoréme (III.4) démontré plus loin nous seront utiles pour démontrer des propriétés de convergence de $(C_nf)^{(k)}$ vers $f^{(k)}$, en particulier la convergence uniforme sur tout compact $[a,b]\subset]0,+\infty]$

II.5 - Evaluation de

$$\phi_{n,m}(x) = \sum_{k=0}^{\infty}\frac{n^{2k+1}}{(k!)^2}x^ke^{-nx}\int_0^{+\infty}t^k(x-t)^me^{-nt}dt$$

Pour n et m entiers nous définissons $\phi_{n,m}$ par

(II.5.1)

$$\phi_{n,m}(x) = \sum_{k=0}^{+\infty} \frac{n^{2k+1}}{(k!)^2} x^k e^{-nx} \int_0^{+\infty} t^k (x-t)^m e^{-nt} dt$$

$$\phi'_{n,m}(x) =$$

$$= \sum_{k=0}^{+\infty} \frac{n^{2k+1}}{(k!)^2} (k-nx) x^{k-1} e^{-nx} \int_0^{+\infty} t^k (x-t)^m e^{-nt} dt + m\phi_{n,m-1}(x)$$

$$= \sum_{k=1}^{\infty} \frac{n^{2k+1}}{(k!)^2} k x^{k-1} e^{-nx} \int_0^{+\infty} t^k (x-t)^m e^{-nt} dt + m\phi_{n,m-1}(x) - n\phi_{n,m}(x)$$

(II.5.2)

$$\phi'_{n,m}(x) + n\phi_{n,m}(x) - m\phi_{n,m-1}(x)$$

$$= \sum_{r=0}^{\infty} \frac{n^{2r+3}}{(r!)^2} \cdot \frac{x^r}{r+1} e^{-nx} \int_0^{+\infty} t^{r+1} e^{-nt} (x-t)^m dt$$

$$= n^2 \sum_{r=0}^{+\infty} \frac{n^{2r+1}}{(r!)^2} x^r e^{-nx} \int_0^{+\infty} \frac{t^{r+1}}{r+1} e^{-nt} (x-t)^m dt$$

Par intégration par parties on obtient (pour $n \geq 1$)

(II.5.3)

$$\phi'_{n,m}(x) + n\phi_{n,m}(x) - m\phi_{n,m-1}(x)$$

$$= n^2 \sum_{r=0}^{+\infty} \frac{n^{2r+1}}{(r!)^2} x^r e^{-nx} \int_0^{+\infty} \left(t^r - \frac{nt^{r+1}}{r+1}\right) \frac{(x-t)^{m+1}}{m+1} e^{-nt} dt$$

(II.5.4)

$$\phi'_{n,m}(x) + n\phi_{n,m}(x) - m\phi_{n,m-1}(x)$$

$$= -\frac{n^3}{m+1} \sum_{r=0}^{\infty} \frac{n^{2r+1}}{(r!)^2} x^r e^{-nx} \int_0^{+\infty} \frac{t^{r+1}}{r+1} e^{-nt} (x-t)^{m+1} dt.$$

En changeant m en m+1 dans (II.5.3) et en combinant avec (II.5.4) on obtient:

(II.5.5)

$$n\phi'_{n,m+1}(x) + (m+1)\phi'_{n,m}(x) = m(m+1)\phi_{n,m-1}(x)$$

On peut aussi écrire à partir de l'expression donnant $\phi'_{n,m}(x)$:

$$- m\phi_{n,m-1}(x) + x\phi'_{n,m} =$$

$$= \int_0^{+\infty} \sum_{k=0}^{+\infty} \frac{n^{2k+1}}{(k!)^2} x^k e^{-nx}\Big(k+1-nt+n(t-x)-1\Big)(x-t)^m t^k e^{-nt} dt$$

$$x\phi'_{n,m}(x) + n\phi_{n,m+1}(x) + \phi_{n,m}(x) - m\phi_{n,m-1}(x)$$

$$= \sum_{k=0}^{+\infty} \frac{n^{2k+1}}{(k!)^2} x^k e^{-nx} \int_0^{+\infty} (x-t)^m \frac{d}{dt}(t^{k+1} e^{-nt}) dt$$

$$= \sum_{k=0}^{\infty} \frac{n^{2k+1}}{(k!)^2} x^k e^{-nx} \int_0^{+\infty} m(x-t)^{m-1} t^{k+1} e^{-nt} dt$$

$$= + mx\phi_{n,m-1}(x) - m\phi_{n,m}(x) \quad \Big(\text{en écrivant } t^{k+1} = t^k[x-(x-t)]\Big)$$

On obtient donc la relation

(II.5.6)

$$n\phi_{n,m+1}(x) + (m+1)\phi_{n,m}(x) + x\phi'_{n,m}(x) = m(1+x)\phi_{n,m-1}(x)$$

Posons dans (II.5.5): $\phi_{n,m}(x) = \frac{(2p)!}{n^p} \psi_{n,m}(x)$ pour $m = 2p$

et $\phi_{n,m}(x) = \frac{(2p-1)!}{n^{(p-1)/2}} \psi_{n,m}(x)$ pour $m = 2p-1$,

c'est-à-dire $\phi_{n,m}(x) = \frac{m!}{n^{m/2}} \psi_{n,m}(x)$.

On obtient donc:

$$n \frac{(m+1)!}{n^{(m+1)/2}} \psi'_{n,m+1}(x) + \frac{(m+1)!}{n^{m/2}} \psi'_{n,m}(x) = \frac{(m+1)!}{n^{(m+1)/2}} \psi_{n,m-1}(x)$$

(II.5.7)

$$\psi'_{n,m+1}(x) + \frac{1}{\sqrt{n}} \psi'_{n,m}(x) = \psi_{n,m-1}(x) \quad .$$

On a

$$\psi_{n,m}(0) = n\int_0^{+\infty}(-t)^m e^{-nt}dt = (-1)^m \frac{m!}{n^m} \quad \text{et donc}$$

$$\psi_{n,m}(0) = \frac{(-1)^m}{n^{m/2}} .$$

En intégrant (II.5.7), on obtient

$$|\psi_{n,m+1}(x)| \le \frac{1}{\sqrt{n}}|\psi_{n,m}(x)| + x \sup_{t\in[0,x]}|\psi_{n,m-1}(t)|$$

Mais lo fonction $t \to |\psi_{n,m-1}(t)|$ est croissante et donc:

(II.5.8)

$$|\psi_{n,m+1}(x)| \le \frac{1}{\sqrt{n}}|\psi_{n,m}(x)| + x|\psi_{n,m-1}(x)|$$

(II.5.9) <u>Théorème</u>

$$|\phi_{n,m}(x)| \le \frac{m!}{n^{m/2}}\left(\sqrt{x} + \frac{1}{\sqrt{n}}\right)^m \quad \text{et pour m \underline{impair}}$$

$$|\phi_{n,m}(x)| \le \frac{m!(\sqrt{x}+1)^m}{n^{(m+1)/2}}$$

Il suffit de montrer que $|\psi_{n,m}(x)| \le \left(\sqrt{x} + \frac{1}{\sqrt{n}}\right)^m$

Cette relation est vraie pour m=0 et m=1. En raisonnant par récurrence:

$$|\psi_{n,m+1}(x)| \le \left[\frac{1}{\sqrt{n}}\left(\sqrt{x} + \frac{1}{\sqrt{n}}\right) + x\right]\left(\sqrt{x} + \frac{1}{\sqrt{n}}\right)^{m-1} \le$$

$$\le \left(\sqrt{x} + \frac{1}{\sqrt{n}}\right)^2 \left(\sqrt{x} + \frac{1}{\sqrt{n}}\right)^{m-1} \le \left(\sqrt{x} + \frac{1}{\sqrt{n}}\right)^{m+1}$$

On peut aussi obtenir une relation de récurrence entre quatre $\phi_{n,m}$ consécutifs. En effet par (II.5.6):

$$x\,\phi'_{n,m}(x) = m(1+x)\phi_{n,m-1}(x) - n\,\phi_{n,m+1}(x) - (m+1)\phi_{n,m}(x)$$

et, en changeant m en m+1:

$$x\,\phi'_{n,m+1}(x) = (m+1)(1+x)\phi_{n,m}(x) - n\,\phi_{n,m+2}(x) - (m+2)\phi_{n,m+1}(x)$$

et un utilisant (II.4.5) on obtient

(II.5.10)

$$n^2\phi_{n,m+2}(x) + n(2m+3)\phi_{n,m+1}(x) + (m+1)(m+1-n-nx)\phi_{n,m}(x) =$$

$$= m(m+1)\phi_{n,m-1}(x)$$

## III - QUELQUES PROPRIETES DE CONVERGENCE DE $C_n$

III.1 - Proposition: $C_n$ est une contraction dans $L^p$ pour $1 \le p \le +\infty$ .

En effet $\frac{t^k e^{-nt}}{k!} \le \frac{k^k e^{-k}}{n^k\ k!}$ sur $[0,+\infty[$ et donc

$$|C_n f(x)| \le \sum_{k=0}^{+\infty} \frac{n^{2k+1}}{(k!)^2} x^k e^{-nx} \times \left(\frac{k}{n}\right)^k e^{-k} \times \|f\|_1$$

$$\le \left(\sum_{k=0}^{+\infty} \frac{n^{k+1}}{k!} x^k e^{-nx}\right)\|f\|_1 \quad \text{car} \quad \frac{k^k}{k!} e^{-k} \le 1$$

On a donc $\| C_n f\|_1 \le \|f\|_1$

On a aussi

$$|C_n f(x)| \le \|f\|_\infty \cdot |C_n e_o(x)| = \|f\|_\infty$$

et donc

$$\| C_n f\|_\infty \le \|f\|_\infty$$

Dans $L^p (1 < p < +\infty)$ on a

$$|C_n f(x)| \le \int_0^{+\infty} K_n(x,t)|f(t)|dt \le$$

$$\le \left(\int_0^{+\infty} K_n(x,t)dt\right)^{1/q} \left(\int_0^{+\infty} K_n(x,t)|f(t)|^p dt\right)^{1/p}$$

$$|C_n f(x)|^p \le \int_0^{+\infty} K_n(x,t)|f(t)|^p dt$$

$$\| C_n f\|_p \le \left(\int_0^{+\infty}\left(\int_0^{+\infty} K_n(x,t)dx\right)|f(t)|^p dt\right)^{1/p} = \|f\|_p \ .$$

III.2 - <u>Proposition</u>: On peut approcher f dans $L^p$ par une fonction $g_\varepsilon$ continue sur $\mathbb{R}$, dérivable presque partout et telle que:

a) $\omega_\infty(g_\varepsilon,h) \leq \frac{2}{\varepsilon^{1/p}} \omega_p(f,h)$ pour $h \geq 0$

b) $\| \tilde{f}-g_\varepsilon \|_{\tilde{p}} \leq \omega_p(f,\varepsilon)$

c) $\| g' \|_{\tilde{p}} < \frac{2}{\varepsilon} \omega_p(f,\varepsilon)$

En effet soit $\phi_\varepsilon$ la fonction paire, nulle pour $|u| \geq \varepsilon$ et telle que $\phi_\varepsilon(u) = \frac{1}{\varepsilon} - \frac{u}{\varepsilon^2}$ pour $0 \leq u \leq \varepsilon$ .

Posons $g_\varepsilon(x) = \int_{\mathbb{R}} \tilde{f}(t) \phi_\varepsilon(x-t)dt = \int_{-\varepsilon}^{+\varepsilon} \tilde{f}(x+u) \phi_\varepsilon(u)du$

$$g_\varepsilon(x) - \tilde{f}(x) = \int_{-\varepsilon}^{+\varepsilon} [\tilde{f}(x+u)-\tilde{f}(x)] \phi_\varepsilon(u)du$$

a) $g_\varepsilon(x+h')-g_\varepsilon(x) = \int_{\mathbb{R}} [\tilde{f}(x+u+h')-\tilde{f}(x+u)] \phi_\varepsilon(u)du$

$$|g_\varepsilon(x+h')=g_\varepsilon(x)| \leq \int_{\mathbb{R}} [\tilde{f}(y+h')-\tilde{f}(y)] \phi_\varepsilon(y-x)dy$$

et $\omega_\infty(g_\varepsilon,h) < \omega_p(f,h) \| \phi_\varepsilon \|_q$ (car $h' \leq h$)

Or $\| \phi_\varepsilon \|_q = \left(\frac{2}{q+1}\right)^{1/q} \times \frac{1}{\varepsilon^{1/p}} \leq \frac{2}{\varepsilon^{1/p}}$

On a donc

$$\omega_\infty(g_\varepsilon,h) \leq \frac{2}{\varepsilon^{1/p}} \omega_p(f,h)$$

b) $|g_\varepsilon(x)-\tilde{f}(x)| \leq \left(\int_{-\varepsilon}^{+\varepsilon} |\tilde{f}(x+u)-f(x)|^p \phi_\varepsilon(u)du\right)^{1/p}$

$$\| g_\varepsilon-\tilde{f} \|_{\tilde{p}} \leq \left(\int_{-\varepsilon}^{+\varepsilon}\left(\int_{\mathbb{R}} |\tilde{f}(x+u)-\tilde{f}(x)|^p dx\right) \phi_\varepsilon(u)du\right)^{1/p}$$

$$\leq \omega_p(f,\varepsilon)$$

c) $$g'_\varepsilon(x) = \int_{\mathbb{R}} \tilde{f}(t)\ \phi'_\varepsilon(x-t)dt = -\int_{-\varepsilon}^{+\varepsilon} \tilde{f}(x+u)\ \phi'_\varepsilon(u)du$$

$$= \int_{-\varepsilon}^{+\varepsilon} [\tilde{f}(x)-\tilde{f}(x+u)]\ \phi'_\varepsilon(u)du \quad \text{car } \phi'_\varepsilon \text{ est impaire}$$

$$|g'_\varepsilon(x)| \le \frac{1}{\varepsilon^2}\int_{-\varepsilon}^{+\varepsilon} |\tilde{f}(x)-\tilde{f}(x+u)|du$$

$$\le \frac{(2\varepsilon)^{1/q}}{\varepsilon^2}\left(\int_{-\varepsilon}^{+\varepsilon} |\tilde{f}(x)-\tilde{f}(x+u)|^p\ du\right)^{1/p}$$

et donc

$$\| g'_\varepsilon\|_p \le \frac{2}{\varepsilon}\ \omega_p(f,\varepsilon)$$

Remarque: Si f est uniformément continue on a:
$|g'_\varepsilon(x)| < \frac{2}{\varepsilon}\ \omega_\infty(f,\varepsilon)$

III.3 - Théorème

Si f est dans $L^\infty$ et est uniformement continue sur $[0,+\infty[$ on a

$$|C_n f(x) - f(x)| \le 2\left(1 + \sqrt{2(x+1)}\right)\ \omega_\infty\left(f,\frac{1}{\sqrt{n}}\right)$$

En effet

$$|C_n f(x)-f(x)| \le |C_n(f-g_\varepsilon)(x)| + |g_\varepsilon(x)-f(x)| + |C_n g_\varepsilon(x)-g_\varepsilon(x)|$$

$$\le 2\| f-g_\varepsilon\|_\infty + |C_n g_\varepsilon(x)-g_\varepsilon(x)|$$

$$|C_n g_\varepsilon(x)-g_\varepsilon(x)| \le \int_0^{+\infty} K_n(x,t)\left|\int_x^t g'_\varepsilon(v)dv\right|\ dt$$

$$\le \frac{2}{\varepsilon}\ \omega_\infty(f,\varepsilon)\int_0^{+\infty} K_n(x,t)\,|x-t|\ dt$$

$$\le \frac{2}{\varepsilon}\ \omega_\infty(f,\varepsilon)\sqrt{C_n e_2(x) - 2x\ C_n e_1(x) + x^2}$$

Mais

$$C_n e_2(x) - 2x\, C_n e_1(x) + x^2 = \frac{2x}{n} + \frac{2}{n^2}$$

$$|C_n g_\varepsilon(x) - g_\varepsilon(x)| \le \frac{2}{\varepsilon\sqrt{n}}\, \omega_\infty(f,\varepsilon)\sqrt{2x + \frac{2}{n}} + 2\omega_\infty(f,\varepsilon)$$

En faisant $\varepsilon = \frac{1}{\sqrt{n}}$ on obtient le résultat.

## III.4 - Théorème

S'il existe N reel $\ge 0$ tel que, pour tout $t \ge 0$ $|f(t)| \le e^{Nt}$, et si f est continue au point x alors

$$\lim_{n \to +\infty} |C_n f(x) - f(x)| = 0 .$$

La démonstration qui suit est inspirée de Korovkin [4] et de Coatmélec [1].

En effet, f étant continue au point x on peut introduire un module de continuité local, pour $\delta$ suffisamment petit:

$$\omega(x,f,\delta) = \sup_{\substack{t \in [0,\infty[ \\ |x-t| \le \delta}} |f(x)-f(t)|$$

$$|C_n f(x) - f(x)| < \int_0^{+\infty} K_n(x,t)\,|f(x)-f(t)|\,dt$$

Mais

$$|f(x)-f(t)| \le \omega(x,f,\delta) + \frac{|f(x)| + |f(t)|}{\delta^2}\,|t-x|^2$$

et donc

$$|C_n f(x) - f(x)| \le \omega(x,f,\delta) + \frac{|f(x)|}{\delta^2} \cdot \frac{(2x+2)}{n}$$

$$+ \frac{1}{\delta^2}\int_0^{+\infty} K_n(x,t)\, e^{Nt}\,|t-x|^2\,dt$$

$$|C_n f(x) - f(x)| \le \omega(x,f,\delta) + \frac{|f(x)|}{n\delta^2}\,(2x+2)$$

$$+ \frac{1}{\delta^2}\left[C_n e_{2,N}(x) - 2x C_n e_{1,N}(x) + x^2 C_n e_{0,N}(x)\right]$$

Mais par (II.2) pour n > N:

$$0 \leq C_n e_{2,N}(x) - 2xC_n e_{1,N}(x) + x^2 C_n e_{0,N}(x) \leq \frac{1}{n} r(N,x)$$

En effet

$$C_n e_{0,N}(x) = \frac{n}{n-N} e^{\frac{nN}{n-N}x} \text{ et } C_n e_{1,N}(x) = \frac{n}{(n-N)^2}\left[1 + \frac{n^2 x}{n-N}\right] e^{\frac{nNx}{n-N}}$$

$$C_n e_{2,N}(x) = \frac{n}{(n-N)^3}\left[2 + \frac{4n^2 x}{n-N} + \frac{n^4 x^2}{(n-N)^2}\right] e^{\frac{nNx}{n-N}}$$

$$C_n e_{2,N}(x) - 2xC_n e_{1,N}(x) + x^2 C_n e_{0,N}(x) = \frac{n\, e^{\frac{nNx}{n-N}}}{(n-N)^3} \times r(x)$$

$$r(x) =$$

$$= \frac{1}{(n-N)^2}\left[2(n-N)^2 - 2(n-N)^3 x + (n-N)^4 x^2 + 4n^2 x(n-N) + n^4 x^2 - 2n^2 x^2 (n-n)^2\right]$$

et donc

$$r(x) \leq \frac{1}{n} r(N,x)$$

il en résulte que $|C_n f(x) - f(x)| \leq \omega\left(x, f, \frac{1}{n^{(1/2)-\varepsilon}}\right) + \frac{R(N,x)}{n^{2\varepsilon}}$

pour tout $\varepsilon$ fixé compris entre 0 et 1/2.

En particulier pour $\varepsilon = \frac{1}{4}$ on obtient:

$$|C_n f(x) - f(x)| \leq \omega\left(x, f, \frac{1}{n^{1/4}}\right) + \frac{R(N,x)}{n^{1/4}}$$

## III.5 - Corollaire

S'il existe N réel $\geq 0$ tel que, pour tout $t \geq 0$ : $|f(t)| \leq e^{Nt}$ et si f est continue sur $[0,+\infty[$ alors $C_n f$ converge uniformément vers f sur tout compact de $[0,+\infty[$.

Il suffit de reprendre la démonstration précédente: on obtient alors sur $[0,\ell] \subset [0,+\infty[$

$$\| C_n f - f \|_{\infty,[0,\ell]} \leq \omega_{\infty,[0,\ell]}\left(f, \frac{1}{n^{1/4}}\right) + \frac{S(N,\ell)}{n^{1/4}}$$

et plus précisément pour tout $\varepsilon$ $\left(0 < \varepsilon < \frac{1}{2}\right)$

$$\| C_n f - f \|_{\infty,[0,\ell]} \leq \omega_{\infty,[0,\ell]}\left(f, \frac{1}{n^{(1/2)-\varepsilon}}\right) + \left(\frac{S(N,\ell)}{n^{2\varepsilon}}\right)$$

III.6 - Théorème

Si $f \in L^p$ $(1 \leq p < +\infty)$ on a

$$| g_\varepsilon(x) - C_n g_\varepsilon(x) | \leq 2(2x+2)^{1/2q}\, \omega_p\left(f, \frac{1}{n^{(p-1)/2p}}\right)$$

et donc

$$\| f - C_n f \|_{p,K} \leq H\, \omega_p\left(f, \frac{1}{n^{(p-1)/2p}}\right)$$

En effet

$$\begin{aligned}
| C_n g_\varepsilon(x) - g_\varepsilon(x) | &\leq \int^{+\infty} K_n(x,t) \left| \int_x^t g'(v)\,dv \right| dt \\
&\leq \| g'_\varepsilon \|_p \int_0^{+\infty} K_n(x,t)\, |x-t|^{1/q} dt \\
&\leq \| g'_\varepsilon \|_p \left( \int_0^{+\infty} K_n(x,t)\, |x-t|^2 dt \right)^{1/2q} \\
&\leq \frac{2\omega_p(f,\varepsilon)}{\varepsilon} \times \frac{1}{n^{1/2q}} \times \left(2x + \frac{2}{n}\right)^{1/2q} \\
&\leq 2(2x+2)^{1/2q}\, \omega_p\left(f, \frac{1}{n^{(p-1)/2p}}\right)
\end{aligned}$$

On a donc

$$\| C_n g_\varepsilon - g_\varepsilon \|_{p,K} \leq H_1\, \omega_p\left(f, \frac{1}{n^{(p-1)/2p}}\right)$$

Mais

$$\| f - g_\varepsilon \|_{p,K} \leq \| \tilde{f} - g_\varepsilon \|^{\sim}_{p,K} \leq \| \tilde{f} - g_\varepsilon \|^{\sim}_p \leq \omega_p(f,\varepsilon)$$

D'où

$$\| C_n f - f \|_{p,K} \le \| C_n (f - g_\varepsilon) \|_{p,K} + \| \tilde{f} - g_\varepsilon \|^{\sim}_{p,K} + \| g_\varepsilon - C_n g_\varepsilon \|^{\sim}_{p,K}$$
$$\le \| C_n (f - g_\varepsilon) \|_p + \| \tilde{f} - g_\varepsilon \|^{\sim}_p + H_1 \omega_p\left(f, \frac{1}{n^{(p-1)/2p}}\right)$$
$$\le \left(H_1 + 2\right)\omega_p \left(f, \frac{1}{n^{(p-1)/2p}}\right)$$

## IV - TRANSFORMEE DE LAPLACE DE $C_n f$

Nous posons $(Lf)(p) = \int_0^{+\infty} e^{-px} f(x)\,dx = F_o(p)$ et nous nous limitons au cas où p est réel et strictement positif.

$$(LC_n f)(p) = \sum_{k=0}^{+\infty} \left(\frac{n^k}{k!}\int_0^{+\infty} t^k e^{-nt} f(t)\,dt\right)\int_0^{+\infty} \frac{n^{k+1} x^k e^{-(n+p)x}}{k!}\,dx$$

$$= \sum_{k=0}^{+\infty} \left(\frac{n}{n+p}\right)^{k+1} \frac{n^k}{k!} \int_0^{+\infty} t^k e^{-nt} f(t)\,dt$$

$$= \frac{n}{n+p} \int_0^{+\infty} \left( \sum_{k=0}^{+\infty} \frac{(n^2 t)^k}{(n+p)^k} \times \frac{1}{k!} \right) e^{-nt} f(t)\,dt$$

$$(LC_n f)(p) = \frac{n}{n+p} \int_0^{+\infty} e^{-\frac{npt}{n+p}} f(t)\,dt$$

On a donc

$$|(LC_n f)(p) - (Lf)(p)| \le \frac{K(p)}{n} |(Lf)(p)|$$

Il en résulte que

### IV.1 <u>Théorème</u>

$(LC_n f)(p)$ converge vers $(Lf)$ $(p)$, pour tout $p > 0$, lorsque n tend verse $+\infty$ et on a sur tout compact $[a,b] \subset ]0,+\infty[$ :

$$\|(LC_n)(f) - (Lf)\|_{\infty,[a,b]} \le \frac{K(a,b)}{n} \|Lf\|_{\infty,[a\ b]}$$

IV.2 <u>Remarque</u>

$$(LC_n f)(p) = \int_0^{+\infty} e^{-pu} f\left((1 + \frac{p}{n})u\right) du$$

et donc, si f est uniformèment continue,

$$|(LC_n f)(p)-(Lf)(p)| \leq$$

$$\leq \int_0^{+\infty} e^{-pu}\omega_\infty\left(f, \frac{pu}{n}\right) du \leq$$

$$\leq \int_0^{+\infty} (1+pu)e^{-pu}\omega_\infty\left(f, \frac{1}{n}\right) du \leq$$

$$\leq \frac{2}{p}\,\omega_\infty\left(f, \frac{1}{n}\right)$$

De plus si $f' \in L^\infty$ : $|(LC_n f)(p)-(Lf)(p)| \leq \frac{1}{pn} \|f'\|_\infty$ .

REFERENCES


1. Coatmelec, Chr., Approximation et interpolation des fonctions differentiables de plusieurs variables, Annales Scientifiques de l'Ecole Normale Supèrieure (1966), pp. 271-341.
2. Derriennic, M. M., Sur l'approximation de fonctions intègrables sur [0,1] par des polynômes de Bernstein modifiès. *J. Approx. Theory (a paraitre).*
3. Durrmeyer, J.L., Une formule d'inversion de la transformée de Laplace. Applications a la théorie des moments. These 3e cycle Paris (1967).
4. Korovkin, P. O., "Linear Operators and Approximation Theory." Delhi 1960.
5. Szasz, O., Generalization of S. Bernstein's polynomials to the infinite interval, Journal of Research of the National Bureau of Standards 45 (1950), pp. 239-245.

# BELL-SHAPED BASIS FUNCTIONS FOR SURFACE FITTING

Nira Dyn[1]
David Levin

Department of Mathematical Sciences
Tel-Aviv University, Israel

## 1. INTRODUCTION

Global methods for surface fitting perform interpolation or smoothing by a linear combination of a set of basis functions. This work is concerned with global methods for data consisting of surface values at a finite set of points in the plane. Several of the successful global methods utilize as basis functions radial functions of the form

$$f_i(x,y) = f(r_i), \quad r_i^2 = (x-x_i)^2 + (y-y_i)^2, \tag{1.1}$$

where the points $(x_i, y_i)$, $i = 1,\dots,n$ are the data points. Such methods are the "Thin plate splines" (TPS) of Duchon and Meinguet [4], [8], [9], [11], Duchon pseudo cubics [5] and Hardy's multiquadrics (HMQ) and reciprocal multiquadrics [6], [7].

The thin plate spline surface is given by:

$$\sum_{i=1}^{n} c_i r_i^{2(m-1)} \log r_i + p_m(x,y), \qquad m \geqslant 2 \tag{1.2}$$

[1]Partially supported by the Israeli Academy of Sciences.




ISBN 0-12-780650-4

where $p_m(x,y)$ is a polynomial of total degree $< m$, $r_i$ is defined as in (1) and $c_1,\ldots,c_n$ are constrained to satisfy:

$$\sum_{i=1}^{n} c_i x_i^{\ell} y_i^{k} = 0 \ , \quad 0 \leq \ell + k < m. \tag{1.3}$$

Hardy multiquadric basis functions are of the form

$$f(r_i) = \sqrt{r_i^2 + h^2} \ , \quad i = 1,\ldots,n, \tag{1.4}$$

with $h$ a small parameter chosen to depend linearly on the mean distance to the nearest neighbor point.

The Hardy's reciprocal multiquadric are:

$$f(r_i) = (r_i^2 + h^2)^{-1/2} \ , \quad i = 1,\ldots,n. \tag{1.5}$$

Duchon's pseudo cubic splines and the more general pseudo splines are of the form

$$f(r_i) = |r_i|^{2m-1} \ , \quad m \geq 2. \tag{1.6}$$

A comprehensive comparison of 29 local and global interpolation methods [6] indicates that these basis functions and in particular the Hardy's multiquadrics and the "thin plate splines" perform well as interpolating methods.

The drawbacks in utilizing these basis functions in interpolation processes is the need to solve a large system of linear equations with a corresponding full matrix. Moreover since these basis functions (except (1.5)) increase with the distance from the central point, the matrix of the system becomes ill-conditioned with the increase in the number of data points [6].

While for the HMQ method there is no mathematical basis to explain its efficacy, the TPS method constitutes a certain generalization of the univariate natural splines [1]. The interpolating TPS (1.2) is the solution of the following minimization problem: Find $u^* \in \chi^m(R^2)$ minimizing

$$J_m(u) = \iint_{R^2} \sum_{i=0}^{m} \binom{m}{i} \left( \frac{\partial^m u}{\partial x^i \partial y^{m-i}} \right)^2 dxdy \tag{1.7}$$

among all functions in $\chi^m(R^2)$ satisfying

$$u(x_i, y_i) = z_i, \quad i = 1, \ldots, n \tag{1.8}$$

where

$$\chi^m(R^2) = \{v \mid \frac{\partial^m v}{\partial x^i \partial y^{m-i}} \in L^2(R^2), \quad 0 \leqslant i \leqslant m\} \tag{1.9}$$

For the case $m = 2$, $J_2(u)$ represents the bending energy of a thin plate of infinite extent, and the surface $u^*$ corresponds to the equilibrium position of a thin plate of infinite extent that deforms in bending only under deflections $z_1, \ldots, z_n$ at the points $(x_i, y_i)$, $i = 1, \ldots, n$. It is known that a unique interpolating TPS exists for any set of points $(x_i, y_i)$, $i = 1, \ldots, n$ such that no polynomial of degree less than $m$ vanishes at these points. Under the same condition also a unique smoothing TPS exists [11].

In the univariate situation basis functions of local support, the B-splines, constitute a very convenient basis for approximations with splines. The corresponding linear system for the computation of the interpolating spline is well-conditioned, and efficient methods for its solution are available, independently of the number of data points [1], [2]. Moreover the existence of such a basis indicates the exponential decay of the influence of the data with its distance [3].

For the TPS method no localized basis functions are known. The difficulty in handling large data sets has been overcome by subdividing the domain into several smaller overlapping subdomains and by taking the global surface as a smooth convex combination of the TPS surfaces for the various subdomains [10]. This ad-hoc method reduces the quality of the fitted surface considerably in the overlapping zones.

In the following we present a method for constructing bell-shaped basis functions with an algebraic decay away from the center of the bell, for the TPS and HMQ methods, in case the data points are situated on a square grid. The same idea can be applied to other radial basis functions as the Duchon pseudo splines. With these new basis functions the system of linear equation for the interpolating surface is well-conditioned.

An iterative method for the solution of this system is also presented, with a possible application to the computation of smoothing surfaces.

The extension of these ideas to general configurations of data points is still under investigation.

## 2. CONSTRUCTION OF BELL-SHAPED BASIS FUNCTIONS

The radial basis functions (1.2), (1.4) and (1.6) share in common the property that their high order derivatives decay algebraically away from the center point, while at the center the high even order derivatives are very large or singular. For the purpose of presentation we limit ourselves to the following two cases:

$$f(r_i) = r_i^2 \log r_i \qquad \text{(TPS } m = 2\text{)}, \tag{2.1}$$

$$f(r_i) = \sqrt{r_i^2 + h^2} \qquad \text{(HMQ)}. \tag{2.2}$$

For $j \geq 3$ all the j'th order derivatives of $r_i^2 \log r_i$ decay as $r_i^{2-j}$ while those of $\sqrt{r_i^2 + h^2}$ as $r_i^{1-j}$. Moreover the $2j^{th}$ order derivatives of $\sqrt{r_i^2 + h^2}$ behave as $h^{1-2j}$ at $r = 0$, and

$$\Delta^2 r_i^2 \log r_i = c\delta(r_i), \tag{2.3}$$

where $\Delta = \frac{\partial^2}{\partial x^2} + \frac{\partial^2}{\partial y^2}$, and $\delta$ is the Dirac function.

In the univariate case the original basis for splines consists of the functions $(x-x_i)_+^{2m-1}$ satisfying

$$\frac{d^{2m}}{dx_i^{2m}} (x-x_i)_+^{2m-1} = \delta(x-x_i) \cdot (-(2m-1)!) \tag{2.4}$$

The B-spline basis is obtained by using a finite difference approximation to $\frac{d^{2m}}{dx_i^{2m}}$, e.g., for m = 2 and equidistant points $x_{i+1} - x_i = h$:

$$B_i(x) = h^{-4}[(x-x_{i-2})_+^3 - 4(x-x_{i-1})_+^3 + 6(x-x_i)_+^3 -4(x-x_{i+1})_+^3 + (x-x_{i+2})_+^3]. \tag{2.5}$$

In the bivariate case we apply a finite difference approximation of $\Delta^2$ to the original basis functions. Here also we obtain bell-shaped basis functions which instead of having a finite support decay algebraically with the distance from their center.

The notion of a finite difference approximation to the operator $\Delta^2$ is well known for a square grid:

$$(x_i, y_j),\ x_i = a + ih,\ y_j = b + jh,\ i,j = 1,\ldots,N \tag{2.6}$$

Here we are going to use the finite difference operator

$$(\Delta_h^2 u)_{ij} = \frac{1}{h^4}\Big\{20u(x_i,y_j) - 8[u(x_{i+1},y_j) + u(x_{i-1},y_j) + u(x_i,y_{j-1}) + u(x_i,y_{j+1})] + 2[u(x_{i+1},y_{j+1}) + u(x_{i+1},y_{j-1}) + u(x_{i-1},y_{j+1}) + u(x_{i-1},y_{j-1})] + [u(x_{i+2},y_j) + u(x_{i-2},y_j) + u(x_i,y_{j+2}) + u(x_i,y_{j-2})]\Big\} \tag{2.7}$$

with an $O(h^2)$ error of the form

$$\Delta^2 u - \Delta_h^2 u = h^2 \sum_{i=0}^{6} \sum_{j=1}^{2} c_{ij} \frac{\partial^6 u}{\partial x^i \partial y^{6-i}} (\xi_{ij}, \eta_{ij}), \qquad (2.8)$$

where $c_{ij}$ are constants $c_{i1} \geqslant 0$, $c_{i2} \leqslant 0$ and $x_{i-2} \leqslant \xi_{ij} \leqslant x_{i+2}$, $y_{i-2} \leqslant \eta_{ij} \leqslant y_{i+2}$, $j = 1,2$, $i = 0,\ldots,6$.

Applying $\Delta_h^2$ to the function

$$\psi(x,y,\xi,\eta) = [(x-\xi)^2 + (y-\eta)^2]\log\sqrt{(x-\xi)^2 + (y-\eta)^2} \qquad (2.9)$$

as a function of $\xi,\eta$, at the point $(x_i,y_j)$ of the grid, we obtain a function of the form:

$$B_{ij}(x,y) = \Big\{20\psi_{ij} - 8[\psi_{i+1,j} + \psi_{i-1,j} + \psi_{i,j-1} + \psi_{i,j+1}] + 2[\psi_{i+1,j+1} + \psi_{i+1,j-1} + \psi_{i-1,j+1} + \psi_{i-1,j-1}] + \psi_{i+2,j} + \psi_{i-2,j} + \psi_{i,j+2} + \psi_{i,j-2}\Big\}\frac{1}{h^4} \qquad (2.10)$$

with $\psi_{ij}(x,y) = \psi(x,y,x_i,y_j)$.

By the radial property of the function (2.9) and the particular form of the coefficients of the linear combination in (2.10), $B_{ij}(ph,qh) = h^{-2}B^*(p-i,q-j)$ where $B^*$ is a standard bell-shaped function independent of h. The algebraic decay of $B_{ij}(x,y)$ as the $4^{th}$ power of the reciprocal of the distance from the point $(x_i,y_j)$ can be deduced from equations (2.8), (2.3) and the rate of decay of the sixth order derivatives of the function (2.1).

A graph of this function for h = .5 is given in Fig. 1.

The same procedure when applied to the Hardy's multiquadric basis functions (2.2) yields a similar bell-shaped basis function decaying as the third power of the reciprocal of the distance from the central point.

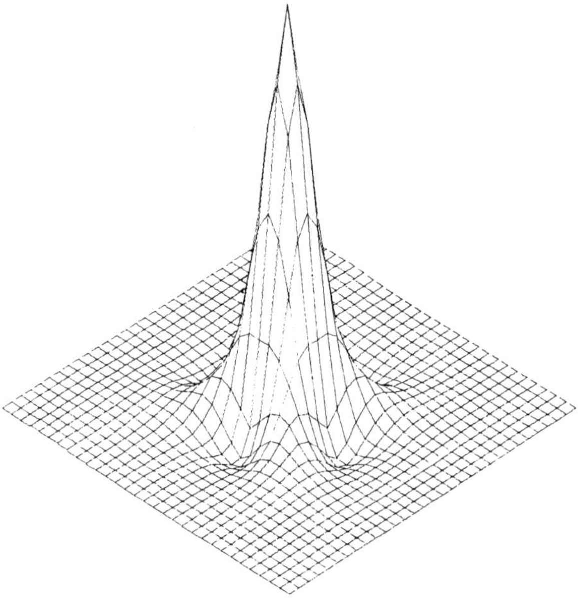

A bell shaped basis function for the thin plate splines.

Figure 1.

## 3. THE USE OF THE BELL-SHAPED FUNCTIONS IN INTERPOLATION AND SMOOTHING

One way of using the bell shaped functions of Section 2 (B-functions) is to consider them as basis functions and to look for an approximating surface of the form $\sum_{i,j=1}^{N} b_{ij}B_{ij}(x,y)$. For the interpolation problem the coefficients are obtained by solving

$$\sum_{i,j=1}^{N} b_{ij}B_{ij}(x_k,y_\ell) = z_{k,\ell}, \quad k,\ell = 1,\ldots,N. \tag{3.1}$$

It turns out that the $B_{ij}$ are so concentrated at $(x_i,y_j)$ that the matrix of the system (3.1) is diagonally dominant for any number of points. This approach is used in the univariate case with the B-splines basis, and it might also be practical for the two dimensional case. However, in this way we immediately lose the polynomial part of the approximation in the TPS method and thus give up the possibility of reproducing planes.

In a second approach we use the bell shaped B-functions to obtain the original TPS solution (1.2), (1.3). Two problems then arise; the construction of adequate basis functions near the boundaries of our domain together with the imposition of the constrains (1.3), and the problem of obtaining the polynomial part of the solution. For the construction of boundary B-functions we used the following guidelines:

(1) Each boundary element is defined to be a linear combination of basis functions $\psi_{ij}$ corresponding to those neighboring points of the original 12 which are in the domain.

(2) The coefficients of the linear combination determining a boundary B-function are chosen so that the corresponding finite difference annihilates as many polynomials as possible from the sequence $1, x, y, xy, x^2, y^2, x^2y, xy^2, \ldots$ and at least the first three of them. The last requirement guarantees that the condition (1.3) is automatically satisfied.

(3) The space spanned by $\{B_{ij}\}_{i,j=1}^{N}$ is of dimension $n - 3 = N^2 - 3$.

(4) The matrix of coefficients relating the B-functions with the original basis functions $\{\psi_{ij}\}$ is symmetric.

The conditions (1) (2) and (4) led us to the following five types of finite differences defining the boundary B-functions:

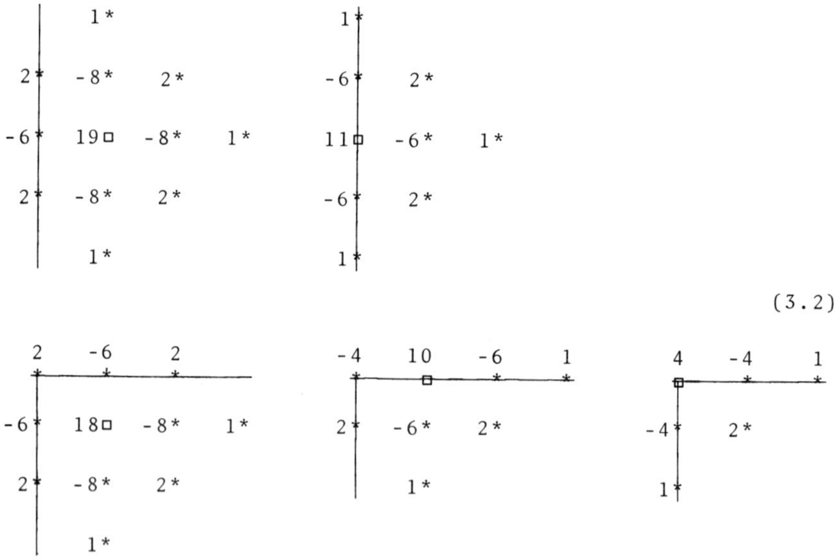


(3.2)

In each array of (3.2) the square denotes the point to which the B-function is assigned, the entries are $h^4$-times the coefficients of the respective functions $\{\psi_{ij}\}$ in the B-function, and the bold lines denote the boundaries.

It can be shown that the space spanned by the above $\{B_{ij}\}_{i,j=1}^{N}$ is of dimension $n - 3 = N^2 - 3$. Having at hand these functions one can represent the interpolating TPS (1.2) as:

$$\sum_{i,j=1}^{N} d_{ij} B_{ij}(x,y) + \alpha + \beta x + \gamma y \tag{3.3}$$

where the coefficients are obtained by solving the system

$$\sum_{i,j=1}^{N} d_{ij} B_{ij}(x_k, y_\ell) + \alpha + \beta x_k + \gamma y_\ell = z_{k,\ell}, \quad 1 \leq k,\ell \leq N. \tag{3.4}$$

These are $N^2$ equations in $N^2 + 3$ unknowns but since $\{B_{ij}(x_k, y_\ell)\}$ is a matrix of rank $N^2 - 3$ three basis functions, e.g. those corresponding to three corners, can be eliminated from the sum. The inclusion of the boundary B-functions disturbs the diagonally dominance of the matrix $B_{ij}(x_k, y_\ell)$. However, most of the rows of this matrix are diagonally dominant and the system is well conditioned and can even be solved iteratively.

Here we adapt an iterative procedure for smoothing and interpolation of data $\{f_i\}_{i=1}^{n}$ given on a set of equally spaced points $\{x_i\}_{i=1}^{n}$ in the univariate situation [2]. This algorithm, which employs the B-splines $\{B_i(x)\}_{i=1}^{n}$ concentrated at the points $\{x_i\}_{i=1}^{n}$, consists of the following steps:

1. Form the initial sum

$$S_0(x) = \sum_{i=1}^{n} f_i B_i(x) \tag{3.5}$$

2. For $j = 1,2,\ldots$ compute

$$e_i^{(j-1)} = f_i - S_{j-1}(x_i), \quad i = 1,\ldots,n, \tag{3.6}$$

$$S_j(x) = S_{j-1}(x) + \sum_{i=1}^{n} e_i^{(j-1)} B_i(x). \tag{3.7}$$

It is shown in [2] that $\lim_{j\to\infty} S_j(x) = S(x)$ where $S(x)$ interpolates the data, while $S_j(x)$, for small values of $j$, perform some kind of data smoothing.

The same process can be employed for the two dimensional interpolation problem using the B-functions:

1. Form the initial surface

$$S_0(x,y) = \sum_{k,\ell=1}^{N} z_{k\ell} B_{k\ell}(x,y) \tag{3.8}$$

2. For $j = 1,2,\ldots$ compute

$$e_{k\ell}^{(j-1)} = z_{k\ell} - S_{j-1}(x_k,y_\ell), \quad k,\ell = 1,\ldots,N, \tag{3.9}$$

$$S_j(x,y) = S_{j-1}(x,y) + \sum_{k,\ell=1}^{N} e_{k\ell}^{(j-1)} B_{k\ell}(x,y). \tag{3.10}$$

For the system (3.1) this process converges since the corresponding matrix of equations is diagonally dominant. On the other hand the matrix of the system (3.4), with the proper boundary B-functions defined by (3.2), is no longer diagonally dominant. The process (3.8)-(3.10) still converges with $S_j(x,y)$ tending to the non-polynomial part of the solution in (3.3), namely to $\sum_{i,j}^{N} d_{ij}B_{ij}(x,y)$. Moreover the residuals $\{e_{k\ell}^{(j)}\}$ tend to limit values $\{e_{k\ell}^{(\infty)}\}$ which can be interpolated by a plane. Therefore an additional step is required for the computation of the coefficients $\alpha,\beta,\gamma$ in (3.3) from the values $\{e_{k\ell}^{(\infty)}\}$.

With a proper definition of the B-functions the above method can be adapted to deal with data given on any rectangular grid and over any domain. In particular, it can handle the case of missing data points in a rectangular grid.

In analogy with the univariate case we expect the iterants $S_j(x,y)$ to perform some kind of data smoothing. In the TPS case $S_j(x,y)$ should be augmented by a least-squares plane fitted to the residuals $\{e_{k\ell}^{(j)}\}$. This point, as well as the generalization of the method to scattered data is yet under investigation.

## 4. NUMERICAL EXAMPLES

We employed the iterative procedure (3.8)-(3.10) to solve the TPS interpolation for several test problems of data given on a square grid. We found out that the convergence is quite fast (convergence factor $\sim 0.5$) and the method can easily handle very large data sets.

<u>Example 4.1</u>. As a data set we take function values at

$$(x_i, y_j)_{i,j=1}^{N}, \quad x_i = y_i = 0.1(i-1), \quad i = 1,\ldots,11, \text{ from}$$

$$F(x,y) = \psi(x,y,0.3,0.3) - \psi(x,y,0.3,0.6)$$

$$- \psi(x,y,0.6,0.3) + \psi(x,y,0.6,0.6)$$

This test function is already a thin plate spline and, therefore, it should be recovered exactly. After 10 iterations the maximum error at grid points was $\sim 0.0002$, and the coefficients of the $\psi_{ij}$'s are displayed in the following array:

| | | | | | | | | | | |
|---|---|---|---|---|---|---|---|---|---|---|
| -.007 | .002 | -.001 | .000 | .000 | .000 | .000 | .000 | .001 | -.002 | .007 |
| .002 | .002 | .000 | .000 | .000 | .000 | .000 | .000 | .000 | -.002 | -.002 |
| -.001 | .000 | .000 | .000 | .000 | .000 | .000 | .000 | .000 | .000 | .001 |
| .000 | .000 | .000 | 1.000 | .000 | .000 | -1.000 | .000 | .000 | .000 | .000 |
| .000 | .000 | .000 | .000 | .000 | .000 | .000 | .000 | .000 | .000 | .000 |
| .000 | .000 | .000 | .000 | .000 | .000 | .000 | .000 | .000 | .000 | .000 |
| .000 | .000 | .000 | -1.000 | .000 | .000 | 1.000 | .000 | .000 | .000 | .000 |
| .000 | .000 | .000 | .000 | .000 | .000 | .000 | .000 | .000 | .000 | .000 |
| .001 | .000 | .000 | .000 | .000 | .000 | .000 | .000 | .000 | .000 | -.001 |
| -.002 | -.002 | .000 | .000 | .000 | .000 | .000 | .000 | .000 | .002 | .002 |
| .007 | -.002 | .001 | .000 | .000 | .000 | .000 | .000 | -.001 | .002 | -.006 |

Example 4.2. Data is taken from $F(x,y) = x \cdot y$ at the points $(0.2(i-1),\ 0.2(j-1))_{i,j=1}^{6}$.

The coefficients of the $\psi_{ij}$'s after 20 iterations with (3.8)-(3.10) are:

| | | | | | |
|---|---|---|---|---|---|
| .585 | -.133 | .043 | -.043 | .133 | -.585 |
| -.133 | -.150 | .006 | -.006 | .150 | .133 |
| .043 | .006 | .013 | -.013 | -.006 | -.043 |
| -.043 | -.006 | -.013 | .013 | .006 | .043 |
| .133 | .150 | -.006 | .006 | -.150 | -.133 |
| -.585 | .133 | -.043 | .043 | -.133 | .585 |

and the residuals $e_{k,\ell}^{(20)}$ are:

| | | | | | |
|---|---|---|---|---|---|
| -.250 | -.150 | -.050 | .050 | .150 | .250 |
| -.150 | -.050 | .050 | .150 | .250 | .350 |
| -.050 | .050 | .150 | .250 | .350 | .450 |
| .050 | .150 | .250 | .350 | .450 | .550 |
| .150 | .250 | .350 | .450 | .550 | .650 |
| .250 | .350 | .450 | .550 | .650 | .750 |

It is easily seen that this residual array can be fitted exactly by $-0.25 + 0.5x + 0.5y$.

Example 4.3. Let the grid points be as in example 4.1 and the data given by

$$z_{ij} = \begin{cases} 1 & i = j = 6 \\ 0 & \text{otherwise:} \quad i,j = 1,\ldots,11. \end{cases}$$

The TPS solution is the fundamental Lagrangian TPS. It turns out that the first iterant $S_0(x,y) = B_{6,6}(x,y)$, is already quite close to this Lagrangian TPS with a maximum error $\sim 0.1$ at grid points.

## REFERENCES


1. de Boor, C., A Practical Guide to Splines, Springer-Verlag, New-York, 1978.
2. de Boor, C., "How does Agee's smoothing method work?" ARO Report 79-3, Proceedings of the 1979 Army Numerical Analysis and Computer Conference, 299-302.
3. Demko, S., "Inverse of band matrices and local convergence of spline projections", SIAM. J. Numer. Anal., vol. 14, no. 4, 1977, 616-619.
4. Duchon, J., "Interpolation des Fonctions de deux variables suivant le principe de la flexion des plaques minces", R.A.I.R.O. Analyse Numérique, Vol. 10, no. 12, 1976, 5-12.
5. Duchon, J., "Splines minimizing rotation-invariant semi-norms in Sobolev spaces", in Constructive Theory of Functions of Several Variables, (W. Schempp and K. Zeller Ed.), Springer-Verlag, 1977, 85-100.

6. Franke, R., "A critical comparison of some methods for interpolation of scattered data", Naval Postgraduate School NPS-53-79-003, March 1979.

7. Hardy, R.L., "Multiquadric equations of topography and other irregular surfaces" J. of Geophysical Research, vol. 76, 1971, 1905-1915.

8. Meinguet, J., "An intrinsic approach to multivariate spline interpolation at arbitrary points" in Polynomial and Spline Approximation-Theory and Applications, (B.N. Sahney ed.) Dordrecht: D. Reidel Publishing Company, 1979; 163-190.

9. Meinguet, J., "Multivariate interpolation at arbitrary points made simple", ZAMP, vol. 30, 1979, 292-304.

10. Paihua Montes, L. and Utreras Diaz, F., "Un ensemble de programmes pour l'interpolation des fonctions, par des fonctions spline du type plaque mince", R.R. no. 140, Mathematiques appliques, Universite Scientifique et Medicale de Grenoble, Oct. 1978.

11. Wahba, G., "How to smooth curves and surfaces with splines and cross-validation", University of Wisconsin-Madison, Statistic Department TR #555, March 1979.

# THE N-WIDTHS OF SETS OF ANALYTIC FUNCTIONS

S. D. Fisher
Northwestern University
Evanston, Illinois

C. A. Micchelli
IBM Thomas Watson Research Center
Yorktown Heights, New York

Let $\Omega$ be a domain in the complex plane, $K$ a compact subset of $\Omega$, and let $A$ be the restriction to $K$ of the closed unit ball of $H^\infty(\Omega)$, the space of bounded holomorphic functions on $\Omega$. We always assume $H^\infty(\Omega)$ contains non-constant functions. In this note we announce some results on the determination of certain n-widths of $A$; extensions, elaborations and proofs will appear elsewhere.

The Kolmogorov n-width of a subset $A$ of a Banach space $X$ is defined by

$$d_n(A;X) = \inf_{X_n} \sup_{x \in A} \inf_{y \in X_n} \|x-y\| \tag{1}$$

where $X_n$ runs over all n-dimensional subspaces of $X$.

The Gel'fand n-width of $A$ in $X$ relative to a Banach space $Y$ is defined by

$$d^n(A;X,Y) = \inf_{Y_n} \sup_{x \in A \cap Y_n} \|x\| \tag{2}$$

where $Y_n$ runs over all closed subspaces of $Y_n$ of codimension $n$. ($Y$ does not necessarily have to be $X$; however, $A$ is a subset of $Y$.)



ISBN 0-12-780650-4

The linear n-width of A in X relative to Y is defined by

$$s_n(A;X,Y) = \inf_{P_n} \sup_{x\varepsilon A} ||x-P_n x|| \tag{3}$$

where $P_n$ runs over all bounded linear operators on Y whose range is a subspace of X of dimension n or less.

Let $\nu$ be a probability measure on K; we take X to be either $L^q(\nu)$, $1\leq q\leq\infty$, or C(K), the space of continuous complex-valued functions on K.

Let $g(z,\zeta)$ be the Green's function of $\Omega$ with pole at $\zeta$. A Blaschke product B(z) of degree d is an analytic function on $\Omega$ of the form

$$B(z) = \lambda\exp\{-\sum_{j=1}^{d}(g(z,\zeta_j)+ih(z,\zeta_j))\} \tag{4}$$

where $\lambda$ is a constant of modulus 1, $\zeta_1,\ldots,\zeta_d$ are points of $\Omega$ (not necessarily distinct) and $h(z,\zeta_j)$ is the harmonic conjugate of $g(z,\zeta_j)$. B(z) may be multiple-valued but its modulus is single-valued. We let $\mathcal{B}_r(\Omega)$ denote the set of Blaschke products of degree r or less on $\Omega$.

Theorem 1. Suppose $\Omega$ has precisely m+1 components in its complement, all of which are larger than a single point. Then all of $d_n(A,X)$, $d^n(A;X,H^\infty(\Omega))$, and $s_n(A;X,H^\infty(\Omega))$ lie in the interval from $\inf\{||B||_X:B\varepsilon\mathcal{B}_n\}$ on the right to $\inf\{||B||_X:B\varepsilon\mathcal{B}_{n+m}\}$ on the left.

Corollary 2. If $\Omega$ is simply-connected, then $d_n, d^n$, and $s_n$ are all equal with value

$$\inf\{||B||_X:B\varepsilon\mathcal{B}_n\}$$

Theorem 3. Suppose $\Omega$ is as in Theorem 1 and that K has the form

$$K=\{z\varepsilon\Omega:|B_0(z)|\leq\rho\} \tag{5}$$

for some $B_0\varepsilon\mathcal{B}_N$ and some $\rho\varepsilon(0,1)$. Then there are positive constants a and b with

$$a \le d_n(A;C(K)) \quad e^{n\gamma} \le b \qquad , n=1,2,3,\ldots \tag{6}$$

where $\gamma$ is a positive constant depending only on $K$ and $\Omega$.

Further comments. (i) The upper bound in Theorem 1 for $d_n$ holds for all domains $\Omega$ (ii) The upper bound in (6) holds also for any domain $\Omega$ provided $K$ is of the given form (iii) the lower bound in (6) holds for any $K$ provided $\Omega$ has $m+1$ complementary components (iv) the asymptotic result, $\lim\ d_n^{1/n}=\exp(-\gamma)$ , which follows from (6), was obtained previously; see [1], [2], and [3]

REFERENCES

1. N.D. Erokhin, Russ. Math. Surveys 23 (1968), 93-135
2. A.L. Levin and V.M. Tikhomirov, appendix to [1]
3. H. Widom, J. Approx. Theory, 5 (1972), 343-361.

# ADMISSIBILITY OF QUADRATURE FORMULAS WITH RANDOM NODES

B. Granovsky

Department of Mathematics
Technion — Israel Institute of Technology
Haifa, Israel

V. Knoh

Leningrad

## I. INTRODUCTION

The general concept of admissibility of a Monte Carlo procedure was stated first by Ermakov [1] who began also to study the problem in application to one common class of such procedures - RQ (quadrature formulas with random nodes or random quadratures) designed for statistical calculation definite integrals. Then the admissibility of RQ was being studied in a number of subsequent papers (see for references [2], [3]). Particularly, in [4] necessary and sufficient conditions for admissibility of RQ with the simplest quadrature sum were established and on its basis some new types of admissible RQ were constructed ([4], [5]).

The conditions obtained in [4] proved to admit the following visual interpretation. A good (i.e., admissible) RQ mustn't be "too random". More precisely, its nodes must be pairwise statistically dependent in such a way that the



ISBN 0-12-780650-4

cumulative density function G(x,y) of any two of them vanishes under all x = y.

The purpose of this paper is to extend these results on a more general class of RQ.

## II. BASIC ASSUMPTIONS AND RESULTS

Here an extension of some basic results in RQ theory ([3], [4]) will be presented.

Let $\{X, \beta\ \mu\}$ be a measurable space and $L_2(\mu)$ be a class of all square integrable functions on $(X, \beta, \mu)$. We'll be concerned throughout with the problem of statistical estimation of the definite integral

$$I = \int_X f\, d\mu = \int f\, d\mu$$

on the class of functions $f \in L_2(\mu)$. Further the following class of linear Monte Carlo procedures will be considered. For a fixed integer N denote $X^{(N)} = X \times \ldots \times X$ (N times), $\Xi = (x_1, \ldots, x_N) \in X^{(N)}, \beta^{(N)} = \beta \times \ldots \times \beta$ (N times) and for any $f \in L_2(\mu)$ define on $\{X^{(N)}, \beta^{(N)}\}$ the quadrature sum $K_N(f) = K_N(f;\Xi) = \Sigma_{i=1}^N A_i(\Xi) f(x_i)$ with coefficients $A_i(\Xi)$ and nodes $x_i \in X$, $i = 1, 2, \ldots, N$. Introduce also a class $\{W\}$ of all probability measures W on $\{X^{(N)}, \beta^{(N)}\}$.

Definition 1. A pair $[K_N(f); W]$ is called a RQ **on** the class of functions $L_2(\mu)$ if the following two conditions hold for all $f \in L_2(\mu)$:

$$E[K_N(f); W] = \int f\, d\mu \tag{1}$$

$$\text{Var.}[K_N(f); W] < \infty \tag{2}$$

It is obvious that (1), (2) impose definite restrictions on both $A_i$, $i = 1, \ldots, N$ and W. Denote further $F \subset L_2(\mu)$ and $K$ the considered classes of functions f and RQ correspondingly.

Taking now dispersion as a measure of an RQ's quality we are led to introduce the following notion of admissibility which is familiar in Mathematical Statistics (see for references, e.g. [6]).

Definition 2. An RQ $[\tilde{K}_N; \tilde{W}]$ is called dominating an RQ $[K_N; W]$ on F if

$$\text{Var}.[\tilde{K}_N(f); \tilde{W}] \leq \text{Var}[K_N(f); W]$$

for all $f \in \mathcal{F}$, with strict inequality somewhere. RQ $[K_N^*; W^*]$ is called admissible on $(\mathcal{F}, \mathcal{K})$ if there exist no RQ $\in \mathcal{K}$ dominating $[K_N^*; W^*]$ on F.

The natural beginning for studying any problem of admissibility is to try to distinguish a sufficiently wide class of nonadmissible procedures.

For this purpose we'll proceed with the following

Definition 3. An RQ $[K_N; W]$ is called symmetric on $\mathcal{F}$ if the quadrature sum $K_N(f) = K_N(f; \Xi)$ is symmetric (mod W) for all $f \in F$ with respect to all its nodes $x_i$, $i = 1, \ldots, N$ and W is a symmetric measure on $\beta^{(N)}$.[1]

The importance of symmetric RQ follows from:

Lemma. For any non-symmetric RQ $[K_N; W]$ on F there exists such a symmetric RQ $[\tilde{K}_N; \tilde{W}]$ on F that Var. $[\tilde{K}_N, \tilde{W}] \leq \text{Var}[K_N, W]$ for all $f \in F$.

Proof is based on the symmetrization method which is common in mathematical statistics ([6]).

---

[1]Measure W is called symmetric on $\mathcal{X}^{(N)}$ if for any set $B \in \beta^{(N)}$, $W(B) = W(B')$ where B' is obtained from B by any of N! permutations of coordinates.

Consider $W(B) = \frac{1}{N!} \Sigma_{i=1}^{N!} W(B^{(i)})$ and $\tilde{K}_N(f ; \Xi) = = \frac{1}{N!} \Sigma_{i=1}^{N!} K_N(f ; \Xi^{(i)}) \frac{dW}{d\tilde{W}}$ where the set $B^{(i)}$ is obtained from the set $B \in \beta^{(N)}$ by one of N! permutations of coordinates $x_j$, and the Radon-Nikodym's derivative exists since the measures W and $\tilde{W}$ are reciprocally absolutely continuous. Then it is easy to show that $[\tilde{K}_N ; \tilde{W}]$ is an RQ and

$$\tilde{K}_N^2 (f ; \Xi) \leq \frac{1}{N!} \sum_{i=1}^{N!} K_N^2 (f ; \Xi^{(i)}) \frac{dW}{d\tilde{W}}$$

whence the assertion of the lemma immediately follows.

Further, throughout the paper only RQ on the class of functions $L_2(\mu)$ are considered. The coefficients $A_i$ of these RQ must possess the following properties:

$A_i(\mu)$ is symmetric with respect to $x_1,\dots,x_{i-1},x_{i+1},\dots,x_N$ ;

$A_i(\Xi) = A_1(\Xi')$, where $\Xi'$ is obtained from $\Xi$ by transposition of coordinates $x_1$ and $x_i$. Both properties are immediate to derive from the symmetry of $K_N(f,\Xi)$ as a functional on $L_2(\mu)$.

Besides the symmetry of W implies that all the nodes $x_1,\dots,x_N$ must have the same marginal distribution law which we shall call $P_W^{(1)}$.

Theorem 1. For $[K_N(f) ; W]$ to be an RQ it is necessary that $\mu \leq P_W^{(1)}$ and

$$E_{x_1} [A_1(\Xi); W] = N^{-1} \frac{d\mu}{dP_W^{(1)}} (x_1) \geq 0, x \in X \quad (\text{mod } \mu) \qquad (3)$$

where $E_{x_1}$ is a version of the conditional expectation operator with respect to $x_1$ and $\frac{d\mu}{dP_W^{(1)}}(x_1)$ is the Radon-Nikodym's derivative. The above conditions and condition

$$\frac{E_{x_1} [A_1^2(\Xi) ; W]}{E_{x_1} [A_1(\Xi) , W]} \leq C = \text{const}, \ x_1 \in X \quad (\text{mod } \mu) \qquad (4)$$

are sufficient for $[K_N ; W]$ to be an RQ.

Proof. Making use of the symmetry of $[K_N(f) ; W]$ and Fubini's theorem we have:

$$E[K_N;W]=NE[f(x_1)A_1(\Xi);W]=N\int_X f(x_1)E_{x_1}[A_1(\Xi);W]\; P_W^{(1)}(dx_1) \tag{5}$$

From here it follows immediately that (3) is necessary and sufficient for (1).

Let now both conditions (3) and (4) hold. Proceeding furthur the same way as when deducing (5) we obtain with the help of the Cauchy-Schwartz inequality and also (4) and (3),

$$\begin{aligned}E[K_N^2(f) ; W] &= NE[f^2(x_1)A_1^2(\Xi); W] + N(N-1) \times \\ &\times E[f(x_1)f(x_2)A_1(\Xi)A_1(\Xi') ; W ] \le \\ &\le N^2E[f^2(x_1)A_1^2(\Xi); W] = \\ &= N^2 \int f^2(x_1)E_{x_1}[A_1^2(\Xi); W]\; P_W^{(1)}(dx_1) \le \\ &\le N^2C\int f^2(x_1)N^{-1} \frac{d\mu}{dP_W^{(1)}} (x_1)P_W^{(1)}(dx_1) = \\ &= CN \int f^2 d\mu < \infty,\; f \in L_2(\mu) .\end{aligned}$$

This completes the proof.

Let $W_1$ and $W_2$ denote respectively the singular and absolutely continuous parts of measure W with respect to measure $\mu^{(N)}$, so as $W = W_1+W_2$. If $W_1 \neq 0$ we come to different kinds of RQ with functionally dependent or partially fixed nodes which were studied in a number of papers (see for references [2]).

Here we restrict ourselves to the case when W is such that $P_W^{(1)}$ has density $p_W^{(1)}$ with respect to measure $\mu$ and the distribution law $P_W^{(2)}$ of any two nodes, say $x_1,x_2$, is absolutely continuous with respect to measure $\mu^{(2)}$, with $p_W^{(2)}(x_1,x_2)$ as the density. Under this assumption and provided the condition (4) holds the dispersion of any RQ $\in K$ admits the following representation:

$$\mathrm{Var}[K_N(f);\ W] = N\int f^2(x)h_W^2(x)\mu(dx) + N(N-1)\iint_{XX} f(x)f(y)q_W(x,y)\mu(dx)\mu(dy) - \left(\int f\, d\mu\right)^2$$

where

$$h_W^2(x_1) = p_W^{(1)}(x_1)E_{x_1}[A_1^2(\Xi);\ W] \le \le \text{const.},\ x_1 \in X$$

and

$$q_W(x_1,x_2) = E_{x_1,x_2}[A_1(\Xi)A_1(\Xi');\ W]\ p_W^{(2)}(x_1,x_2).$$

## III. ADMISSIBILITY CONDITIONS

We proceed now to a direct study of the admissibility problem. For this purpose we have to adopt certain assumptions on topological properties of the space $X$ and corresponding to them restrictions on the whole class $K$ of the considered RQ. The restrictions will be as follows. $X$ is a compact set of a finite-dimensional Euclidean space, the measure $\mu$ is reciprocally absolutely continuous with Lebesgue measure on $X$, the condition (4) holds and the function $q_W(x_1,x_2)$ is continuous (mod $\mu$) on the set of points $(x,x)$, $x \in X$.

Distinguish now the class $K_0 = \{[K_N^{(0)};\ W]\} \subset K$ of RQ with coefficients $A_i$ of the form $A_i(\Xi)=A_i(x_i)=A(x_i)$, $i=1,\ldots,N$ and distribution W having density $G(\Xi)$ (with respect to measure $\mu^{(N)}$), which is continuous (mod $\mu^{(N)}$) on $X^{(N)}$ and equal to zero at all points of its discontinuity. Note that for the class $K_0$ the condition (4) reduces to the condition $A(x) \leq$ $\leq$ const, $x \in X$ (mod $\mu$).

Our goal is to extend the admissibility condition obtained in [4] on the above defined class $K_0$ of RQ.

Theorem 2. For an RQ $\in K_0$ to be admissible in the class $K$ of RQ it is necessary and sufficient that

$$A(x)p_W^{(1)}(x) = N^{-1} , x \in X \qquad (\text{mod } \mu) \tag{7}$$

and

$$p_W^{(2)}(x,x) = 0 , x \in X. \tag{8}$$

Proof. The statement in (7) is a straight consequence of the condition (3) when applied to an RQ $\in K_0$. So only (8) is left for us to prove. First observe that for an RQ $\in K_0$

$$q_W(x_1,x_2) = A(x_1)A(x_2)p_W^{(2)}(x_1,x_2) \tag{9}$$

where $A(x) > 0$ (mod $\mu$) because of (7) and the density function $p_W^{(2)}$ is continuous (mod $\mu^{(2)}$) on $X^{(2)}$ according to our assumptions on G.

To establish the necessity of the condition (8) we employ the construction similar to [4]. Consider an arbitrary RQ $\in K_0$ which doesn't satisfy (7) and let $(\overline{x},\overline{x}) \in \overline{X}^{(2)}$ be such a point that $p_W^{(2)}(\overline{x},\overline{x}) > 0$. Then the above assumptions on G imply the existence of such point $\overline{\Xi} = (\overline{x},\overline{x},\overline{x}_3,\ldots,\overline{x}_N) \in X^{(N)}$ with $\overline{x}_j \neq \overline{x}_i \neq \overline{x}$, $j \neq i$, $i,j = 3,\ldots, N$ that $G(\Xi) > 0$ in some neighborhood of the point $\overline{\Xi}$ and the function G is continuous in this neighborhood. Choose the above neighborhood, say $\Phi^{(N)}$, in the form $\overline{X} \times \overline{X} \times \overline{X}_3 \times \cdots \times \overline{X}_N$ where $\overline{X},\overline{X}_3,\ldots, \overline{X}_N$ are mutually disjoint neighborhoods of the points $\overline{x},\overline{x}_3,\ldots, \overline{x}_N$

respectively and denote by symbol $U\Phi^{(N)}$ the union of all $2^{-1}N!$ mutually disjoint sets which may be obtained from $\Phi^{(N)}$ by all possible permutations of the coordinates $x_i, i=1,2,\dots,N$.

Define now on $X^{(N)}$ a symmetric function

$$\Psi(\Xi) = \eta(\Xi) \sum_{1\leq i<j\leq N} \ell(x_i)\ell(x_j)\eta_1(x_i,x_j) \qquad (10)$$

where $\eta$ and $\eta_1$ are the characteristic functions of the sets $U\Phi^{(N)}$ and $\overline{X}^{(2)} = \overline{X} \times \overline{X}$ respectively and $0 \neq \ell \in L_2(\mu)$ is a continuous on $\overline{X}$ function obeying the condition

$$\int_{\overline{X}} \ell \, d\mu = 0 \qquad (11)$$

Then there exists such $\varepsilon > 0$ that the function $\widetilde{G} = G - \varepsilon\Psi$ is nonnegative on $U\Phi^{(N)}$. In addition (10), (11) imply that $\widetilde{G}$ is a symmetric density function with marginal density $p_{\widetilde{W}} = p_W$. According to Theorem 1 this means that $[K_N ; \widetilde{W}]$ with $\widetilde{W}$ corresponding to $\widetilde{G}$ is also an RQ $\in K_0$.

Here on the basis of (9) - (11) we obtain

$$\mathrm{Var.}[K_N^{(0)}(f);\widetilde{W}] = \mathrm{Var.}[K_N^{(0)}(f);W] - \varepsilon N! \prod_{k=3}^{N} \mu(\overline{X}_k)\left(\int_{\overline{X}} Af \, \ell d\mu\right)^2.$$

This shows the RQ $[K_N^{(0)}(f) ; W]$ is inadmissible, which indeed proves the necessity part of our assertion. To prove sufficiency of the condition (8) we first observe that Cauchy-Schwartz' inequality and (3) imply for any RQ $\in K$

$$h_W^2(x) \geq [N^2 p_W^{(1)}(x)]^{-1}, \quad x \in X \qquad (\mathrm{mod}\ \mu). \qquad (12)$$

In particular if RQ $\in K_0$ then according to (7) $h_W^2(x) = A^2(x)p_W^{(1)}(x)$ and (12) becomes an equality. Consider now $[K_N^{(0)}(f) ; W^*] \in K_0$ satisfying all the conditions of the given theorem. The above arguments allow us to conclude that for any RQ $\in K$ with $h_W^2 \neq h_{W*}^2$ there exists such a set $\overline{X} \subset X$ of non-zero measure that $h_{W*}^2(x) < h_W^2(x)$, $x \in \overline{X}$. Hence for any function $\overline{f} \neq 0$ which vanishes outside $\overline{X}$ we have

$$\mathrm{Var}[K_N^{(0)}(\overline{f}) \; ; W^*] - \mathrm{Var}[K_N(\overline{f}) \; ; W] \leq$$

$$\leq (N-1)N \iint\limits_{\overline{X}\,\overline{X}} \overline{f}(x)\overline{f}(y)[q_{W*}(x,y)-q_W(x,y)] \; \mu(dx) \; \mu(dy)$$

with equality and $\overline{X} = X$ if only $h_{W*}^2(x)=h_W^2(x)$, $x \in X$ (mod $\mu$). From the condition (8) and the above assumptions on $q_W$ it follows that the set $\overline{X}$ can be assumed such that the kernel $\delta(x,y) = q_{W*}(x,y)-q_W(x,y)$ is continuous on $\overline{X} \times \overline{X}$ and $\delta(x,x) \leq 0$, $x \in \overline{X}$. Now it is easy to show the kernel $\delta$ cannot be non-negative definite from whence there follows the admissibility of the RQ $[K_N^{(0)}(f) \; ; W^*]$. The proof is completed.

The Theorem shows in particular the nodes of an admissible RQ cannot be pairwise independent.

Corollary 1. The method of importance sampling (see, e.g. [3]) is inadmissible. We observe in this connection that the inadmissibility of the rude Monte-Carlo method which can be treated as a particular case of the method of importance sampling was established as far back as, 1956, by Gelfand, Frolov and Schenzov.

Corollary 2. The method of stratified sampling (or modified Monte Carlo method) (see [3], [7]) is admissible.

This result was obtained in [4].

## IV. EXAMPLE OF ADMISSIBLE RQ

In this section we make use of Theorem 2 as a tool for construction of a new family of admissible RQ. First we observe that as it is seen from (7) we may construct an RQ $\in K_0$ with any given density function G suitable for simulation. This is just what stipulates the importance for application of the considered class of RQ.

From (7) it also follows that for any RQ $\in K_0$ all the coefficients $A_i(x_i) \geq 0$ , $x_i \in X$ , $i = 1,2,\ldots, N$.

Now we will construct an admissible RQ combining the properties of both Monte Carlo methods as in corollaries 1,2, above.

Let $X_1,\ldots, X_N$ be a partition of $X$ into N non-intersecting subsets and $p(x)$ be a density function on $(X,\beta,\mu)$ such that $p^{-1}(x) \leq \text{const}$, $x \in X$ and

$$\int_{X_j} p(x)\mu(dx) = N^{-1} , \quad j = 1,\ldots, N.$$

Consider

$$K_N^{(0)}(f) = \sum_{i=1}^{N} f(x_i)\ p^{-1}(x_i)$$

and

$$G(\Xi)= \begin{cases} \frac{N^N}{N!}\ p(x_1)\ p(x_2)\ \ldots\ p(x_N), & \Xi \in U\Phi^{(N)} \\ 0 , & \text{otherwise} \end{cases}$$

where $\Phi^{(N)} = X_1 \times \cdots \times X_N$ and the symbol $U\Phi^{(N)}$ has the same meaning as in Section 3. It is easy to verify that the corresponding $[K_N^{(0)}; W]$ is an RQ $\in K_0$ with $p_w^{(1)}(x) = p(x)$ and

$$p_w^{(2)}(x_1,x_2) = \begin{cases} \frac{N}{N-1}\, p(x_1)\,p(x_2), & (x_1,x_2) \in X_i \times X_j, \\ & i \neq j,\ 1 \leq i,j \leq N \\ 0, & \text{otherwise} \end{cases}.$$

So the considered RQ obeys all the conditions of Theorem 2 and is therefore admissible.

From (6) we obtain that the dispersion $D_1$ of our quadrature is equal to

$$D_1(f) = N^{-1}\int_X f^2 p^{-1}\, d\mu - \sum_{i=1}^{N}\left(\int_{X_i} f\, d\mu\right)^2 . \tag{13}$$

To analyse the expression denote $p\, d\mu = d\mu_p$ and $\eta_j$ the characteristic function of the set $X_j$, $j = 1,\ldots, N$. Then we may rewrite (13) in the form:

$$D_1(f) = N^{-1}R_N(fp^{-1})$$

where $R_N(fp^{-1})$ is the distance, in the metric $L_2(\mu_p)$, of the function $fp^{-1}$ from the linear subspace spanned by the orthonormal system of N functions $\sqrt{N}\,\eta_j$, $j = 1,2,\ldots, N$. Consider now the RQ $\in K_0$ corresponding to the classical method of importance sampling with the same density function $p_w^{(1)} = p$. In this case we have $p_w^{(2)}(x,y) = p(x)p(y)$ whence on the basis of (13) the dispersion, say $D_2$, of this RQ is equal to:

$$D_2(f) = N^{-1}\left[\int_X f^2 p^{-1} d\mu - \int_X f\, d\mu\right]^2 = N^{-1}R_1(fp^{-1})$$

where $R_1(fp^{-1})$ denotes the distance in the metric $L_2(\mu_p)$ of the function $fp^{-1}$ from the function $f_0 \equiv 1$, $x \in X$ which indeed belongs to the linear subspace of functions $\eta_j$, $j = 1,\ldots, N$ above.

This shows our RQ dominates the method of importance sampling.

REFERENCES

[1] Ermakov, S. M. (1968), On the Admissibility of Monte Carlo method procedures, *Soviet Math. Dokl., v.8,No. 1*, 66-67.

[2] Granovsky, B.L., Ermakov, S.M. (1977), The Monte Carlo Method, *Journal Soviet Mathematics*, 161-192, Publ. of the Amer. Math. Society.

[3] Ermakov, S. M. (1975), Monte Carlo Methods and some adjacent problems , *Nauka*, Moscow (in Russian).

[4] Granovsky, B.L. (1974), Admissibility conditions for a class of quadrature formulas with random nodes, Academy of Sciences of the USSR, Mathematical Notes, 16,2,763-767.

[5] Granovsky, B.L., Stolyarov, G.V. (1974), On the problem of admissiblity of random quadratures, in the collection, "Monte Carlo Methods in Comput. Math. & Physics", Novosibirsk. (in Russian).

[6] Lehmann, E.L. (1959), Testing statistical hypotheses, Wiley, N. Y.

[7] Haber, S. (1970), Numerical evaluation of multiple integrals, *SIAM Rev. 12*, 4, 481-526.

# CONVERGENCE FOR OPERATORS OF HYPERBOLIC TYPE

Joseph W. Jerome

Department of Mathematics
Northwestern University
Evanston, Illinois

## 1. INTRODUCTION

Consider the Cauchy problem

$$\text{(i)} \qquad \frac{du}{dt} + A(t)u = F(t), \tag{1.1}$$

$$\text{(ii)} \qquad u(0) = u_0,$$

on a Banach space X, where $\{-A(t)\}$ is a stable family of generators of strongly continuous semigroups on X. Kato [8,9] and Dorroh [2] have given a comprehensive analysis of the evolution operators for (1.1). A summary of a special case of that theory is given in §2.

The importance of (1.1) lies in the applications to a wide class of linear hyperbolic problems. Thus, for example, the symmetric linear hyperbolic system of first order in x,t on a space-time domain $D = \Omega \times (0,T)$ given by

$$u_t + \sum_{j=1}^{n} a_j(x,t) \frac{\partial u}{\partial x_j} + b(x,t)u = f(x,t), \tag{1.2}$$

where $a_j$, $1 \leqslant j \leqslant n$, are $m \times m$ symmetric matrices, b is an $m \times m$ matrix and $u = (u_1, \ldots, u_m)$ is a special case of (1.1) with

$$A(t)u = \sum_{j=1}^{n} a_j(\cdot,t) \frac{\partial u}{\partial x_j} + b(\cdot,t)u . \tag{1.3}$$




ISBN 0-12-780650-4

If, for example,

$$a_j, b \in C\left(0,T;\ C^1(\mathbb{R}^n)\right) \ ,$$

where $C^1(\mathbb{R}^n)$ is the set of all $m \times m$ matrix-valued functions a such that a and $\frac{\partial a}{\partial x_j}$, $1 \leq j \leq n$, are in the space $C(\mathbb{R}^n)$ of continuous, bounded functions on $\mathbb{R}^n$, then the operators of (1.3) may be identified with a family of closed, linear operators satisfying an $L^2$ energy inequality; hence, by well-known ideas, comprise a stable family of generators of strongly continuous semigroups on $L^2(\mathbb{R}^n)$ (cf. Friedrichs [3,4]). This permits the solution of the Cauchy problem for $u_0 \in H^1(\mathbb{R}^n)$ and suitable f. Maxwell's equations are well-known to define a system of type (1.2).

One speed neutron transport theory in a homogeneous slab, bounded by the planes $x = -a$ and $x = a$, leads to the integro-differential system

$$\frac{\partial}{\partial t} n(x,y,t) = -vy \frac{\partial}{\partial x} n(x,y,t) - v\Sigma n(x,y,t) + v\gamma \int_{-1}^{1} n(x,y',t)dy' \qquad (1.4)$$

on $(-a,a) \times (-1,1) \times (0,T)$. Equation (1.4) is augmented by homogeneous boundary conditions along $x = -a$, $x = a$, and a standard initial condition. Here v, $\Sigma$ and $\gamma$ are positive constants and $n(x,y,t)$ is the neutron density corresponding to position x, velocity cosine y and time t (see Ref. [1] for discussion and references). The operator of (1.4) may be realized as a bounded perturbation of a stable family of generators in $L^2\left((-a,a) \times (-1,1)\right)$, hence defines a stable family, independent of t.

In this note, we carry out a semidiscretization analysis, defined by applying a backward Euler approximation method to (1.1i). We obtain the following new results: (i) a convergence rate of $O(\Delta t)$ in the $L^2(0,T;Y^*)$ norm for the step function sequence defined by the semidiscretization; and (ii) the verification that the semidiscrete, time independent equations on X resulting from the discretization have solutions derivable from a contraction, hence can be solved by successive approximations. Here $Y^*$ is the dual of a smooth space Y and is the abstract counterpart of $H^{-1}(\Omega)$; Y is the natural state space for $u_0$. Property (i) is obtained in Theorem 3.3 and (ii) in Proposition 3.1.

We shall not explicitly formulate the applications of our general results in the interest of economy. The existing literature closest in spirit to the approach of this paper is the paper of Hersh and Kato [5] who treat a general class of time semidiscretizations for operators A which are not time dependent. They obtain estimates in stronger norms under hypotheses on $u_0$ far more stringent than in this paper. Their results are essentially higher order accuracy results. The reader may consult Ref. [5] for related references. We note only the early, informative paper of Weinberger [10].

We emphasize the expository character of much of this paper. It is our belief that the real interest in the techniques of §3 lies in the potential application to nonlinear problems (*cf.* [7] for a convergence argument in the case of quasi-linear systems).

## 2. OPERATORS OF HYPERBOLIC TYPE

This section reviews well-known concepts and results.

Definition 2.1: Let V be a closed linear operator with domain and range dense in a Banach space X. Denote by $R(\lambda,V)$ the resolveng $(\lambda I-V)^{-1}$ for $\lambda$ in the resolvent set $\rho(V)$. For $M > 0$ and $\omega \in \mathbb{R}$ denote by $G(X,M,\omega)$ the set of all operators $A = -V$ such that

$$\| [R(\lambda,V)]^k \| \leq M(\lambda-\omega)^{-k}, \quad k \geq 1, \ \lambda > \omega \tag{2.1}$$

Write $(G(X,M) = \bigcup_{-\infty < \omega < \infty} G(X,M,\omega)$ and $G(X) = \bigcup_{M>0} G(X,M)$.

Definition 2.2: Let X be a Banach space and suppose that a family $\{A(t)\}_{0\leqslant t\leqslant T} \subset G(X)$ is given. $\{A(t)\}$ is said to be stable if there are (stability) constants M and $\omega$ such that

$$\| \prod_{j=1}^{k} [A(t_j)+\lambda_j]^{-1} \| \leq M \prod_{j=1}^{k} (\lambda_j-\omega)^{-1}, \quad \lambda_j > \omega \tag{2.2}$$

for any finite families $\{t_j\}_{j=1}^k$ and $\{\lambda_j\}_{j=1}^k \subset \mathbb{R}$ with $0 \leq t_1 \leq \cdots \leq t_k \leq T$, $k=1,2,$ . Moreover, $\Pi$ is time-ordered, i.e., $[A(t_j)+\lambda_j]^{-1}$ is to the left of $[A(t_{j'})+\lambda_{j'}]^{-1}$ if $j > j'$.

We recall the first basic results in Ref.[8].

Proposition 2.1: Let Y be a Banach space densely and continuously embedded in a Banach space X such that the norm of Y is determined by an isomorphism S of Y onto X:

$$\|u\|_Y = \|Su\|_X . \tag{2.3}$$

Let $\{A(t)\} \subset G(X)$ be a given stable family such that $Y \subset D_{A(t)}$ and suppose that $\{A_1(t)\} \subset G(X)$ is defined by

$$\begin{aligned} &(\text{i}) \quad A_1(t) = SA(t)S^{-1} \\ &(\text{ii}) \quad D_{A_1(t)} = \{v \in X : A(t)S^{-1} v \in Y\} \end{aligned} \tag{2.4}$$

and that $\{A_1(t)\}$ is stable, with constants $M_1$ and $\omega_1$. Then the restriction of $A(t)$ to $S^{-1}D_{A_1(t)}$ is in $G(Y)$ with stability constants $\omega_1$ and $M_1$.

Remark 2.1: The proposition thus gives sufficient conditions for resolvent estimates for $A(t)$, as in (2.2), to be valid on Y. Such resolvent estimates are essential in deriving fixed point properties associated with the time semidiscretization of the next section. The next result is quoted from Ref. [9].

Proposition 2.2: Suppose that the hypotheses of Proposition 2.1 are satisfied and that $t \longrightarrow A(t)$ is norm continuous $(Y,X)$. Then there exists a unique family of operators $\{U(t,s)\} \subset B(X)$, defined on $D$: $0 \leqslant s \leqslant t \leqslant T$, with the following properties:

$$\begin{aligned}
&\text{(i)} && U(t,s) \text{ is strongly continuous on } D \text{ to } B(X) \text{ with } U(s,s) = I;\\
&\text{(ii)} && U(t,r) = U(t,s)U(s,r),\quad r \leqslant s \leqslant t;\\
&\text{(iii)} && U(t,s)Y \subset Y \text{ and } U(t,s) \text{ is strongly continuous on } D \text{ to } B(Y);\\
&\text{(iv)} && \frac{dU(t,s)}{dt} = -A(t)U(t,s),\quad \frac{dU(t,s)}{ds} = U(t,s)A(s),\\
& && \text{which exist in the strong sense in } B(Y,X) \text{ and are strongly continuous on } D \text{ to } B(Y,X).
\end{aligned} \tag{2.5}$$

Remark 2.2: The existence of evolution operators satisfying (2.5) permits the solution of the Cauchy problem (1.1) in the case $u_0 \in Y$ and $F \in C[0,T;X] \cap L^1(0,T;Y)$. Indeed $u$ is given explicitly by

$$u(t) = U(t,0)u_0 + \int_0^t U(t,s)\, F(s)ds, \tag{2.6}$$

as is well known, and satisfies $u \in C[0,T;Y] \cap C^1[0,T;X]$.

Remark 2.3: If $\frac{dF}{dt} \in L^2(0,T;X)$, $\lambda \geq \max(\omega,\omega_1) + 1$, and $\tau \in [0,T]$, then $(\frac{d^2}{dt^2})R(\lambda,-A(\tau))u \in L^2(0,T;X)$ and we have the estimate

$$\int_{t_1}^{t_2} \left\| \left(\frac{d^2}{dt^2}\right) R(\lambda,-A(\tau))u \right\|_X^2 dt \leq C_\lambda \int_{t_1}^{t_2} \{\|u_t\|_X^2 + \|F_t\|_X^2\} dt \qquad (2.7)$$

for $0 \leq t_1 \leq t_2 \leq T$, where $C_\lambda$ does not depend upon $\tau$. This regularity result and corresponding estimate follow directly from the equation (1.1). Estimate (2.7) will be used in the proof of Theorem 3.3. We note also that additional regularity of u can be obtained, say $u \in C[0,T;S^{-1}Y] \cap C[0,T;Y]$, if $u_0 \in S^{-1}Y$ and $t \to A(t)$ is norm continuous $(S^{-1}Y,Y)$.

## 3. THE TIME SEMIDISCRETIZATION AND RATES OF CONVERGENCE

In this section we assume throughout that X and Y are real Hilbert spaces related by Proposition 2.1 with $\|u\|_X \leq \|u\|_Y$. In the standard way, the dual of Y can be associated with a space with so-called negative norm:

$$(f,f)_{Y^*} = (S^{-1}f, S^{-1}f)_X, \quad f \in X, \qquad (3.1)$$

and (3.1) is extended by continuity to $Y^*$. X thus assumes the role of a pivot space. We also assume that $\{A(t)\}$, $\{A_1(t)\}$ in $G(X)$ are stable, with $M = M_1 = 1$.

Now let $N \geq 1$ be given and set $\Delta t = T/N$. Given $u_0 \in Y$, we are interested in the recursive solution of

$$A(t_n)u_n^N + \frac{1}{\Delta t}u_n^N = F_n^N + \frac{1}{\Delta t}u_{n-1}^N, \quad n = 1,\ldots,N, \qquad (3.2)$$

for $t_n = n\Delta t$, $n = 0,\ldots,N$ and $F_n^N = \frac{1}{\Delta t}\int_{t_{n-1}}^{t_n} F(t)dt$.

**Proposition 3.1:** Let r be defined by

$$r = r_\delta = (1+\delta)(\|u_0\|_Y + \|F\|_{L^1(0,T;Y)})e^{\gamma T}, \tag{3.3i}$$

$$\gamma = \max[1,\ 1 + \max(\omega,\omega_1)] \tag{3.3ii}$$

and $\delta > 0$ is arbitrary. Then, for $N \geq T[(1+\delta^{-1}) + \gamma]$, there exist unique solutions of (3.2) in the ball $\overline{B_Y(0,r)} \subset Y$. The $u_n^N$ may be realized as fixed points of the mappings, where $\mu^2 = 1/\Delta t$,

$$v \to Qv = -R(\mu^2-1,\ -A(t_n))v +$$

$$+ \mu^2 R(\mu^2-1,\ -A(t_n)) \prod_{j=1}^{n-1} \mu^2 R(\mu^2,\ -A(t_n))u_0 + \tag{3.4}$$

$$+ \sum_{i=1}^{n} \frac{1}{\mu^2} R(\mu^2-1,\ -A(t_n))\left\{\prod_{j=i}^{n-1} \mu^2 R(\mu^2,\ -A(t_j))\right\}F_i^N.$$

Q is a contraction with contraction constant $\frac{\delta}{1+\delta}$, on the complete metric subset of X defined by $\overline{B_Y(0,r)}$.

**Proof:** The contraction property of Q is seen from the estimate

$$\|Qv - Qw\|_X \leq \frac{1}{\mu^2-1-\omega}\|v-w\|_X \leq \frac{\delta}{\delta+1}\|v-w\|_X\ , \tag{3.5}$$

which follows from (2.2) and the restriction on N. Note that, by assumption, M = 1. By taking w = 0 we have an estimate for the first term on the r.h.s. of (3.4) in the norm of X, viz., $(\frac{\delta}{\delta+1})r$. The second term is bounded from above by

$$\left(\frac{\mu^2}{\mu^2-1-\omega}\right)\left(\frac{\mu^2}{\mu^2-\omega}\right)^{n-1} \leq \left(\frac{\mu^2}{\mu^2-1-\omega}\right)^{N} \leq e^{(1+\omega)T}\|u_0\|_X \tag{3.6}$$

and the third term by

$$\sum_{i=1}^{n} \frac{\mu^2}{\mu^2-1-\omega}\left(\frac{\mu^2}{\mu^2-\omega}\right)^{n-1}\|F\|_{L^1(t_{i-1},t_i;X)} \leq e^{(1+\omega)T}\|F\|_{L^1(0,T;X)} \tag{3.7}$$

Again, we have used (2.2) and the relation $\|\cdot\|_Y \geq \|\cdot\|_X$. It follows from the three previous estimates that

$$\| Qv \|_X \leq \left(\frac{\delta}{1+\delta}\right) r + \left(\frac{1}{1+\delta}\right) r \leq r. \tag{3.8}$$

A similar analysis, either using the relation (2.9i) or the direct resolvent estimates on Y noted in Proposition 2.1, yields

$$\| Qv \|_Y \leq r. \tag{3.9}$$

The fixed point property of Q on $\overline{B_Y(0,r)}$ follows from (3.5), (3.8) and (3.9). By using the resolvent formulation of (3.2) (cf. (3.10) to follow), coupled with induction, we see that such fixed points satisfy (3.2).

Remark 3.1: One of course does not determine $u_n^N$ by means of Q as defined in (3.4), but rather by the recursive solution of

$$\begin{aligned} u_n^N &= -R(\mu^2-1, -A(t_n))u_n^N + \mu^2 R(\mu^2-1, -A(t_n))u_{n-1}^N + \\ &\quad + R(\mu^2-1, -A(t_n))F_n^N = Qu_n^N, \end{aligned} \tag{3.10}$$

which is simply the resolvent formulation of (3.2).

Definition 3.1: For $0 \leq t \leq T$ and $\lambda \geq \gamma$ define the bilinear form $E(t,\lambda;\cdot,\cdot)$ on X by

$$E(t,\lambda;f,g) = (R(\lambda, -A(t))f,g)_X. \tag{3.11}$$

The following result is a slight generalization of a result of [7]. It makes use of the energy inequality.

$$(A(t)f,f)_X \geq -\omega(f,f)_X \tag{3.12}$$

satisfied by $A(t) \in G(1,\omega,X)$ ([1, p.142]).

Proposition 3.2: $E(t,\lambda;\cdot,\cdot)$ is a positive definite (non-symmetric) bilinear form on X; the corresponding form $\{E(t,\lambda;\cdot,\cdot)\}^{1/2}$ dominates the Y* norm, with constant independent of t and depending reciprocally upon $\lambda^{1/2}$.

Remark 3.2: We are now prepared to state the fundamental convergence theorem of this section. u denotes the solution of the Cauchy problem (1.1). The regularity of u is asserted in Remark 2.2. The hypothesis

$$\frac{dF}{dt} \in L^2(0,T;X) \tag{3.13}$$

will be employed in the theorem.

Theorem 3.3: Suppose (3.13) holds. There is a constant C such that

$$\sum_{n=1}^{N-1} \| u(t_n) - u_n^N \|_{Y*}^2 \Delta t \le C(\Delta t)^2 \tag{3.14}$$

for each N defined in Proposition 3.1. In particular, if $\{u^N\}$ denotes the step function sequence defined by

$$u^N(t) = u_n^N, \; t_{n-1} \le t < t_n, \; n=1,\ldots, N-1, \tag{3.15}$$

then

$$\| u-u^N \|_{L^2(0,T;Y*)} \le C_1(\Delta t) \tag{3.16}$$

for some constant $C_1$.

*Proof:* Since $u_t \in L^2(0,T;x)$, and hence, 'a fortiori' to $L^2(0,T;Y*)$, it follows that (3.15) is a simple consequence of (3.14) and the triangle inequality. To verify (3.14) we write

$$\text{(i)} \quad \frac{u(t_n) - u(t_{n-1})}{\Delta t} + A(t_n)\, u(t_n) =$$

$$= \left[\frac{u(t_n) - u(t_{n-1})}{\Delta t} - \frac{\partial u}{\partial t}(t_n)\right] + F(t_n) \; , \tag{3.17}$$

$$\text{(ii)} \quad \frac{u_n - u_{n-1}}{\Delta t} + A(t_n)u_n = F_n,$$

where (3.17i) represents (1.1) evaluated at $t_n$ and where (3.17ii) is a notationally simplified version of (3.2).

Operating on (3.17) with $R(\lambda,-A(t_n))$ ($\lambda \geq \gamma$; cf. (3.3ii)), followed by inner product with $R(\lambda,-A(t_n))(u(t_n)-u_n)$, gives, upon subtraction,

$$\begin{aligned}
&\frac{1}{\Delta t}(R_\lambda(t_n)(u(t_n)-u_n),\ R_\lambda(t_n)(u(t_n)-u_n))_X \\
&- \frac{1}{\Delta t}(R_\lambda(t_n)(u(t_{n-1})-u_{n-1}),\ R_\lambda(t_n)(u(t_n)-u_n))_X \\
&+ (R_\lambda(t_n)(u(t_n)-u_n),\ u(t_n)-u_n)_X \\
&= \lambda(R_\lambda(t_n)(u(t_n)-u_n),\ R_\lambda(t_n)(u(t_n)-u_n))_X \\
&+ (R_\lambda(t_n)(F(t_n)-F_n),\ R_\lambda(t_n)(u(t_n)-u_n))_X \\
&+ \left(\frac{1}{\Delta t}\int_{t_{n-1}}^{t_n} R_\lambda(t_n)\left[\frac{du}{dt}(t)-\frac{du}{dt}(t_n)\right]dt,\ R_\lambda(t_n)(u(t_n)-u_n)\right)_X,
\end{aligned} \tag{3.18}$$

where we have written $R_\lambda(t_n) = R(\lambda,-A(t_n))$ and have interchanged $\int_{t_{n-1}}^{t_n}$ and $R_\lambda(t_n)$. For conciseness, we simply summarize the estimation. Apply the inequality

$$|(f,g)_X| \leq \frac{1}{2}\left\{\varepsilon\|f\|_X^2 + \varepsilon^{-1}\|g\|_X^2\right\} \tag{3.19}$$

to the second term on the l.h.s. of (3.18) with $\varepsilon=1$ and sum from $k=1$ to $k=n$. Follow this by an application of (3.19) to the second and third terms on the r.h.s. of the resultant sum with $\varepsilon=4$; make a choice of $\Delta t$ sufficiently small so that $\lambda\Delta t < 1/8$. Altogether, then, we have, for $1 \leq n \leq N-1$,

$$\begin{aligned}
&\frac{1}{8}\|R_\lambda(t_n)(u(t_n)-u_n)\|_X^2 + \sum_{k=1}^{n} E(t_k,\lambda;u(t_k)-u_k,u(t_k)-u_k)\Delta t \\
&\leq (4+\lambda)\sum_{k=1}^{n-1}\|R_\lambda(t_k)(u(t_k)-u_k)\|_X^2\Delta t + \left(\frac{c}{\lambda^2}\right)\sum_{k=1}^{n}\|F(t_k)-F_k\|_X^2\,\Delta t \\
&+ 2\sum_{k=1}^{n}\left\|\left(\frac{d^2}{dt^2}\right)R_\lambda(t_k)u\right\|_{L^2(t_{k-1},t_k;X)}^2(\Delta t)^2.
\end{aligned} \tag{3.20}$$

Notice that we have multiplied through by $\Delta t$, used the fact that $u(0) = u_0$, and used the estimate

$$\left\| \frac{1}{\Delta t} \int_{t_{k-1}}^{t_k} R_\lambda(t_k)\left[\frac{du}{dt}(t) - \frac{du}{dt}(t_k)\right]dt \right\|_X^2$$

$$\leq \left(\frac{1}{\Delta t}\right)^2 \left\{ \int_{t_{k-1}}^{t_k} \int_t^{t_k} \left\|\left(\frac{d^2}{dt^2}\right) R(t_k)u\right\|_X ds \right\}^2$$

$$\leq (\Delta t)^2 \left\| \left(\frac{d^2}{dt^2}\right) R(t_k)u \right\|^2_{L^2(t_{k-1}, t_k; X)}$$

together with the permutation of $\frac{d}{dt}$ and $R(t_k)$, and the (time) differentiability property of $R(t_k)u$ (cf. Remark 2.3). An application of the discrete Gronwall inequality (cf.[6]) to (3.20) yields

$$\sum_{k=1}^{N-1} E(t_k, \lambda;\ u(t_k) - u_k,\ u(t_k - u_k)) \leq C(\Delta t)^2$$

upon use of (2.7) and (3.13). Proposition 3.2 and (3.21) now give (3.14). □

## REFERENCES


1. A. Bellini-Morante, Applied Semigroups and Evolution Equations, Oxford Mathematical Monographs, Clarendon Press, Oxford, 1979.
2. J. Dorroh, A Simplified Proof of a Theorem of Kato on Linear Evolution Equations, *J. Math. Soc. Japan, 27* (1975), 474-478.
3. K. Friedrichs, The Identity of Weak and Strong Extensions of Differential Operators, *Trans. Amer. Math. Soc. 55* (1944), 132-151.
4. K. Friedrichs, Symmetric Hyperbolic Linear Differential Equations, *Comm. Pure Appl. Math. 7* (1954), 345-392.

5. R. Hersh and T. Kato, High-Accuracy Stable Difference Schemes for Well-Posed Initial-Value Problems, *SIAM J. Numer. Anal. 16* (1979), 670-682.

6. J.Jerome, Nonlinear Equations of Evolution and a Generalized Stefan Problem, *J. Differential Equations, 26* (1977), 240-261.

7. J. Jerome, A Global Existence Theorem for Quasi-Linear Symmetric Hyperbolic Systems, manuscript.

8. T. Kato, Linear Evolution Equations of Hyperbolic Type, *J. Fac. Science Univ. Tokyo, 17* (1970), 241-258.

9. T. Kato, Linear Evolution Equations of Hyperbolic Type, II, *J. Math. Soc. Japan 25* (1973), 648-666.

10. H. Weinberger, Error Bounds in Finite Difference Approximation to Solutions of Symmetric Hyperbolic Systems, *J. Soc. Ind. Appl. Math. 7* (1959), 49-75.

# EXPLICIT $\mathcal{C}^n$-EXTENSIONS OF FUNCTIONS OF TWO VARIABLES IN A STRIP BETWEEN TWO CURVES, OR IN A CORNER, IN $\mathbb{R}^2$

Alain Le Mehaute
Laboratoire d'Analyse Numérique
Institut National des Sciences Appliquees
Rennes, France

## I. INTRODUCTION AND NOTATIONS

In this note we present some explicit methods to construct $\mathcal{C}^n$-extension of a function, in a strip, called $\mathcal{B}$, between two curves, $\gamma_1$ and $\gamma_2$ even they intersect in a corner.

First, a function f and its Taylor field of order 2n is assumed to be known on $\gamma_1$ and $\gamma_2$, and we construct two $\mathcal{C}^n$ extensions of f to $\mathcal{B}$ and then, using a reflection principe, we construct another $\mathcal{C}^n$-extension of f to $\mathcal{B}$.

We use the following notations :

* $\alpha = (\alpha_1, \alpha_2) \in \mathbb{N} \times \mathbb{N}$ $\qquad |\alpha| = \alpha_1 + \alpha_2 \in \mathbb{N}$

$\alpha! = \alpha_1!\ \alpha_2!$ $\qquad \binom{\alpha}{\beta} = \binom{\alpha_1}{\beta_1}\binom{\alpha_2}{\beta_2}$

$\beta \leqslant \alpha$ iff $\beta_1 \leqslant \alpha_1$ and $\beta_2 \leqslant \alpha_2$

$$D^{\alpha} f(M) = \frac{\partial^{\alpha_1+\alpha_2} f}{\partial x^{\alpha_1}\ \partial y^{\alpha_2}}(M)$$

If $A = (x_A, y_A)$ and $B = (x_B, y_B)$ are two points in $\mathbb{R}^2$,

$$AB = ((x_A - x_B)^2 + (y_A - y_B)^2)^{1/2}$$

$$AB^{\alpha} = (x_B - x_A)^{\alpha_1}\ (y_B - y_A)^{\alpha_2}$$




ISBN 0-12-780650-4

* If f is m-times continuously differentiable on a compact E, $T_A^m f$ is the Taylor expansion of f of order m at A, ie :

$$T_A^m f \; : \quad M \to T_A^m f(M) = \sum_{|\alpha| \leq k} \frac{1}{\alpha!} D^\alpha f(A).AM^\alpha$$

* Whitney expansion theorem (WHITNEY [5])

A polynomial field $A \to T_A f$ of order m is the restriction to a compact $E \subset R^n$ of the Taylor field of a function $f \in \mathcal{C}^n(E)$ if and only if there exists a modulus of continuity $\omega_1$ such that

$$(W) \qquad \forall A \in E \quad \forall B \in E \qquad \sup_{0 \leq |\alpha| \leq m} \frac{|D^\alpha T_B f(A) - D^\alpha T_A f(A)|}{AB^{m-|\alpha|}} \leq \omega_1(AB)$$

It is often more convenient to use the equivalent formulations of (W) (TOUGERON [4], COATMELEC [1])

$$\forall \alpha, \quad |\alpha| \leq m, \quad \forall A \in E \quad \forall B \in E \quad \forall M \in \mathbb{R}^n$$

$$|D^\alpha(T_A f)(M) - D^\alpha(T_B f)(M)| \leq 2^{m-|\alpha|} e^{n/2} \omega_1(AB) \left\{ AM^{m-|\alpha|} + BM^{m-|\alpha|} \right\} \qquad (1)$$

or

$$|D^\alpha(T_A f)(M) - D^\alpha(T_B f)(M)| \leq AM^{m-|\alpha|} \omega_2(AM) + BM^{m-|\alpha|} \omega_2(BM) \qquad (2)$$

(when $\omega_2$ is another modulus of continuity).

* FAA DE BRUNO Formula (LE MEHAUTE [3])

Let $\phi : \mathbb{R}^s \to \mathbb{R}$, $f : \mathbb{R} \to \mathbb{R}$, m times continuously differentiables. There, for $|\nu| \leq m$, for all $M \in \mathbb{R}^s$,

$$\frac{1}{\nu!} D^\nu (f \circ \phi)(M) = \sum_\sigma \frac{1}{\alpha!} f^{(|\alpha|)} (\phi(M)) \prod_{1 \leq |\beta| \leq m} \left( \frac{1}{\beta!} D^\beta \phi(M) \right)^\beta$$

where $\sigma$ means that

$$\alpha \in \mathbb{N}^{\binom{n+s}{s}-1} \quad \text{and} \quad \sum_{\substack{1 \leq |\gamma| \\ \gamma \leq \nu}} \gamma_i \alpha_\gamma = \nu_i \quad ; \quad i = 1,2,\ldots,s$$

Let $\gamma_1$ and $\gamma_2$ two curves in $\mathbb{R}^2$, represented by $y = \gamma_1(x)$ and $y = \gamma_2(x)$, we suppose that :

a) $\gamma_1$ and $\gamma_2$ are $\mathcal{C}^n$ continuously differentiables

b) $\gamma_1(x) \leqslant \gamma_2(x)$ or $\gamma_2(x) \leqslant \gamma_1(x)$ , for all $x \in E$

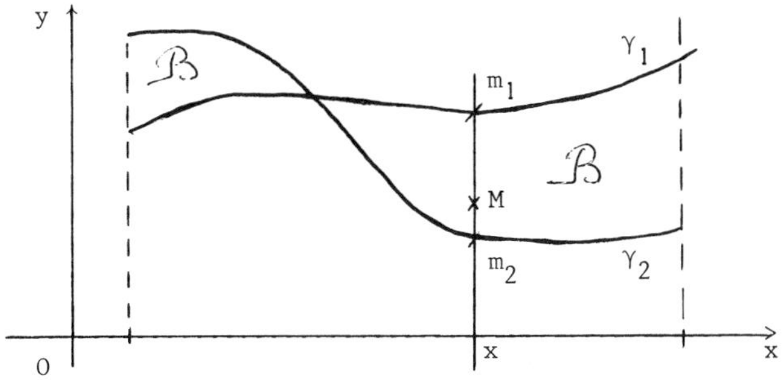


we call

$$\mathcal{B} = \{ (x,y) \in \mathbb{R}^2 \mid \gamma_1(x) \leqslant y \leqslant \gamma_2(x) \quad \text{or} \quad \gamma_2(x) \leqslant y \leqslant \gamma_1(x) \}$$

and, henceforth, we assume that $\operatorname{diam} \mathcal{B} \leqslant 1$ (for the simplicity of formulas).

Given $M = (x,y)$ in $\mathcal{B}$ , let

$$m_1 = (x_1, \gamma_1(x)) \qquad m_2 = (x, \gamma_2(x)) \tag{3}$$

$$X = X(M) = \frac{y-\gamma_1(x)}{\gamma_2(x)-\gamma_1(x)} \qquad 1 - X = \frac{y - \gamma_2(x)}{\gamma_1(x) - \gamma_2(x)} \tag{4}$$

Using FAA DE BRUNO Formula, we obtain the two lemmas :

<u>Lemma 1</u> : $\forall\, p \in \mathbb{N}$, $M \to X^{p+1}$ is p-flat on $\gamma_1$ (and $(1-X)^{p+1}$ is p flat on $\gamma_2$)

<u>Lemma 2</u> : For $\alpha$ , $|\alpha| \leqslant n$, if $P(X)$ is a polynomial in X, for all $M \in \mathcal{B}$,

$$|D^\alpha(P(X)(M))| \leqslant \frac{C_1}{(m_1 m_2)^{|\alpha|}} \quad ; \quad |D^\alpha(P(1-X)(M))| \leqslant \frac{C_2}{(m_1 m_2)^{|\alpha|}} ,$$

where $C_1$ and $C_2$ are real constants.

Problem P1 : Given f and its Taylor polynomial of order 2n on $\gamma_1 \cup \gamma_2$, construct $\hat{F}$ such that

1) $\hat{F} \in \mathcal{C}^n(\bar{\mathcal{B}})$

2) $D^\alpha \hat{F} \equiv D^\alpha f$ on $\gamma_1 \cup \gamma_2$, $|\alpha| \leqslant n$

3) $f \to \hat{F}$ is linear.

(we say that $\hat{F}$ is a $\mathcal{C}^n$-extension of f to $\bar{\mathcal{B}}$)

## II. AN EXTENSION OF DEGREE 3n+1

We construct first a partition of unity, of two terms, $s_1$ and $s_2$ such that

$$\begin{cases} s_1 + s_2 \equiv 1 \\ s_i \geqslant 0 \quad ; \quad i = 1,2 \\ s_i \text{ is n-flat on } \gamma_i \quad ; \quad i = 1,2 . \end{cases}$$

It is easy to see that we can take :

$$s_1(M) = X^{n+1} \sum_{q=0}^{n} \frac{(n+q)!}{n!q!} (1-X)^q \tag{5}$$

and

$$s_2(M) = (1-X)^{n+1} \sum_{q=0}^{n} \frac{(n+q)!}{n!q!} X^q \tag{6}$$

Let $\hat{F}$ define on $\mathcal{B}$ by :

$$\hat{F}(M) = s_1(M)\, T^n_{m_2} f(M) + s_2(M)\, T^n_{m_1} f(M) \tag{7}$$

Proposition 1 : If $\gamma_1$ and $\gamma_2$ are $\mathcal{C}^n$, if $f \in \mathcal{C}^{2n}$, then $\hat{F}$ is a $\mathcal{C}^n$-extension of f to $\mathcal{B}$, solution of problem P1.

Proof : * If $M \in \gamma_1 \cup \gamma_2$, $\hat{F}(M) = T^n_M f(M) = f(M)$

* $E : f \to \hat{F}$ is linear.

* Let $m_o \in \gamma_1 \cup \gamma_2$ . To prove that $D^\alpha f(m_o) - D^\alpha F(M) \to 0$ as $Mm_o \to 0$ , we prove that $(D^\alpha F(M) - T^n_{m_o} f(M)) \to 0$ as $Mm_o \to 0$ .

$$D^{\alpha}(\hat{F}(M)) - D^{\alpha}(T^n_{m_o} f(M)) = s_2(M)\left\{D^{\alpha}T^n_{m_1} f(M) - D^{\alpha}T^n_{m_o} f(M)\right\}$$

$$+ s_1(M)\left\{D^{\alpha}T^n_{m_2} f(M) - D^{\alpha}T^n_{m_o} f(M)\right\}$$

$$+ \sum_{\substack{\beta\leqslant\alpha \\ \beta\neq(o,o)}} \binom{\alpha}{\beta} D^{\beta}(s_2)\left\{D^{\alpha-\beta}(T^n_{m_1} f(M) - T^n_{m_2} f(M))\right\}$$

1st case : $m_o = \gamma_1 \cap \gamma_2$. With Lemma 2 and Whitney Theorem,

$$|D^{\alpha}\hat{F}(M) - D^{\alpha}(T^n_{m_o} f(M))| \leqslant m_1m_o^{n-|\alpha|}\,\omega(m_1m_o) + m_2m_o^{n-|\alpha|}\,\omega(m_2m_o) +$$

$$+ \sum_{\substack{\beta\leqslant\alpha \\ \beta\neq(o,o)}} \binom{\alpha}{\beta} \frac{C}{(m_1m_2)^{|\beta|}} (m_1m_2)^{n-|\alpha|+|\beta|}\,\omega(m_1m_2)$$

Hence, there exists a constant K such that

$$|D^{\alpha}\hat{F}(M) - D^{\alpha}(T^n_{m_o} f(M))| \leqslant K\, m_oM^{n-|\alpha|}\omega(m_oM)$$

2nd case : $m_o \in \gamma_1$ (or $\gamma_2$) only. Using the facts that $s_1$ is n-flat on $\gamma_1$, $s_2 \equiv 1$ on $\gamma_1$, and, with Whitney Theorem,

$$|D^{\alpha}(T^n_{m_2} f(M)) - D^{\alpha}(T^n_{m_o} f(M))| \leqslant m_2M^{n-|\alpha|}\omega(m_2M) + m_oM^{n-|\alpha|}\,\omega(m_oM) \leqslant$$

$$\leqslant C\, m_o\, M^{n-|\alpha|}\,\omega(m_oM)\ .$$

Thus, in the two cases, we can concluded that $D^{\alpha}\hat{F}$ is continuous on $\bar{\mathcal{B}}$, $|\alpha| \leqslant n$ ■

Remark : Is is necessary that $f \in \mathcal{C}^{2n}(\gamma_1 \cup \gamma_2)$, as we can see when calculating explicitly $D^{\alpha}(T^n_{m_i} f(M))$ ; $i = 1,2$ .

Examples : → $n = 0$ : $s_1 = X$ ; $s_2 = 1-X$

$$\hat{F}_0(M) = \hat{F}_0(x,y) = X\, f(m_2) + (1-X)\, f(m_1) =$$

$$= \frac{y-\gamma_1(x)}{\gamma_2(x) - \gamma_1(x)}\, f(x,\gamma_2(x)) + \frac{y-\gamma_2(x)}{\gamma_1(x) - \gamma_2(x)}\, f(x,\gamma_1(x))$$

→ $n = 1$ : $s_1 = X^2(1 + 2(1-X))$ ; $s_2 = (1-X)^2(1+2X)$

$$\hat{F}_1(M) = \left(\frac{y-\gamma_1}{\gamma_2-\gamma_1}\right)^2 \left(1 + 2\,\frac{y-\gamma_2}{\gamma_1-\gamma_2}\right)\left(f(m_2) + (y-\gamma_2)\,\frac{\partial f}{\partial y}(m_2)\right) +$$

$$+ \left(\frac{y-\gamma_2}{\gamma_1-\gamma_2}\right)^2 \left(1 + 2\,\frac{y-\gamma_1}{\gamma_2-\gamma_1}\right)\left(f(m_1) + (y-\gamma_1)\,\frac{\partial f}{\partial y}(m_1)\right)$$

→ In particular :

$\gamma_2 : y = x$ ; $\gamma_1 : y = 0$ ; $x > 0$

. $n=0$ : $\hat{F}_0(x,y) = \frac{y}{x}\, f(x,x) + \frac{x-y}{x}\, f(x,o)$

. $n=1$ : $$\hat{F}_1(x,y) = \left(\frac{y}{x}\right)^2 \left(1+2\,\frac{x-y}{x}\right)\left[f(x,x) + (y-x)\,\frac{\partial f}{\partial y}(x,x)\right]$$
$$+ \left(\frac{x-y}{x}\right)^2 \left(1+2\,\frac{y}{x}\right)\left[f(x,o) + y\,\frac{\partial f}{\partial y}(x,o)\right]$$

. $n=2$ : $$\hat{F}_2(x,y) = \left(\frac{y}{x}\right)^3 \left[1+3\,\frac{x-y}{x} + 6\left(\frac{x-y}{x}\right)\right]^2$$
$$\left[f(x,x) + (y-x)\,\frac{\partial f}{\partial y}(x,x) + \frac{1}{2}(y-x)^2\,\frac{\partial^2 f}{\partial y^2}(x,x)\right]$$
$$+ \frac{x-y}{x}^3 \left[1+3\,\frac{y}{x} + 6\left(\frac{y}{x}\right)^2\right]$$
$$\left[f(x,o) + y\,\frac{\partial f}{\partial y}(x,o) + \frac{1}{2}\,y^2\,\frac{\partial^2 f}{\partial y^2}(x,o)\right]$$

## III. AN EXTENSION OF DEGREE 2n+1

Definition : We define $\hat{f}$ on $\mathcal{B}$ by :

$$\hat{f}(M) = \sum_{q=0}^{n} \frac{(n+q)!}{n!q!} \left\{ X^{n+1}(1-X)^q \, T_{m_2}^{n-q} f(M) \right\} + \sum_{q=0}^{n} \frac{(n+q)!}{n!q!} \left\{ (1-X)^{n+1} X^q \, T_{m_1}^{n-q} f(M) \right\} \quad (8)$$

Proposition 2 : If $f \in \mathcal{C}^{2n}(\gamma_1 \cup \gamma_2)$, if $\bar{\gamma}_1$ and $\gamma_2$ are $\mathcal{C}^n$, then $\hat{f}$ is a $\mathcal{C}^n$-extension of f to $\mathcal{B}$ , solution of problem P1.

Proof : The same arguments as in proposition 1 hold. We had to write $\hat{f}$ as :

$$\hat{f}(M) = T_{m_1}^n f(M) + \left\{ T_{m_2}^n f(M) - T_{m_1}^n f(M) \right\} \left\{ \sum_{q=0}^{n} \frac{(n+q)!}{n!q!} X^{n+1} (1-X)^q \right\}$$
$$- \left\{ \sum_{q=1}^{n} \frac{(n+q)!}{n!q!} X^{n+1}(1-X)^q \left[T_{m_2}^n f(M) - T_{m_2}^{n-q} f(M)\right] + \sum_{q=1}^{n} \frac{(n+q)!}{n!q!} (1-X)^{n+1} X^q \left[T_{m_1}^n f(M) - T_{m_1}^{n-q} f(M)\right] \right\}$$

and to proof that the last term $\{ \Sigma \ldots + \Sigma \}$ is n-flat on $\gamma_1 \cup \gamma_2$ ■

Examples :

* n=1 :
$$\hat{f}_1(M) = f(m_2) \left\{ \left(\frac{y-\gamma_1}{\gamma_2-\gamma_1}\right)^2 \left(1+2\,\frac{y-\gamma_2}{\gamma_1-\gamma_2}\right) \right\} + \left(\frac{y-\gamma_1}{\gamma_2-\gamma_2}\right)^2 (y-\gamma_2) \frac{\partial f}{\partial y}(m_2)$$
$$+ f(m_1) \left\{ \left(\frac{y-\gamma_2}{\gamma_1-\gamma_2}\right)^2 \left(1+2\,\frac{y-\gamma_1}{\gamma_2-\gamma_1}\right) \right\} + \left(\frac{y-\gamma_2}{\gamma_1-\gamma_2}\right)^2 (y-\gamma_1) \frac{\partial f}{\partial y}(m_1)$$

* n=2 :
$$X^3 \left\{1+3(1-X) + 6(1-X)^2\right\} f(m_2) + X^3 \left\{1+3(1-X)\right\} (y-\gamma_2) \frac{\partial f}{\partial y}(m_2)$$
$$+ X^3 \frac{1}{2} (y-\gamma_2)^2 \frac{\partial^2 f}{\partial y^2}(m_2) + (1-X)^3 \left\{1+3X+6X^2\right\} f(m_1)$$
$$+ (1-X)^3 (1+3X) (y-\gamma_1) \frac{\partial f}{\partial y}(m_1) + (1-X)^3 \frac{1}{2} (y-\gamma_1)^2 \frac{\partial^2 f}{\partial y^2}(m_1)$$

* In particular : $\gamma_2 : y = x \quad ; \quad \gamma_1 : y = 0 \quad ; \quad x > 0$

$$\cdot \quad \hat{f}_1(x,y) = \frac{y^2}{x^2} \left\{ \left(1+2\,\frac{x-y}{x}\right) f(x,x) + (y-x)\,\frac{\partial f}{\partial y}(x,x) \right\}$$

$$+ \left(\frac{x-y}{x}\right)^2 \left\{ \left(1+2\,\frac{y}{x}\right) f(x,0) + y\,\frac{\partial f}{\partial y}(x,0) \right\}$$

$$\cdot \quad \hat{f}_2(x,y) = \left(\frac{y}{x}\right)^3 \left\{ \left[1+3\left(\frac{x-y}{x}\right) + 6\left(\frac{x-y}{x}\right)^2\right] f(x,x) + \left[1+3\left(\frac{x-y}{x}\right)\right](y-x)\frac{\partial f}{\partial y}(x,x) + \frac{1}{2}(y-x)^2 \frac{{}^2 f}{x^2}(x,x) \right\}$$

$$+ \left(\frac{x-y}{x}\right)^3 \left\{ \left[1+3\,\frac{y}{x} + 6\left(\frac{y}{x}\right)^2\right] f(x,0) + \left[1+3\,\frac{y}{x}\right] y\,\frac{\partial f}{\partial y}(x,0) + \frac{1}{2}\,y^2\,\frac{\partial^2 f}{\partial y^2}(x,0) \right\}$$

## IV. $\mathcal{C}^n$-EXTENSION, USING A REFLECTION PRINCIPE

Consider now another problem :

Problem P2 : Given a function f defined and $\mathcal{C}^n$ outside $\bar{\mathcal{B}}$, and all of whose derivatives up to order n have continuous limits at the boundary $\gamma_1 \cup \gamma_2$ of $\mathcal{B}$, can f be extended to a $\mathcal{C}^n$ function over

$$\mathcal{B} = \{ (x,y) \mid \gamma_1(x) < y < \gamma_2(x) \ \text{ or } \ \gamma_2(x) < y < \gamma_1(x) \} \ ?$$

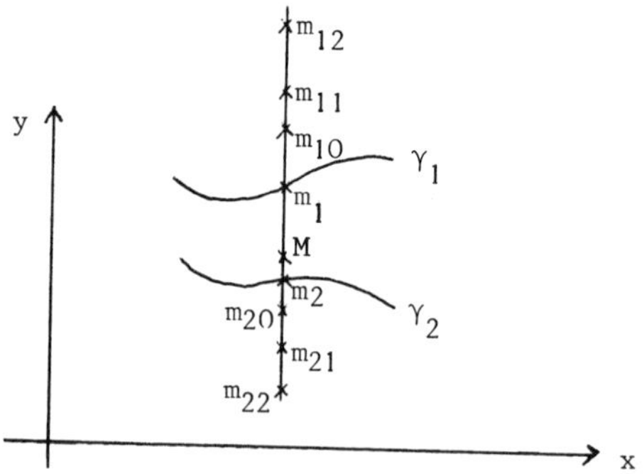

Let $b_0, b_1, \ldots, b_n$ , (n+1) reals, all differents, and all strictly positives. Given $M = (x,y) \in \mathcal{B}$, let

$$m_1 = (x,\gamma_1(x)) \qquad m_2 = (x,\gamma_2(x))$$

$$m_{1k} = \Big(x,\gamma_1(x) + b_k(\gamma_1(x) - y)\Big) = (x,y_{1k})$$

$$m_{2k} = \Big(x,\gamma_2(x) + b_k(\gamma_2(x) - y)\Big) = (x,y_{2k})$$

Consider $P_1$, the polynomial of Lagrange interpolation (in one variable y) constructed over $m_{10}, m_{11}, \ldots m_{1n}$, and $P_2$, the one constructed over $m_{20}, m_{21}, \ldots, m_{22}$.

From Lagrange formula, $P_1(M) = \sum_{k=0}^{n} \ell_k(M)\ f(m_{1k})$ ,

$$\text{where} \quad \ell_k(M) = \prod_{\substack{j=0 \\ j\neq k}}^{n} \frac{y-y_j}{y_k-y_j} \tag{9}$$

We find that $\ell_k = \prod_{\substack{j=0 \\ j\neq k}}^{n} \frac{1+b_j}{b_j-b_k}$ , which is independant of M.

Hence :

$$P_i(M) = \sum_{k=0}^{n} \ell_k f(m_{ik}) \quad ; \quad i = 1,2 \ . \tag{10}$$

<u>Lemma 3</u> : For all p = 0,1,...,n , $\sum_{k=0}^{n} \ell_k\ (b_k)^p = (-1)^p$

<u>Definition</u> : For all $M \in \mathcal{B}$, let

$$E[f](M) = s_1(M)\ P_2(M) + s_2(M)\ P_1(M) \tag{11}$$

$$\text{ie} \quad E[f](M) = s_1(M)\left(\sum_{k=0}^{n} \ell_k f(m_{2k})\right) + s_2(M)\left(\sum_{k=0}^{n} \ell_k f(m_{1k})\right)$$

where $s_1$ and $s_2$ are defined by (5) and (6).

<u>Proposition 3</u> : If $\gamma_1$ and $\gamma_2$ are $\mathcal{C}^n$, if f is $\mathcal{C}^n$, then E[f] is a $\mathcal{C}^n$-extension of f to $\mathcal{B}$.

Proof : * $f \to E[f]$ is linear

* let $m_o \in \gamma_1 \cup \gamma_2$ $D^\alpha f(m_o) = \lim D^\alpha f(N)$, $N \to m_o$ , $N \in \mathbb{R}^2 \setminus \bar{\mathcal{B}}$

Then $m_o m_{ik} \leqslant (m_o M) \quad ( ||\gamma_i||_1 \quad (1+b_k) + b_k )$

As $s_1 + s_2 \equiv 1$, and $\sum_{k=0}^{n} \ell_k = 1$ ,

$$D^\alpha f(m_o) - D^\alpha(E[f](M)) = s_2(M) \sum_{k=0}^{n} \ell_k (D^\alpha f(m_o) - D^\alpha f(m_{1k}))$$

$$+ s_1(M) \sum_{k=0}^{n} \ell_k (D^\alpha f(m_o) - D^\alpha f(m_{2k}))$$

$$+ \sum_{\substack{\beta \leqslant \alpha \\ \beta \neq (0,0)}} \binom{\alpha}{\beta} D^\beta s_2(M) \sum_{k=0}^{n} \ell_k (D^{\alpha-\beta} f(m_{1k}) - D^{\alpha-\beta} f(m_{2k}))$$

Using similar arguments than in the above propositions, we can concluded that $D^\alpha E[f](M)$ is continuous.

Remark : By the reflection principe, we use the values of f in a strip outside $\mathcal{B}$ which is parallel to $\gamma_1$ or $\gamma_2$. It is thus interesting to be allowed to use $b_i$ **not fixed.** We can choose them in connection with the geometry of the domain of knowledge of f.

## REFERENCES

[1] COATMELEC, Chr. : Approximation et interpolation de fonctions différentiables de plusieurs variables. Ann. Sc. Ecole Normale Sup. 3e série T. 83 (1966)

[2] HESTENES, M.R. : Extension of the range of a differentiable function. Duke Math Journal 8 (1941), 183-192.

[3] LE MEHAUTE, A. : Quelques méthodes explicites de prolongements de fonctions numériques de plusieurs variables. Thèse 3e cycle, Rennes (1976).

[4] TOUGERON, J.C. : Idéaux de fonctions différentiables. Springer Verlag (1972).

[5] WHITNEY, H. : Analytic extensions of differentiable functions defined in closed sets. Trans. Amer. Math Soc. Vol. 36, N° 1 (1934).

# TAYLOR INTERPOLATION OF ORDER n AT THE VERTICES OF A TRIANGLE. APPLICATIONS FOR HERMITE INTERPOLATION AND FINITE ELEMENTS

Alain Le Mehaute
Laboratoire d'Analyse Numérique
Institut National des Sciences Appliquees
Rennes, France

## I - INTRODUCTION

In this note we give first a constructive method to solve the problem of Taylor interpolation at three points, not on the same line, i.e given $A_1\ A_2\ A_3$, we construct a polynomial P, of degree 2n+1, solution of the interpolation problem :

$$D^{\alpha}P(A_i) = D^{\alpha}f(A_i) \quad ; \quad i=1,2,3 \quad ; \quad |\alpha| \leqslant n \ .$$

Then, a method for constructing Hermite interpolation polynomials of degree 2n+1 is given, using the effective solution of a linear system of $\frac{n(n+1)}{2}$ equations with $\frac{n(n+1)}{2}$ unknowns. In this way, we can construct triangular Hermite finite elements of order $2n+1, n \in \mathbb{N}$, generalizing classical elements of order 3 or 5.

The method is also used to construct the unique polynomial of degree 4n+1 realizing $\mathcal{C}^n$-continuity between two triangles. In each case, we obtain explicit upper bounds for the interpolating error on a triangle, which generalize the ones used in the usual cases.



ISBN 0-12-780650-4

## II - NOTATION

The following notations are used :

* $\alpha = (\alpha_1, \alpha_2) \in \mathbb{N} \times \mathbb{N}$ $\quad |\alpha| = \alpha_1 + \alpha_2 \in \mathbb{N}$

$\alpha! = \alpha_1!\ \alpha_2!$ $\quad \binom{\alpha}{\beta} = \binom{\alpha_1}{\beta_1}\binom{\alpha_2}{\beta_2}$

$\beta \leqslant \alpha$ iff $\beta_1 \leqslant \alpha_1$ and $\beta_2 \leqslant \alpha_2$

$$D^{\alpha} f(M) = \frac{\partial^{|\alpha|} f}{\partial x^{\alpha_1}\, \partial y^{\alpha_2}} (M)$$

if $A = (x_A, y_A)$ and $B = (x_B, y_B)$ are two points in $\mathbb{R}^2$,

$$AB = \left((x_B - x_A)^2 + (y_B - y_A)^2\right)^{1/2} , \quad AB^{\alpha} = (x_B - x_A)^{\alpha_1} (y_B - y_A)^{\alpha_2}$$

* Let $A_1$, $A_2$, $A_3$ three points in $\mathbb{R}^2$; $\lambda_i(M)$ is the barycentric coordinate of a point $M$ with respect to $A_i$, $i = 1,2,3$. (when it is possible, we write only $\lambda_i$ for $\lambda_i(M)$ ).

* $T_A^k f$ denotes, when it exists, the Taylor expansion of order $k$ of a function $f$ at $A$, ie

$$T_A^k f : M \rightarrow T_A^k f(M) = \sum_{|\alpha| \leqslant k} \frac{1}{\alpha!} D^{\alpha} f(A) . AM^{\alpha}$$

## III - TAYLOR INTERPOLATION AT THE THREE VERTICES OF THE TRIANGLE $A_1A_2A_3$

We say that a polynomial $p$ realizes Taylor interpolation of order $n$ of a given function $f$ at a point $A$ iff

$$D^{\alpha} p(A) = D^{\alpha} f(A) \qquad \text{for all } \alpha \ , \quad |\alpha| \leqslant n$$

Problem P1 : Given 3 points $A_1$, $A_2$, $A_3$, not on the same line, construct a polynomial $p$, of degree $2n+1$, that realizes Taylor interpolation of order $n$ at the three vertices of the triangle $A_1A_2A_3$.

Proposition III.1 : The polynomial $P_{2n+1}[f]$ , explicitly given by

$$P_{2n+1}[f](M) = \sum_{i=1}^{3} \lambda_i^{n+1} \sum_{j=0}^{n} \frac{(n+j)!}{n!j!} (1-\lambda_i)^j T_{A_i}^{n-j} f(M) \qquad (1)$$

is a solution of problem P1.

Proof : * As barycentric coordinates are polynomials of degree 1, $P_{2n+1}[f]$ is exactly of degree 2n+1.

* If $k \in \{1,2,3\}$, $\lambda_i(A_k) = 0$ if $i \neq k$, and $\lambda_k(A_k) = 1$ thus

$$P_{2n+1}[f](A_k) = (\lambda_k(A_k))^{n+1} \; T_{A_k}^{n} f(A_k) = f(A_k)$$

* For the derivatives, if $|\alpha| \leq n$ , using Leibniz formula and the notation

$$\nabla_i = \left(\frac{\partial \lambda_i}{\partial x}, \frac{\partial \lambda_i}{\partial y}\right) , \quad \nabla_i^{\beta} = \left(\frac{\partial \lambda_i}{\partial x}\right)^{\beta_1} \left(\frac{\partial \lambda_i}{\partial y}\right)^{\beta_2}$$ , we obtain :

$$D^{\alpha} P_{2n+1}[f](M) = \sum_{i=1}^{3} \sum_{\beta \leq \alpha} \binom{\alpha}{\beta} \frac{(n+1)!}{(n+1-|\beta|)!} \nabla_i^{\beta} \lambda_i^{n+1-|\beta|} \times$$

$$\times \sum_{j=0}^{n} \frac{(n+j)!}{n!j!} (-1)^j \sum_{\substack{\gamma \leq \alpha-\beta \\ |\gamma| \leq j}} \binom{\alpha-\beta}{\gamma} \frac{j!}{(j-|\gamma|)!} \nabla_i^{\gamma} (\lambda_i - 1)^{j-|\gamma|} \times \qquad (2)$$

$$\times \; T_{A_i}^{n-j-|\alpha-\beta-\gamma|} \; D^{\alpha-\beta-\gamma} f \; (M)$$

For $M = A_k$, $D^{\alpha} P_{2n+1}[f](A_k) = D^{\alpha} f(A_k)$. ■

Remark : The result is valid in $\mathbb{R}^p$, $p \geq 1$ ( change $\sum_{i=1}^{3}$ in $\sum_{i=1}^{p+1}$ )

we have (2n+1) unisolvency only in two cases :

- if p = 1, for all n (and then we have a more usuable form that the one of DAVIS [ 3])
- if p > 1, for n = 0 only (corresponding to Lagrange-interpolation).

Examples : * n = 0 : $P_1(M) = \sum_{i=1}^{3} \lambda_i(M) \; f(A_i)$

* n = 1 : $P_3(M) = \sum_{i=1}^{3} \lambda_1^2 \left\{ T_{A_i}^1 f(M) + 2(1-\lambda_i) \; f(A_i) \right\} =$

$$= \sum_{i=1}^{3} \lambda_i^2 \left\{ \lambda_{i-1} T_{A_i}^1 f(A_{i-1}) + \lambda_{i+1} T_{A_i}^1 f(A_{i+1}) + 2(1-\lambda_i) f(A_i) \right\}$$

$$* \quad n = 2 \quad : \quad P_5(M) = \sum_{i=1}^{3} \lambda_i^3 \left\{ T^2_{A_i} f(M) + 3(1-\lambda_i) T^1_{A_i} f(M) + 6(1-\lambda_i^2) f(A_i) \right\}$$

IV - APPLICATION : HERMITE INTERPOLATION IN 4 POINTS

Problem P2 : Let $A_1A_2A_3$ a triangle, and $A_o$ not on the edges of this triangle. We want to construct a polynomial $\Pi_{2n+1}$ of degree $2n+1$ solution of the interpolation problem :

$$D^{\alpha}\Pi_{2n+1}(A_i) = D^{\alpha}f(A_i) \qquad |\alpha| \leqslant n, \quad i = 1,2,3$$

$$D^{\beta}\Pi_{2n+1}(A_o) = D^{\beta}f(A_o) \qquad |\beta| \leqslant n-1$$

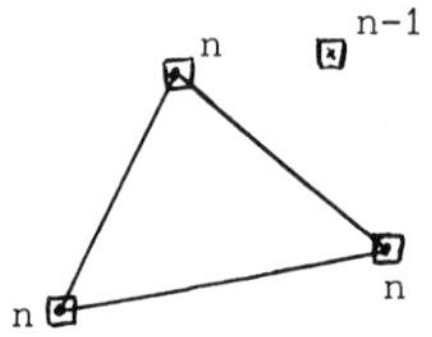


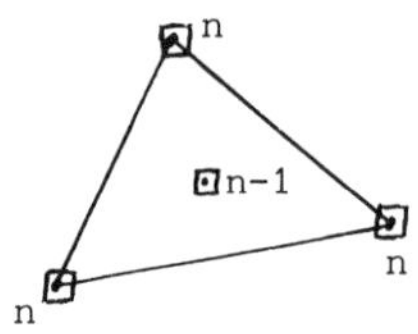


First we have to prove the unisolvency of Problem P2, ie there is one and only one solution.

A. UNISOLVENCY

Proposition IV.1 : Problem P2 is (2n+1) unisolvent.

Proof : It is equivalent to prove that any polynomial $\Pi_{2n+1}$ of degree $2n+1$, such that

$$D^{\alpha}\Pi_{2n+1}(A_i) = 0 \qquad |\alpha| \leqslant n \quad ; \quad i = 1,2,3$$

$$D^{\alpha}\Pi_{2n+1}(A_o) = 0 \qquad |\beta| \leqslant n-1$$

is identically zero.

Let $M \in \mathbb{R}^2$ → if $M = A_i$, $i = 1,2$ or $3$, we have already $\Pi_{2n+1}(M)=0$.

→ If $M \in a_i$, line containing the side $[A_{i-1}, A_{i+1}]$ of the triangle $A_1A_2A_3$. Then the restriction $p_{2n+1}$ of $\Pi_{2n+1}$ to $a_i$ is a polynomial of degree $2n+1$ in one variable which satisfies :

$$p_{2n+1}(A_k) = p'_{2n+1}(A_k) = \dots = p^{(n)}_{2n+1}(A_k) = 0 \quad , \quad k = i+1 \text{ or } i-1.$$

Thus $p_{2n+1}$ is identically zero.

Therefore $\Pi_{2n+1}(M) = 0$, and similar arguments hold for the other sides.

→ $M \notin a_1 \cup a_2 \cup a_3$ . Through the five points $A_0, A_1, A_2, A_3, M$ there is one and only one conic $\Gamma$ .

* $\Gamma$ is degenerate in two lines. As $A_0$ and M are not on $a_1 \cup a_2 \cup a_3$, $\Gamma$ is necessarily the union of $a_i$ and the line $A_0A_i$, i = 1, 2 or 3.

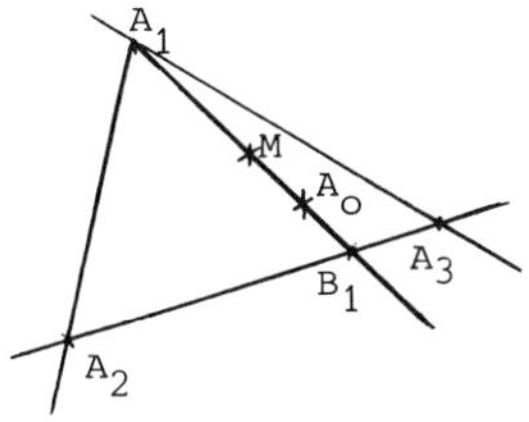


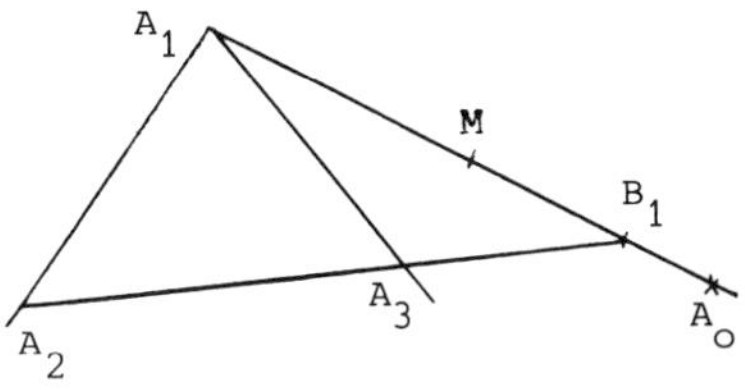


for example, $\Gamma = a_1 \cup A_0A_1$ .

Let $B_1 = A_0A_1 \cap A_2A_3$. Then $\Pi_{2n+1}(B_1) = 0$ . The restriction $q_{2n+1}$ of $\Pi_{2n+1}$ to $A_0A_1$ is a polynomial of degree 2n+1 in one variable which satisfies :

$$q_{2n+1}(A_1) = q'_{2n+1}(A_1) = \dots = q^{(n)}_{2n+1}(A_1) = 0$$

$$q_{2n+1}(A_0) = q'_{2n+1}(A_0) = \dots = q^{(n-1)}_{2n+1}(A_0) = 0$$

$$q_{2n+1}(B_1) = 0$$

thus $q_{2n+1}$ is identically zero and $\Pi_{2n+1}(M) = 0$ .

* $\Gamma$ is a proper conic : for example, $\Gamma$ is an hyperbola. There is a system of axis where $\Gamma$ has a parametric equation of the form :

$$x(t) = a\,\frac{1+t^2}{1-t^2} \quad ; \quad y(t) = b\,\frac{2t}{1-t^2} \quad ; \quad t \in \mathbb{R} - \{-1,+1\}$$

then $$\Pi_{2n+1}(M) = \pi(t) = \sum_{i+j=0}^{2n+1} a_{ij}x^iy^j = \sum_{i+j=0}^{2n+1} a_{ij}a^ib^j2^j\,\frac{(1+t^2)^i\,t^j}{(1+t^2)^{i+j}} ,$$

$$t \notin \{-1,+1\}$$

Let $q(t) = (1-t^2)^{2n+1}\pi(t) = \sum_{i+j=0}^{2n+1} a_{ij} a^i b^j 2^j (1+t^2)^i \, t^j (1-t^2)^{2n+1-(i+j)}$, $t \in \mathbb{R}$

q is a polynomial in t of degree 4n+2 and

$$\frac{d^k}{dt^k}\left(q(t)\right) = \sum_{\ell \leq k} \binom{k}{\ell} \frac{d^\ell}{dt^\ell}\left((1-t^2)^{2n+1}\right) \frac{d^{k-\ell}}{dt^{k-\ell}}\left(\pi(t)\right)$$

Using Faa de Bruno formula [4] we obtain that

$$\frac{d^{k-\ell}}{dt^{k-\ell}}\left(\pi(t_1)\right) = 0 \quad , \quad \begin{array}{ll} i = 1,2,3 & k-\ell \leq n \\ i = 0 & k-\ell \leq n-1 \end{array} \qquad (t_i \sim A_i)$$

Therefore $q(t_i) = q'(t_i) = \ldots. = q^{(n)}(t_i) = 0 \quad , \quad i = 1,2,3$

$q(t_o) = q'(t_o) = \ldots. = q^{(n-i)}(t_o) = 0$

Thus q is identically zero, and $\Pi_{2n+1}(M) = 0$

Similar arguments hold for the other cases and we find that $\Pi_{2n+1}$ is identically zero. ■

As a particular case, we mention Hermite triangle of type (3) (CIARLET [1]).

B. CONSTRUCTION OF $\Pi_{2n+1}$

Let $P_{2n+1}[f](M) = \sum_{i=1}^{3} \lambda_i^{n+1} \sum_{j=0}^{n} \frac{(n+q)!}{n!q!} (1-\lambda_i)^j \, T_{A_i}^{n-j} f(M)$

Let $\Pi_{2n+1}[f] = P_{2n+1}[f] + R_{2n+1}[f]$ (3)

where $R_{2n+1}[f](M) = \lambda_1\lambda_2\lambda_3 \sum_{1+p+q=n-1} \alpha_{pq} (\lambda_1\lambda_2)^p (\lambda_2\lambda_3)^q (\lambda_3\lambda_1)^r$ (4)

where $\alpha_{pq}$ are real constants.

Lemma 1 : $R_{2n+1}$ is a polynomial of degree 2n+1 and is n-flat at $A_1$, $A_2$ and $A_3$ (ie $D^\alpha R_{2n+1}(A_k) = 0$ , $k = 1,2,3$ , $|\alpha| \leq n$)

Explicitly, we have, for $|\alpha| \leq n$ ,

$$D^{\alpha}R_{2n+1}[f](M) = \sum_{p+q+r=n-1} \alpha_{pq} \left\{ \sum_{\beta\leqslant\alpha} \sum_{\gamma\leqslant\beta} \binom{\alpha}{\beta}\binom{\beta}{\gamma} |\alpha-\beta|!\,|\beta-\gamma|!\,|\gamma|! \times \right.$$

$$\times \binom{p+q+1}{|\beta-\gamma|}\binom{p+r+1}{|\alpha-\beta|}\binom{r+q+1}{|\gamma|} \nabla_1^{\alpha-\beta}\ \nabla_2^{\beta-\gamma}\ \nabla_3^{\gamma} \times \tag{5}$$

$$\left. \times\ \lambda_1^{p+r+1-|\alpha-\beta|}\ \lambda_2^{p+q+1-|\beta-\gamma|}\ \lambda_3^{r+q+1-|\gamma|} \right\}$$

for brevity, we write $D^{\alpha}R_{2n+1}[f](M) = \sum_{p+q=0}^{n-1} \alpha_{pq}\ B^{\alpha}_{pq}(M)$

It is obvious that $B^{\alpha}_{pq}(A_k) = 0$ ; $|\alpha| \leqslant n$ ; $k = 1,2,3$ ∎

Thus $D^{\alpha}\Pi_{2n+1}[f](A_k) = D^{\alpha}P_{2n+1}[f](A_k) = D^{\alpha}f(A_k)$ ; $|\alpha| \leqslant n$ ; $k = 1,2,3.$

<u>Proposition IV.2</u> : There exist $\frac{n(n+1)}{2}$ constants $\alpha_{pq}$ such that

$$D^{\beta}\Pi_{2n+1}[f](A_o) = D^{\beta} f(A_o) \qquad |\beta| \leqslant n-1$$

(and $\Pi_{2n+1}$ is then the unique solution of problem P2)

<u>Proof</u> : It is now easy to see that the $\alpha_{pq}$ must be the solutions of the linear system :

$$(\mathcal{S}) \left\{ \sum_{p+q=0}^{n-1} \alpha_{pq}\ B^{\alpha}_{pq}(A_o) = D^{\alpha}f(A_o) - D^{\alpha}P_{2n+1}(A_o) \right.$$

where $D^{\alpha}P_{2n+1}(A_o)$ is obtain from (2) and $B^{\alpha}_{pq}(A_o)$ from (5).

<u>Remark</u> : Hermite finite elements of type (2n+1).

If $A_o$ is strictly interior to the triangle $A_1A_2A_3$, we have construct a finite element, which we can call Hermite finite element of type (2n+1), following Ciarlet [1]

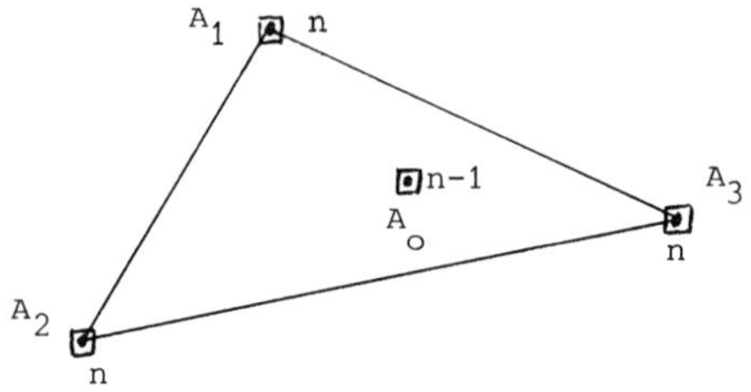


Example : n=1 : Hermite element of type (3).

$A_o$ = centroid of $A_1A_2A_3$. $\lambda_i(A_o) = \frac{1}{3}$ , $i = 1,2,3$ .

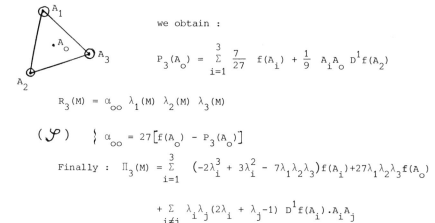

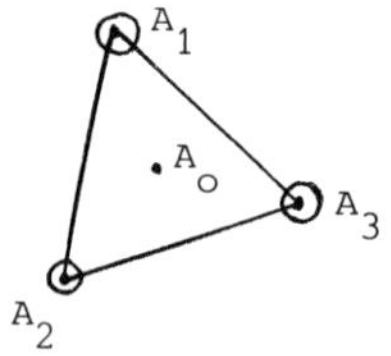


we obtain :

$$P_3(A_o) = \sum_{i=1}^{3} \frac{7}{27} f(A_i) + \frac{1}{9} A_iA_o D^1 f(A_2)$$

$$R_3(M) = \alpha_{oo} \lambda_1(M) \lambda_2(M) \lambda_3(M)$$

$$(\mathcal{S}) \quad \{ \alpha_{oo} = 27[f(A_o) - P_3(A_o)]$$

Finally : $\Pi_3(M) = \sum_{i=1}^{3} (-2\lambda_i^3 + 3\lambda_i^2 - 7\lambda_1\lambda_2\lambda_3) f(A_i) + 27\lambda_1\lambda_2\lambda_3 f(A_o)$

$$+ \sum_{i \neq j} \lambda_i\lambda_j(2\lambda_i + \lambda_j - 1) D^1 f(A_i).A_iA_j$$

C. BOUNDS FOR ERROR INTERPOLATION

If $\Omega$ is a compact which contains $A_1, A_2, A_3, A_o$, and if $M \in \Omega$, it is possible to obtain bounds for error interpolation. We do this here only if $\Omega$ is the triangle $A_1A_2A_3$, and $A_o \in \overset{\circ}{\Omega}$ (ie, for Hermite finite element). In the general case, similar results may be obtained.

Let h = diameter of $\Omega$, $\rho$ = diameter of the sphere inscribed in $\Omega$

$$|f|_k = \sup_{|\alpha| = k} ||D^\alpha f||_\infty = \sup_{|\alpha|=k} \sup_{M \in \bar{\Omega}} |D^\alpha f(M)| \ , \quad 0 \leqslant j \leqslant n$$

Proposition IV.3 : If $\frac{h}{\rho} < \sigma$, when $\sigma$ is a constant, if $|\alpha| \leqslant n$, there exists a constant $C(\alpha,\Omega)$ such that, for all $f \in \mathcal{C}^{2n+2}(\bar{\Omega})$,

$$\| D^\alpha f - D^\alpha \Pi_{2n+1}[f] \|_\infty \leqslant C(\alpha,\Omega)\, h^{2n+2-|\alpha|}\, |f|_{2n+2}$$

Proof : Let pf a polynomial of best approximation of f, of degree 2n+1 . Because of (2n+1) unisolvency, $\Pi_{2n+1}[pf] \equiv pf$

Let u = f - pf . Using a result of MEINGUET [5]

$$|D^\alpha u(M)| \leqslant h^{2n+2-|\alpha|} \frac{(\sqrt{2})^{2n+2-|\alpha|}}{(2n+2-|\alpha|)!} |f|_{2n+2}$$

As $|D^\alpha f(M) - D^\alpha \Pi_{2n+1}[f](M)| \leqslant |D^\alpha u(M)| + |D^\alpha \Pi_{2n+1}[u](M)|$

we have only to bound $D^\alpha \Pi_{2n+1}[u](M)$.

$$|D^\alpha \Pi_{2n+1}[u](M)| \leqslant |D^\alpha P_{2n+1}[u](M)| + |D^\alpha R_{2n+1}[u](M)|$$

* $|T_{A_i}^{n-|\delta|} D^\alpha u(M)| \leqslant \sum_{|\beta| \leqslant 2n+1-|\delta|} \frac{1}{\beta!} |MA_i|^{|\beta|} |D^{\alpha+\beta} u(A_i)|$

$$\leqslant h^{2n+2-|\delta|} (\sqrt{2})^{2n+2-|\delta|} |f|_{2n+2} \sum_{|\beta| \leqslant 2n+1-|\delta|} \frac{1}{\beta!} \frac{(\sqrt{2})^{-|\beta|}}{(2n+2-|\beta+\delta|)!}$$

* $\frac{\partial \lambda_i}{\partial x} = \frac{1}{\Delta}(y_{i+1} - y_{i-1}) \qquad \frac{\partial \lambda_i}{\partial y} = \frac{1}{\Delta}(x_{i-1} - x_{i+1})$

where $\Delta = 2S$ , S = area of triangle $\Omega$

Thus $\nabla_i^\gamma = \frac{1}{\Delta^{|\gamma|}} (y_{i+1} - y_{i-1})^{\gamma_1} (x_{i-1} - x_{i+1})^{\gamma_2}$

and $|\nabla_i^\gamma| \leqslant \frac{h^{|\gamma|}}{\Delta^{|\gamma|}}$

* With (2), we obtain

$$|D^{\alpha}P_{2n+1}[u](M)| \leqslant h^{2n+2-|\alpha|}|f|_{2n+2}\,(\sqrt{2})^{2n+2-|\alpha|}\left\{\sum_{i=1}^{3}\sum_{\beta\leqslant\alpha}\binom{\alpha}{\beta}\frac{(n+1)!}{(n+1-|\beta|)!}\times\right.$$

$$\times\sum_{j=0}^{n}\frac{(n+j)!}{n!j!}\sum_{\substack{\gamma\leqslant\alpha-\beta\\|\gamma|\leqslant j}}\binom{\alpha-\beta}{\gamma}\frac{j!}{(j-|\gamma|)!}(\sqrt{2})^{|\beta|+|\gamma|}\frac{h^{2(|\beta|+|\gamma|)}}{\Delta^{|\beta|+|\gamma|}}$$

$$\left.\times\sum_{|\delta|\leqslant n-j-|\alpha-\beta-\gamma|}\frac{1}{\delta!}\frac{(\sqrt{2})^{-|\delta|}}{((2n+2)-|\alpha-\beta-\gamma-\delta|)!}\right\}$$

$$\frac{h^{2(|\beta|+|\gamma|)}}{\Delta^{|\beta|+|\gamma|}} \leqslant \left(\frac{h}{\rho}\right)^{2(|\beta|+|\gamma|)} \leqslant \left(\frac{h}{\rho}\right)^{2|\alpha|} \leqslant \sigma^{2|\alpha|}$$

$$(\sqrt{2})^{2n+2-|\alpha|}\,(\sqrt{2})^{|\beta|+|\gamma|} \leqslant (\sqrt{2})^{2n+2}$$

Hence, there exists a constant $C_1$ such that

$$|D^{\alpha}P_{2n+1}[u](M)| \leqslant h^{2n+2-|\alpha|}\,|f|_{2n+2}\,(\sqrt{2})^{2n+2}\left(\frac{h}{\rho}\right)^{2|\alpha|} C_1$$

* From (5), we easily find that each $B^{\alpha}_{pq}$ has $\frac{1}{\Delta^{|\alpha|}}$ in factor, and that

$$|B_{pq}(M)| \leqslant \frac{|\alpha|}{\Delta^{|\alpha|}}\sum_{\beta\leqslant\alpha}\sum_{\gamma\leqslant\beta}\binom{\alpha}{\beta}\binom{\beta}{\gamma}|\alpha-\beta|!\,|\beta-\gamma|!\,|\gamma|!\binom{p+r+1}{|\alpha-\beta|}\binom{p+q+1}{|\beta-\gamma|}\binom{r+q+1}{|\gamma|}$$

or $$|B^{\alpha}_{pq}(M)| \leqslant \frac{h^{|\alpha|}}{\Delta^{|\alpha|}}\,C(p,q,\alpha)$$

* Using Cramer formula for the solution of the system $(\mathcal{S})$,

$$\alpha_{pq} = \frac{\det D_{pq}}{\det D}$$

$$\det D = \sum_{\sigma\in\mathfrak{S}_p} p(\sigma)\prod_{i=1}^{p} B^{\alpha_i}_{i,\sigma(i)}$$ when $\mathfrak{S}_p$ = set of permutations of $\{1,2,\ldots p\}$ ; $|\alpha_i| = i$

In each term of this sum, there is one and only one term of each column and each line of the determinant. As $B^{\alpha}_{pq}$ has $\frac{1}{\Delta^{|\alpha|}}$ in factor , det D has $\frac{1}{\Delta^r}$ in factor,

where $r = 1\times 0 + 2\times 1 + \ldots + n(n-1) = \frac{1}{3} n(n-1)(n+1)$.

Thus $\det D = \frac{1}{\Delta^r} C$ , where C is a constant (depending on $\Omega$).

With the same argument,

$$|\det D_{pq}| \leqslant \sum_{i+j=0}^{n-1} h^{2n+2-(i+j)} \; |f|_{2n+2} \; C_{ij} \; \left(\frac{h}{\Delta}\right)^{r-(i+j)}$$

where $C_{ij}$ are constants for M.

$$|\alpha_{pq}| \leqslant h^{2n+2} \; |f|_{2n+2} \sum_{i+j=0}^{n-1} \frac{1}{C} \; h^{r-2(i+j)} \; \Delta^{i+j} \; C_{ij} \; .$$

As $r-2(i+j) \geqslant 0$, for all $n \in \mathbb{N}$, there exists a constant $K_{pq}$ such that

$$|\alpha_{pq}| \leqslant h^{2n+2} \; |f|_{2n+1} \; K_{pq} \; .$$

Therefore, with (4), there exists a constant K such that

$$|R_{2n+1}[u](M)| \leqslant h^{2n+2} \; |f|_{2n+2} \; K$$

Directly, or with a Markoff inequality (COATMELEC [2]), we find that there is a constant K' such that

$$|D^{\alpha} R_{2n+1}[u](M)| \leqslant h^{2n+2-|\alpha|} \; |f|_{2n+2} \; \left(\frac{h}{\rho}\right)^{2|\alpha|} \; K'$$

and finally,

$$|D^{\alpha} \Pi_{2n+1}[u](M)| \leqslant h^{2n+2-|\alpha|} \; |f|_{2n+2} \; \left(\frac{h}{\rho}\right)^{2|\alpha|} C \; ,$$

where C is a constant. ■

In the Sobolev spaces, with the usual norms and semi-norms, using the same arguments (and also a result of MEINGUET [5]), we can proof

<u>Proposition IV.4</u> : If $\frac{h}{\rho} < \sigma$ , there exists a constant C such that for all $f \in H^{2n+2}(\bar{\Omega})$,

$$||f - \Pi_{2n+1}[f]||_{m,\Omega} \leqslant C \, h^{2n+2-m} \; |f|_{2n+2,\Omega} \quad , \quad m \leqslant n$$

## V - HERMITE - ŽENIŠEK FINITE ELEMENT

A. ŽENIŠEK [6] prooved the following theorem :

Theorem : The simplest polynomial $Z(x,y)$, which generates piecewise polynomial and n-times continuously differentiable functions in an arbitrarily triangulated domain is of degree $4n+1$ and is on the triangle uniquely determined by the values :

$$D^{\alpha}Z(A_i) \quad ; \quad i = 1,2,3 \quad ; \quad |\alpha| \leqslant 2n$$

$$D^{\alpha}Z(A_o) \qquad |\alpha| \leqslant n-2$$

$$\frac{\partial^k Z}{\partial \nu_i^k}(Q_{ri}^{(k)}) \quad k = 1,2,\dots,n \quad ; \quad r = 1,2,\dots,k \quad ; \quad i = 1,2,3$$

where $A_1, A_2, A_3$ are the vertices of the triangle, $A_o$ its centroid, $Q_{ri}^{(k)}$ the points dividing the side $A_{i+1}A_{i-1}$ into $(k+1)$ equal parts and $\frac{\partial Z}{\partial \nu_i}$ the normal derivative.

Problem P3 : How to construct this polynomial Z ?

Consider first a rotation of the coordinate axes through an angle $\alpha$ , and let the new axes be denoted by $\tau$ and $\nu$ .
The relationship between the two systems may be expressed as

$\tau = cx + sy$ , $\nu = -sx + cy$ , where $s = \sin\alpha$, $c = \cos\alpha$

Differentiations with respect to $\tau$ and $\nu$ give

$$\frac{\partial^p}{\partial \nu^p} = \sum_{k=0}^{p} \binom{p}{k} (-1)^k s^k c^{p-k} \frac{\partial^p}{\partial x^k \partial y^{p-k}}$$

$$\frac{\partial^p}{\partial \tau^p} = \sum_{k=0}^{p} \binom{p}{k} c^k s^{p-k} \frac{\partial^p}{\partial x^k \partial y^{p-k}}$$

we have first the

Lemma : if M is on the side $A_2 A_3$, then

$$T_{A_2}^n f(M) = \lambda_3^n T_{A_2}^n f(M) + (1-\lambda_3) \sum_{i=0}^{n-1} \lambda_3^i T_{A_2}^i f(A_3)$$

and same results for the other subscripts by circular permutation.

A. CONSTRUCTION OF THE POLYNOMIAL

Let $Z = Z_{4n+1}[f]$ the polynomial of degree 4n+1 unique solution of the problem.

Let $Z_{4n+1}[f] = P_{4n+1}[f] + R_{4n+1}[f]$

where $P_{4n+1}[f]$ realizes Taylor interpolation of order 2n at the three points $A_1, A_2, A_3$

$$P_{4n+1}[f](M) = \sum_{i=1}^{3} \lambda_i^{2n+1} \sum_{j=0}^{2n} \frac{(2n+j)!}{(2n)!j!} (1-\lambda_i)^j \, T_{A_i}^{2n-j} f(M) \tag{6}$$

and

$$R_{4n+1}[f](M) = \lambda_1\lambda_2\lambda_3 \sum_{p+q+r=2n-1} \alpha_{pq} (\lambda_1\lambda_2)^p (\lambda_2\lambda_3)^q (\lambda_3\lambda_1)^r \tag{7}$$

($\alpha_{pq}$ real constants)

It is obvious that $R_{4n+1}$ is 2n-flat at $A_1, A_2, A_3$.

Explicitly, we obtain

$$\frac{\partial^k}{\partial\nu^k} (P_{4n+1}[f](M)) = \sum_{i=1}^{3} \sum_{j=0}^{2n} \frac{(2n+j)!}{(2n)!j!} \sum_{p\leq k} \sum_{q\leq p} \binom{k}{p}\binom{p}{q} \frac{(2n+1)!}{(2n+1-p)!} \frac{j!(-1)^j}{(j-q)!} \tag{8}$$

$$\times \lambda_i^{2n+1-p} \quad (\lambda_i - 1)^{j-q} \left(\frac{\partial\lambda_i}{\partial\nu}\right)^{p+q} \quad T_{A_i}^{2n-j-(k-p-q)} \frac{\partial^{k-p-q}}{\partial\nu^{k-p-q}} f(M)$$

$$\frac{\partial^k}{\partial\nu^k} (R_{4n+1}[f](M)) = \sum_{p+q+r=2n-1} \alpha_{pq} \, B_{pq}^{k,\nu}(M) \tag{9}$$

where

$$B_{p,q}^{k,\nu}(M) = \sum_{i\leq k} \sum_{j\leq i} \binom{k}{i}\binom{i}{j} \frac{(p+r+1)!(q+p+1)!(r+p+1)!}{(p+q+1-k+i)!(q+p+1-i+j)!(q+r+1-j)!} \times \tag{10}$$

$$\times \left(\frac{\partial\lambda_1}{\partial\nu}\right)^{k-i} \left(\frac{\partial\lambda_2}{\partial\nu}\right)^{i-j} \left(\frac{\partial\lambda_3}{\partial\nu}\right)^{j} \lambda_1^{p+r+1-k+i} \quad \lambda_2^{q+p+1-i+j} \quad \lambda_3^{q+r+1-j}$$

<u>Proposition V.1</u> : There exists $\frac{2n(2n+1)}{2}$ constants $\alpha_{pq}$ such that $Z_{4n+1}$ is the solution of Problem P3.

They are solutions of the linear system :

$$\mathcal{S}' \left\{ \begin{array}{ll} \sum_{p+q=0}^{2n-1} \alpha_{pq} B_{pq}^{\alpha}(A_o) = D^{\alpha}f(A_o) - D^{\alpha}P_{4n+1}(A_o) & |\alpha| \leqslant n-2 \\ \sum_{p+q=0}^{2n-1} \alpha_{pq} B_{pq}^{k,\nu_i}(Q_{ri}^{(k)}) = \frac{\partial^k}{\partial\nu_i^k} f(Q_{ri}^{(k)}) - \frac{\partial^k}{\partial\nu_i^k}(P_{4n+1}(Q_{ri}^{(k)})) & \end{array} \right.$$

$k=1,2,..,n$ ; $r=1,2,..,k$ ; $i=1,2,3$

B. BOUNDS OF ERROR INTERPOLATION

Using the same arguments as before, we can proof :

Proposition V.2 : If $\frac{h}{\rho} < \sigma$, if $|\alpha| \leqslant 2n$, there exists a constant $C(\alpha,\Omega)$ such that, for all $f \in \mathcal{C}^{4n+2}(\bar{\Omega})$,

$$||D^{\alpha}f - D^{\alpha}Z_{4n+1}[f]||_{\infty} \leqslant C(\alpha,\Omega)h^{4n+2-|\alpha|} |f|_{4n+2}$$

Proposition V.3 : If $\frac{h}{\rho} < \sigma$, if $|\alpha| \leqslant 2n$, there exists a constant K such that, for all $f \in H^{4n+2}(\Omega)$,

$$||f - Z_{4n+1}[f]||_{m,\Omega} \leqslant K\, h^{4n+2-m} |f|_{4n+2,\Omega}$$

Remark

It is interesting to compare the number of degrees of freedom of the elements (ie : the order of the linear system we have theorically to solve), and the order of system ($\mathcal{S}$) we have in fact to solve

| n | degrees of polynomial Z | degrees of freedom | order of $\mathcal{S}$ |
|---|---|---|---|
| 1 | 5 | 21 | 3 |
| 2 | 9 | 55 | 10 |
| 3 | 13 | 105 | 21 |

## REFERENCES


[1] CIARLET, P.G. : The finite Element Method for Elliptic Problems. North Holland (1978).

[2] COATMELEC, Chr. : Approximation et interpolation de fonctions différentiables de plusieurs variables. Ann. Sc. Ecole Normale Sup. 3e série, T. 83 (1966).

[3] DAVIS, D.J : Interpolation and Approximation, Blaisdell Publishing Company (1963).

[4] LE MEHAUTE, A. : Quelques méthodes explicites de prolongements de fonctions numériques de plusieurs variables. Thèse 3e cycle, Université de Rennes (1976).

[5] MEINGUET, J. : Structure et estimations de coefficients d'erreurs. Journées Eléments Finis, Université de Rennes (1977). RAIRO, Anal.Num. Vol.11, 355-368 (1977)

[6] ZENISEK, A. : Interpolation Polynomials on the Triangle. Numer. Math 15 (1970), 283-296.

# JACOBI PROJECTIONS

William A. Light

Department of Mathematics
University of Lancaster
Lancaster, England

## I. INTRODUCTION

Let $C[-1,1]$ be the space of continuous functions on $[-1,1]$, with norm defined by $||x|| = \max_{t\in[-1,1]} |x(t)|$. $P_n$ will denote the subspace of $C[-1,1]$ comprising algebraic polynomials of degree at most n. A projection $L_n : C[-1,1] \to P_n[-1,1]$ is a linear operator which is also idempotent. By virtue of the elementary inequality

$$||x-L_n x|| \le (1 + ||L_n||) \operatorname{dist}(x,P_n)$$

attention is naturally focussed on those $L_n$ which have small norm. A good introduction to this area may be found in [4], although several important questions posed there have since been resolved.

In this paper we concentrate on a family of projections called the Jacobi projections. The classical Jacobi polynomials $P_n(\alpha,\beta)$ are orthogonal on $[-1,1]$ with respect to the weight function $(1-t)^{\alpha}(1+t)^{\beta}$ where $\alpha,\beta > -1$. Throughout the paper we shall adopt the normalisation $P_n^{(\alpha,\beta)}(1) = 1$. The Jacobi projection $L_n^{(\alpha,\beta)} : C[-1,1] \to P_n$ is then obtained by truncating the expansion of x in Jacobi polynomials after n terms, i.e.



ISBN 0-12-780650-4

$$L_n^{(\alpha,\beta)}x = \sum_{i=0}^{n} a_i(\alpha,\beta)\, P_i^{(\alpha,\beta)}$$

where $a_i(\alpha,\beta) = h_i(\alpha,\beta) \int_{-1}^{1} (1-s)^{\alpha}(1+s)^{\beta} x(s) P_i^{(\alpha,\beta)}(s)ds,$

and $h_i(\alpha,\beta)$ is the appropriate normalisation factor. $L_n^{(\alpha,\beta)}$ clearly has the alternative representation

$$(L_n^{(\alpha,\beta)}x)(t) = \int_{-1}^{1} (1-s)^{\alpha}(1+s)^{\beta} x(s) K_n^{(\alpha,\beta)}(s,t)ds,$$

where $K_n^{(\alpha,\beta)}(s,t) = \sum_{i=0}^{n} h_i(\alpha,\beta)\, P_i^{(\alpha,\beta)}(s) P_i^{(\alpha,\beta)}(t).$

It is well-known and trivial to verify that

$$||L_n^{(\alpha,\beta)}|| = \max_{t\in[-1,1]} \int_{-1}^{1} (1-s)^{\alpha}(1+s)^{\beta} |K_n^{(\alpha,\beta)}(s,t)|ds.$$

During the last two decades there has been considerable progress in the understanding of classical orthogonal polynomials. The best reference remains the book by Askey [2]. The following theorem is a simple consequence of some deeper work due initially to Askey and Wainger [3] and in a refined form to Gasper [6]

Theorem 1.1

For $\alpha \geq \beta \geq -\frac{1}{2}$ we have

$$||L_n^{(\alpha,\beta)}|| = \int_{-1}^{1} (1-s)^{\alpha}(1+s)^{\beta} |K_n^{(\alpha,\beta)}(s,1)|ds$$

$$= A(\alpha,\beta,n) \int_{-1}^{1} (1-s)^{\alpha}(1+s)^{\beta}\, |P_n^{(\alpha+1,\beta)}(s)|ds.$$

Proof

Gasper's result is that if $\sum_{m=0}^{\infty} a_m P_m^{(\alpha,\beta)}$ is a convergent Jacobi series, then

$$\int_{-1}^{1} \left| \sum_{m=0}^{\infty} a_m P_m^{(\alpha,\beta)}(s) P_m^{(\alpha,\beta)}(t) \right| (1-s)^{\alpha}(1+s)^{\beta} ds$$

$$\leq \int_{-1}^{1} \left| \sum_{m=0}^{\infty} a_m P_m^{(\alpha,\beta)}(s) \right| (1-s)^{\alpha}(1+s)^{\beta} ds$$

$$\forall t \in [-1,1].$$

The first assertion of our theorem is obtained by setting

$$a_m = \begin{cases} 0 & m > n \\ 1 & m \leq n \end{cases}.$$

The second assertion is given in Szego [9]. ∎

The asymptotic behaviour of $||L_n^{(\alpha,\beta)}||$ is known [8] to be

$$||L_n^{(\alpha,\beta)}|| \sim A(\alpha,\beta) n^{\alpha+\frac{1}{2}} + O(n^{\alpha+\frac{1}{2}})$$

where
$$A(\alpha,\beta) = \frac{2}{\pi^{\frac{3}{2}}} \frac{\Gamma(\frac{\alpha}{2}+\frac{1}{4})\Gamma(\frac{\beta}{2}+\frac{3}{4})}{\Gamma(\alpha+1)\Gamma(\frac{\alpha+\beta}{2}+1)}.$$

From this it can be seen that it is reasonable to enquire into the location of the minimum of $||L_n^{(\alpha,\beta)}||$ for $\alpha \geq -\frac{1}{2}$, $\beta \geq -\frac{1}{2}$. Of course $\alpha = \beta = -\frac{1}{2}$ is the projection which associates each $x \in C[-1,1]$ with its truncated Chebyshev expansion. This has been a form of approximation which has long been favoured by numerical analysts - see [5] for details. From numerical evidence it would seem that $\alpha = \beta = -\frac{1}{2}$ is the location of the required minimum. In [7], it was shown that if $-\frac{1}{2} \leq \alpha = \beta \leq \frac{1}{2}$ then $||L_n^{(\alpha,\beta)}|| > ||L_n^{(-\frac{1}{2},-\frac{1}{2})}||$. In this paper we shall extend these results, and, as a by-product, obtain simpler proofs of them. Since we shall be working off the ultraspherical line $\alpha = \beta$, it is worth noting that the symmetry properties of the Jacobi polynomials in the region of $\alpha,\beta \geq -\frac{1}{2}$ ensure that $||L_n^{(\alpha,\beta)}|| = ||L_n^{(\beta,\alpha)}||$. Thus, for example, although Theorem 1.1 appears to be giving us only half the picture, we can by the symmetry extend the result to $\alpha<\beta$, when the maximum is attained at $t = -1$.

## 2. Preliminary Results

In this section we collect those results which are basic to all that follows. They are either easy to establish, well-known, or consequences of work by other mathematicians. Therefore proofs are omitted. We denote by $\hat{x}_n$ the function defined by

$$\hat{x}_n(t) = \mathrm{sgn}\{K_n^{(\frac{1}{2},-\frac{1}{2})}(t,1)\}.$$

It is then easy to see that $(L_n^{(-\frac{1}{2},-\frac{1}{2})}\hat{x}_n)(1) = ||L_n^{(-\frac{1}{2},\frac{1}{2})}||$.

<u>Lemma 2.1</u>

If $\hat{x}_n = \sum_{k=o}^{\infty} d_k C_k^{(-\frac{1}{2},-\frac{1}{2})}$ then

(i) $d_k = \dfrac{2}{\pi k} \tan \dfrac{\pi k}{2n+1}$, $K > 0$, $d_o = \dfrac{1}{2n+1}$.

(ii) $d_{p(n+\frac{1}{2})} = 0$,

$|d_{p(n+\frac{1}{2})-r}| > |d_{p(n+\frac{1}{2})+r}|$, $1 \le r \le n-1$, $p = 2,4,6,\ldots$ .

(iii) The sums $\sum_{j=1}^{\infty} d_{n+2j}^{(n)}$, $\sum_{j=1}^{\infty} d_{n+j}$ and $\sum_{j=1}^{\infty} d_{n+2j-1}$ all converge to negative limits.

<u>Proof</u>

See [7] for details.

<u>Theorem 2.2</u>

If $P_n^{(\alpha,\beta)} = \sum_{k=o}^{n} a_k(\alpha,\gamma,\beta,n) P_k^{(\gamma,\beta)}$

then $$a_k(\alpha,\gamma,\beta,n) = \frac{n!(\alpha+1)_k(n+\gamma+\beta+1)_k(\alpha-\gamma)_{n-k}(k+\beta+1)_{n-k}}{k!(n-k)!(\gamma+1)_k(k+\alpha+\beta+1)_k(k+\gamma+1)_{n-k}(2k+\alpha+\beta+2)_{n-k}}$$

<u>Proof</u>

See [2] for example, recalling that the normalisation here is $P_n^{(\alpha,\beta)}(1) = P_k^{(\gamma,\beta)}(1) = 1$, and $(\alpha)_k = \alpha(\alpha+1)(\alpha+2)..(\alpha+k-1)$.

Theorem 2.3

If $P_n^{(\alpha,\beta)} = \sum_{k=o}^{n} b_k(\alpha,\beta,\delta,n) P_k^{(\alpha,\delta)}$

then $b_k(\alpha,\beta,\delta,n) = \dfrac{n!(n+\alpha+\delta+1)_k(-1)^{n-k}(\beta-\delta)n-k}{k!(n-k)!(k+\alpha+\beta+1)_k(2k+\alpha+\beta+2)_{n-k}}.$

Proof

As for Theorem 2.2.

The $a_k(\alpha,\gamma,\beta,n)$ and $b_k(\alpha,\beta,\delta,n)$ are called connection coefficients. Where the parameters are obvious they will be dropped to give $a_k(n)$, $b_k(n)$. One of the main purposes behind obtaining results in the same vein as these theorems is that the signs of the connection coefficients can now be read off. In most of the previous applications attention has been focussed on the case when all the connection coefficients are non-negative. In our case we are interested in the case when they are nearly all non-positive. Here "nearly all" means all except $a_n(n)$.

Lemma 2.4

If $-1 \le \alpha-\gamma \le 1$ then

(i) $a_k(\alpha,\gamma,\beta,n) \le 0$ for $k < n$.

(ii) $\left|\dfrac{a_k(n)}{a_k(n+1)}\right| < 1, \quad \left|\dfrac{a_k(n)}{a_{k+1}(n)}\right| < 1.$

(iii) $\left|\dfrac{a_{n+1}(n+j)}{a_n(n+j)}\right| > \left|\dfrac{a_{n+1}(n+j+1)}{a_n(n+j+1)}\right|$ for $j \ge 1$.

Proof

(i) This is obvious - the only term controlling the sign of $a_k(\alpha,\gamma,\beta,n)$ is $(\alpha-\gamma)_{n-k}$, which is non-positive for $k < n$.

(ii) + (iii) None of these inequalities is obvious. The ratio in each case needs to be obtained, when some careful algebraic manipulations provide the results. The proofs are omitted to avoid the use of an excessive space for such tedious calculations. However, a prototypical case is done in [7].

## 3. Main Theorems

All the theorems given here rest on the following single elementary observation : if $L_n : C[-1,1] \to P_n$ and $H_n : C[-1,1] \to P_n$ have the property $||L_n|| = (L_n x)(t) < (H_n x)(t)$ then $||H_n|| > ||L_n||$.

<u>Theorem 3.1</u>

If $-\frac{1}{2} \le \gamma \le \frac{1}{2}$ then $||L_n^{(-\frac{1}{2},-\frac{1}{2})}|| < ||L_n^{(\gamma,-\frac{1}{2})}||$, for $n \ge 1$.

<u>Proof</u>

We know that $||L_n^{(-\frac{1}{2},-\frac{1}{2})}|| = (L_n \hat{x}_n)(1)$ where

$$\hat{x}_n = \sum_{k=0}^{\infty} d_k C_k^{(-\frac{1}{2},-\frac{1}{2})} \qquad \text{(Lemma 2.1).}$$

Now a simple computation shows that

$$(L_n^{(-\frac{1}{2},\gamma)} \hat{x}_n)(1) = (L_n^{(-\frac{1}{2},-\frac{1}{2})} \hat{x}_n)(1) + \sum_{r=0}^{n} b_r$$

where $b_r = \sum_{j=1}^{\infty} a_r(n+j) d_{n+j}$.

Now by Lemmas 2.1 (ii), 2.4 (ii) we must have

$$|d_{p(n+\frac{1}{2})-r}\, a_r(p(n+\tfrac{1}{2})-r)| > |d_{p(n+\frac{1}{2})+r}\, a_r(p(n+\tfrac{1}{2})+r)|$$

$$p = 2,4,6,\ldots\ldots$$

$$1 \le r \le n-1.$$

Furthermore for $\gamma < \frac{1}{2}$, from Lemmas 2.1(i) and 2.4(ii) we have that

$$d_{p(n+\frac{1}{2})-r}\ a_{r(p(n+\frac{1}{2})-r)} > 0$$

$$d_{p(n+\frac{1}{2})+r}\ a_{r}(p(n+\tfrac{1}{2})+r) < 0.$$

Now rewriting

$$b_r = \sum_{j=1}^{\infty} a_r(n+j)d_{n+j} = \sum_{p=1}^{\infty}\ \sum_{k=-n+1}^{n-1} d_{p(2n+1)+k}\ a_r(p(2n+1)+k$$

and observing that

$$\sum_{j=-n+1}^{-n-1} d_{p(2n+1)+k}a_r(p(2n+1)+k) = \sum_{k=1}^{n-1} d_{p(2n+1)-k}a_r(p(2n+1)-k)$$

$$+ \sum_{k=1}^{n-1} d_{p(2n+1)+k}a_r(p(2n+1)-k$$

we can see that the sum on the left-hand side of the last equality is positive and hence the $b_r$ are positive. We therefore have the inequalities

$$||L_n^{(-\frac{1}{2},-\frac{1}{2})}|| = (L_n^{(-\frac{1}{2},-\frac{1}{2})}\hat{x}_n(1) = (L_n^{(-\frac{1}{2},\gamma)}\hat{x}_n)(1) - \sum_{r=0}^{n} b_r$$

$$< (L_n^{(-\frac{1}{2},\gamma)}\hat{x}_n)(1) \le ||L_n^{(-\frac{1}{2},\gamma)}||. \blacksquare$$

Corollary 3.2

If $-\frac{1}{2} \le \delta \le \frac{1}{2}$ then $||L_n^{(-\frac{1}{2},-\frac{1}{2})}|| \le ||L_n^{(-\frac{1}{2},\delta)}||$, for $n \ge 1$.

Note that the proof of this corollary directly, without the aid of Theorem 3.1, is rather tricky - at least if we try to mimick the techniques employed in the proof of Theorem 3.1. The alternation in sign of the coefficients $b_k(\alpha,\beta,\delta,n)$ makes the whole argument very much more difficult. We can, with difficulty, extend the range of Theorem 3.1 to $-\frac{1}{2} \le \gamma \le \frac{3}{2}$.

Theorem 3.3

For $-\frac{1}{2} \le \gamma \le \frac{3}{2}$ then $||L_n^{(-\frac{1}{2},-\frac{1}{2})}|| < ||L_n^{(\gamma,-\frac{1}{2})}||$, for $n \ge 1$.

Proof

Writing $\hat{x}_n = \sum_{k=0}^{n} d_k P_k^{(-\frac{1}{2},-\frac{1}{2})} + \sum_{k=n+1}^{\infty} d_k P_k^{(-\frac{1}{2},-\frac{1}{2})}$

$= p_n + q_n,$

Again formally, we have $q_n = \sum_{k=n+1}^{\infty} d_k \sum_{i=0}^{k} a_i(k) P_i^{(\gamma,-\frac{1}{2})}$

Now we may observe that, in Theorem 2.2, if $\gamma = \alpha+1$ then $a_k(\alpha,\alpha+1,\beta,n) = 0$ for $k \le n-2$. In fact the formula is

$$P_n^{(\alpha-1,\beta)} = \frac{\alpha(n+\alpha+\beta)}{(\alpha+n)2n+\alpha+\beta)} P_n^{(\alpha,\beta)} - \frac{\alpha(n+\ \ )}{n(2n\ \ +\alpha+\beta)} P_{n-1}^{(\alpha,\beta)}$$

(Here the normalisation has been retained, $P_n^{(\alpha,\beta)}(1) = 1$. The formula may also be found in [1] with the normalisation $P_n^{(\alpha,\beta)}(1) = \frac{(\alpha+1)n}{n!}$ ). Now let $\hat{x}_n = \sum_{k=0}^{\infty} e_k P_k^{(\alpha,-\frac{1}{2})}$ where $-\frac{1}{2} \le \alpha \le \frac{1}{2}$. Then from the above relationship between $P_k^{(\alpha-1,-\frac{1}{2})}$ and $P_k^{(\alpha,-\frac{1}{2})}$ we can see that

$$L_n^{(\alpha+1,-\frac{1}{2})}\hat{x}_n = \sum_{k=0}^{n} e_k P_k^{(\alpha,-\frac{1}{2})} + \lambda(\alpha,n) e_{n+1} P_n^{(\alpha,-\frac{1}{2})}$$

where $\lambda(\alpha,n) = \dfrac{-\alpha(n+\frac{1}{2})}{(n+1)(2n+\alpha+\frac{3}{2})}$ .

Now by reference to the proof of Theorem 3.1 we can write

$$(L_n^{(\alpha+1,-\frac{1}{2})}\hat{x}_n)(1) = ||L_n^{(-\frac{1}{2},-\frac{1}{2})}|| + \sum_{r=0}^{n} b_r + \lambda(\alpha,n)e_{n+1}$$

$$= ||L_n^{(-\frac{1}{2},-\frac{1}{2})}|| + \sum_{r=0}^{n-1} b_r + b_n + \lambda(\alpha,n)e_{n+1} .$$

The term $b_n + \lambda(\alpha,n)e_{n+1}$ is easily computed as

$$b_n + \lambda(\alpha,n)e_{n+1} = \sum_{j=1}^{\infty} a_n(n+j)d_{n+j} + \lambda(\alpha,n) \sum_{j=0}^{\infty} a_{n+1}(n+1+j)d_{n+1+j}$$

$$= \sum_{j=1}^{\infty} \{a_n(n+j) + \lambda(\alpha,n)a_{n+1}(n+j)\}d_{n+j}.$$

We shall now seek to establish the two points

(i) $a_n(n+j) + \lambda(\alpha,n)\, a_{n+1}(n+j) < 0, \quad j \geq 1$

(ii) $|a_n(n+j) + \lambda(\alpha,n) a_{n+1}(n+j)|$ is a decreasing function of $j$.

The first is a consequence of the fact that $|\lambda(\alpha,n)| < 1$ and $|a_{n+1}(n+j)| < |a_n(n+j)|$, as given by Lemma 2.4 with the exception of the term for $j = 1$ which is clearly negative. Now from Lemma 2.4 we have

$$\frac{a_{n+1}(n+j)}{a_n(n+j)} > \frac{a_{n+1}(n+j+1)}{a_n(n+j+1)} .$$

Hence $$\frac{\lambda(\alpha,n) a_{n+1}(n+j)}{a_n(n+j)} + 1 < \frac{\lambda(\alpha,n) a_{n+1}(n+j+1)}{a_n(n+j+1)} + 1,$$

the reversal of the inequality being forced by $\lambda(\alpha,n) < 0$.

So

$$-a_n(n+j)-\lambda(\alpha,n)a_{n+1}(n+j) < \frac{a_n(n+j)}{a_n(n+j+1)}\left[-a_n(n+j+1)-\lambda(\alpha,n)\, a_{n+1}(n+j+1)\right]$$

or, since $\left|\dfrac{a_n(n+j)}{a_n(n+j+1)}\right| < 1$ by Lemma 2.4

$$|a_n(n+j) + \lambda(\alpha,n)a_{n+1}(n+j)| < |a_n(n+j+1)+\lambda(\alpha,n)a_{n+1}(n+j+1)|.$$

Now, precisely those arguments which secured Theorem 3.1 can be applied to $b_n + \lambda(\alpha,n)e_{n+1}$ showing that the sum

$$\sum_{j=1}^{\infty}\{a_n(n+j) + \lambda(\alpha,n)a_{n+1}(n+j)\}\, d_{n+j}$$

is positive. We already know from the same Theorem 3.1 that $\sum_{r=0}^{n-1} b_r > 0$. Thus we may conclude that

$$||L_n^{(\alpha+1,-\frac{1}{2})}|| \geq (L_n^{(\alpha+1,-\frac{1}{2})}\hat{x}_n(1) = ||L_n^{(-\frac{1}{2},-\frac{1}{2})}|| + \sum_{r=0}^{n} b_r + \lambda(\alpha,n)e_{n+1} > ||L_n^{(-\frac{1}{2},-\frac{1}{2})}||, \text{ where } n \geq 1.$$

<u>Corollary 3.4</u>

For $-\frac{1}{2} \le \beta \le \frac{3}{2}$, $||L_n^{(-\frac{1}{2},-\frac{1}{2})}|| < ||L_n^{(-\frac{1}{2},\beta)}||$, for $n \ge 1$.
By applying the same line of argument to the ultraspherical case and using the results of [7], one may then show

<u>Theorem 3.5</u>

If $-\frac{1}{2} \le \alpha \le \frac{3}{2}$ then $||L_n^{(-\frac{1}{2},-\frac{1}{2})}|| < ||L_n^{(\alpha,\alpha)}||$, for $n \ge 1$.
It is conceivable that by more careful analysis of these results one could extend the theorems to intervals of length greater than two. However, ideally one would like the results say for all $\alpha \ge -\frac{1}{2}$. Thus it seems pointless to pursue the arguments developed here.

## 4. Extension Results

In this section we use the theorems from the previous section along with a simple connection formula between the norms of three distinct projections. This allows us to "fill in" some of the missing results in the region $-\frac{1}{2} \le \alpha,\beta \le \frac{3}{2}$. Section three may be summarised by saying that in this region we have the desired results on the coordinate axes (recalling that the origin is shifted to the point $(-\frac{1}{2},-\frac{1}{2})$) and on the line $\alpha = \beta$. We begin by proving the connection result. It becomes convenient in this section to alter the normalisation to $P_n^{(\alpha,\beta)}(1) = \frac{(\alpha+1)_n}{n!}$. In general the normalisation of unity at the right-hand end point is always more convenient for rearrangement of series, but seldom for any other purpose.

<u>Theorem 4.1</u>

For all $\alpha,\beta, \ge -\frac{1}{2}$ the following relationship holds:

$$||L_n^{(\alpha,\beta)}|| \le \frac{1}{\Gamma(n+\alpha+\beta+2)}\{ \Gamma(\alpha+1)||L_n^{(\alpha+1,\beta)}|| + \Gamma(n+\beta+1)||L_n^{(\alpha,\beta+1)}||\}.$$

<u>Proof</u>

We have $||L_n^{(\alpha,\beta)}|| = \frac{\Gamma(n+\alpha+\beta+2)2^{-\alpha-\beta-1}}{\Gamma(\alpha+1)\Gamma(n+\beta+1)} \int_{-1}^{1} (1-s)^{\alpha}(1+s)^{\beta}.$

$$|P_n^{(\alpha+1,\beta)}(s)|ds$$

from Theorem 1.1. Now we use the relationship [1, pg. 782]

$$(1-t)P_n^{(\alpha+1,\beta)}(t) + (1+t)P_n^{(\alpha,\beta+1)}(t) = 2P_n^{(\alpha,\beta)}(t)$$

giving

$$||L_n^{(\alpha,\beta)}|| \le \frac{\Gamma(n+\alpha+\beta+2)2^{-\alpha-\beta-2}}{\Gamma(\alpha+1)\Gamma(n+\beta+1)} \int_{-1}^{1}\{(1-s)^{\alpha+1}(1+s)^{\beta}|P_n^{\alpha+2,\beta}(s)|$$

$$+ (1-s)^{\alpha}(1+s)^{\beta+1}|P_n^{(\alpha+1,\beta+1)}(s)|\}\, ds.$$

Again by reference to Theorem 1.1 we may rewrite the right-hand side of this inequality as

$$\frac{\Gamma(\alpha+1)}{\Gamma(n+\alpha+\beta+2)} ||L_n^{(\alpha+1,\beta)}|| + \frac{\Gamma(n+\beta+1)}{\Gamma(n+\alpha+\beta+2)} ||L_n^{(\alpha,\beta+1)}||.$$

<u>Theorem 4.2</u>

For all $\alpha \ge -\tfrac{1}{2}$, $||L_n^{(\alpha,\alpha)}|| \le ||L_n^{(\alpha+1,\alpha)}||$.

<u>Proof</u>

From the remarks following Theorem 1.1 we know that $||L_n^{(\alpha+1,\alpha)}|| = ||L_n^{(\alpha,\alpha+1)}||$ so that Theorem 4.1 gives

$$||L_n^{(\alpha,\alpha)}|| \le \frac{||L_n^{(\alpha+1,\alpha)}||}{\Gamma(n+2\alpha+2)} \{\Gamma(\alpha+1) + \Gamma(n+\alpha+1)\}.$$

Now for $n \ge 2$ we can write

$$\frac{\Gamma(\alpha+1) + \Gamma(n+\alpha+1)}{\Gamma(n+2\alpha+2)} = \frac{\Gamma(\alpha+1) + \Gamma(n+\alpha+1)}{(n+2\alpha+1)\Gamma(n+2\alpha+1)} \le 1$$

and so for such n $||L_n^{(\alpha,\alpha)}|| \le ||L_n^{(\alpha+1,\alpha)}||$.

For $n = 1$ and $\alpha \ge 0$ we have

$$\frac{\Gamma(\alpha+1) + \Gamma(n+\alpha+1)}{\Gamma(n+2\alpha+2)} = \frac{\Gamma(\alpha+1) + \Gamma(\alpha+2)}{(2\alpha+2)\ \Gamma(2\alpha+2)} \le 1.$$

For $n = 1$ and $-\frac{1}{2} \leq \alpha \leq 0$ we have

$$\frac{\Gamma(\alpha+1) + \Gamma(n+\alpha+1)}{\Gamma(n+2 +2)} = \frac{\Gamma(\alpha+1) + \Gamma(\alpha+2)}{\Gamma(2\alpha+3)}$$

$$= \frac{\Gamma(\alpha+1) + (\alpha+1)\Gamma(\alpha+1)}{\Gamma(2\alpha+3)}$$

$$= \frac{(\alpha+2)\ \Gamma(\alpha+1)}{\Gamma(2\alpha+3)}$$

$$\leq \frac{2\Gamma(1)}{\Gamma(2)}, \qquad \text{since } -\tfrac{1}{2} \leq \alpha \leq 0$$

$$\leq 1.$$

Corollary 4.3

For $0 \leq \alpha \leq \frac{3}{2}$ we have $||L_n^{(\alpha+1,\alpha)}|| > ||L_n^{(-\frac{1}{2},-\frac{1}{2})}||$.

Proof

We simply combine Theorems 4.2 and 3.5.

Corollary 4.4

Also for $0 \leq \alpha \leq \frac{3}{2}$ we have $||L_n^{(\alpha,\alpha+1)}|| > ||L_n^{(-\frac{1}{2},-\frac{1}{2})}||$.

Proof

This is simply the symmetry result.

Corollary 4.5

For $\beta = \frac{1}{2}$ and $-\frac{1}{2} \leq \alpha \leq \frac{1}{2}$ or $\alpha = \frac{1}{2}$, $-\frac{1}{2} \leq \beta \leq \frac{1}{2}$, we have $||L_n^{(\alpha,\beta)}|| > ||L_n^{(-\frac{1}{2},-\frac{1}{2})}||$.

Proof

Use Theorem 4.1 and Theorem 3.5.

## 5. Summary

Numerical evidence suggests extremely strongly that the minimum of $||L_n^{(\alpha,\beta)}||$ for each n and for $\alpha,\beta \geq -\frac{1}{2}$ occurs when $\alpha = \beta = -\frac{1}{2}$. This is clearly true asymptotically, but for low values of n, which are of interest in practice, the analytic proof remains an open question. There are many other interesting open questions in this area. For example it is again numerically well-known that $||L_n^{(0,0)}|| > ||L_{n-1}^{(0,0)}||$ for $n = 1,2,\ldots$ . However again the proof of this result has evaded several able mathematicians. However, the main purpose of this section is to summarise in diagrammatic form the various results established in sections three and four. In the following figure the full lines represent those values of the pair $(\alpha,\beta)$ for which it is known that $||L_n^{(\alpha,\beta)}|| > ||L_n^{(-\frac{1}{2},-\frac{1}{2})}||$. The dotted lines represent the two coordinate axes (with the usual shifted origin $(-\frac{1}{2},-\frac{1}{2})$ and the ultraspherical line $\alpha = \beta$.

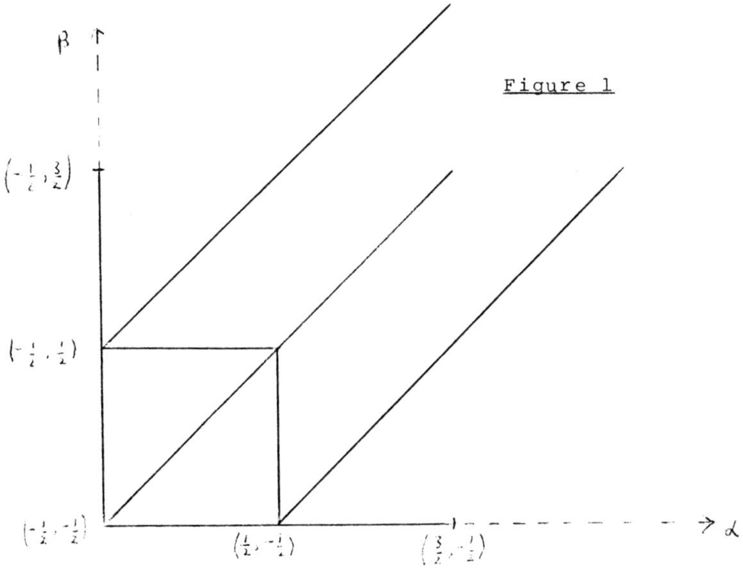


Figure 1

## 6. References

1. Abramowitz, M. and Stegun, I.A. "Handbook of Mathematical Functions", Dover, 1965.
2. Askey, R.A. "Orthogonal Polynomials and Special Functions" Regional Conference Lectures in Applied Maths. Vol. 21, SIAM. Philadelphia, 1975.
3. Askey, R.A. and Wainger, S. "A Convolution Structure for Jacobi Series", Amer. J. Math., 91 (1969) pp.463-485.
4. Cheney, E.W. and Price, K.H. "Minimal Projections", in "Approximation Theory", Talbot, A. (Ed.) Academic Press, 1970.
5. Fox, L. and Parker, I.B. "Chebyshev Polynomials in Numerical Analysis", O.U.P. 1968.
6. Gasper, G. "Positivity and the Convolution Structure for Jacobi Series", Ann. of Math. 93 (1971), pp.112-118.
7. Light, W.A. "Norms of some Projections on C[a,b]", J. Approx. Th. 20 (No. 1, 1977) pp.51-56.
8. Lorch, L. "Lebesgue Constants for Jacobi Series", Proc. Amer. Math. Soc. 10 (1959), pp. 756-761.
9. Szegö, G. "Orthogonal Polynomials", Amer. Math. Soc. Colloquium Pub. Vol. 23, 1959.

# OSCILLATING MONOSPLINES OF LEAST UNIFORM NORM

R.B. Barrar

H.L. Loeb*

Department of Mathematics
University of Oregon
Eugene, Oregon

In this article we announce several new results concerning polynomial and extended totally positive, monosplines which oscillate in a prescribed manner. Among such families we characterize the unique function of minimal uniform norm. Indeed, these results are corollaries of more general theorems about these oscillating families.

These results extend the classical development in two directions. Firstly, they generalize the corresponding properties of Tchebycheff polynomials [6] and of Tchebycheff systems [4]. Secondly, they generalize the characterization theorem for best uniform monosplines with multiple nodes and simple zeros to the complete multiple node and multiple zero setting [1,2,7,9,10].

We first consider the case of extended totally positive monosplines. This means that the underlying kernel K(x,y) is extended totally positive in the sense that

$$0 \leqslant x_1 \leqslant x_2 \leqslant \dots \leqslant x_r \leqslant 1 \tag{1}$$

$$0 \leqslant y_1 \leqslant y_2 \leqslant \dots \leqslant y_r \leqslant 1$$

---

* Partially written while visiting the Technion, Haifa, Israel.




ISBN 0-12-780650-4

implies that

$$K^*\begin{pmatrix}x_1,\dots,x_r\\ y_1,\dots,y_r\end{pmatrix} > 0 . \tag{2}$$

In the situation where strict inequality occurs in (1) within each group, $\{x_i\}$ and $\{y_i\}$,

$$K^*\begin{pmatrix}x_1,\dots,x_r\\ y_r,\dots,y_r\end{pmatrix} = \det\left\{K(x_i,y_j);\ i,j = 1,\dots,r\right\}. \tag{3}$$

When equality occurs in (1), the appropriate partial derivatives of $K(x,y)$ with respect to x and y are employed in the modified definition (3). (For more details see [8, p.5]).

For a given set of odd integers $\{m_i\}_{i=1}^{p}$, and non-negative integers $M_0$ and $M_1$, set

$$N = M_0 + \sum_{i=1}^{n}(m_i + 1) + M_1 .$$

Then consider all monosplines of the form,

$$M(x) = \int_0^1 K(x,y)\,dy + \sum_{i=1}^{M_0-1} a_i K^{(i)}(x,0) + \sum_{i=1}^{p}\sum_{j=0}^{m_i-1} a_{ij}K^{(j)}(x,\xi_i) + \sum_{i=0}^{M_1-1} b_i K^{(i)}(x,1) \tag{4}$$

where

$$K(x,0) \equiv 1,\ K^{(j)}(x,\xi) \doteq \frac{\partial^j}{\partial\xi^j}K(x,\xi) ,$$

and $0 < \xi_1 < \dots < \xi_p < 1$.

Further we are given a set of positive integers $\{n_i\}_{i=1}^s$ with $N = \sum_{i=1}^s n_i$. If

$$\Delta_s = \{\underline{x} = (x_1,\ldots,x_s) : 0 < x_1 < \ldots < x_s < 1\}$$

for each $\underline{x} \in \Delta_s$, by the results of Karlin, Pinkus [11] and Barrow [5], there is a unique monospline $M(x,\underline{x})$ of the form (4) such that

$$\left.\frac{dM^j(x,\underline{x})}{dx^j}\right|_{x=x_i} = 0 \qquad \begin{pmatrix} j = 0,1,\ldots,n_i-1 \\ i = 1,\ldots,s \end{pmatrix} \tag{5}$$

Then we seek to minimize the uniform norm of $M(\cdot,\underline{x})$ $\left(\|M(\cdot,\underline{x})\| = \max_{x \in [0,1]} |M(x,\underline{x})|\right)$ as $\underline{x}$ varies over $\Delta_s$. Using some non-linear techniques and our differential equation approach [2], we are able to demonstrate

Theorem 1: The problem

$$\min_{\underline{x} \in \Delta_s} \|M(\cdot,\underline{x})\|,$$

has exactly one solution. The solution $M(\cdot,\underline{x}^*)$ is uniquely defined by the properties:

a. There exists a set of s+1 points $\{t_i\}_{i=0}^s$ such that

(i) $0 = t_0 < x_1^* < t_1 < \ldots < t_{s-1} < x_s^* < t_s = 1$

(ii) $\|M(\cdot,\underline{x}^*)\| = (-1)^{N-\sum_{j=1}^i n_j} M(t_i,\underline{x}^*) \qquad (i = 0,1,\ldots,s)$

Our results for the polynomial monospline problem depend very heavily on a recent extension of the Fundamental Theorem of Algebra for Polynomial Monosplines which encompasses both multiple knots and multiple zeros [3].

We consider the class of polynomial monosplines of the form,

$$M(x) = \sum_{j=0}^{n-1} a_j K^{(j)}(x,0) + \sum_{i=1}^{p} \sum_{j=0}^{m_i} a_{ij} K^{(j)}(x,\xi_i) , \qquad (6)$$

where $K(x,\xi) = \frac{(x-\xi)_+^{n-1}}{(n-1)!}$ is the spline kernel and $0 < \xi_1 < \ldots < \xi_p < 1$, $\{m_i\}_{i=1}^{p}$ are of course a fixed set of odd integers with $N = n + \sum_{i=1}^{p} (m_i + 1)$, and $m_i \leqslant n \quad (i = 1,\ldots,p)$.

Theorem 2: (Fundamental theorem of Algebra for polynomial monosplines with multiple knots and multiple zeros, [3]). Given a set of positive integers $\{n_i\}_{i=1}^{s}$ where $N = \sum_{i=1}^{s} n_i$ , $n \geqslant 2$ and $n_i \leqslant n$ $(i = 1,\ldots,s)$. Then for each $\underline{x} \in \Delta_s$ there exists at most one $M(x,\underline{x})$ at the form (6) such that

$$M^{(j)}(x_i) = 0 \qquad (j = 0,1,\ldots,n_i-1) \qquad (7)$$
$$(i = 1,\ldots,s).$$

Further if

$$n \geqslant 2 \quad \text{and} \quad \max_{1\leqslant i\leqslant p} \{m_i\} + \max_{1\leqslant i\leqslant s} \{n_i\} \leqslant n \qquad (8)$$

there exists exactly one $M(x,\underline{x})$ which satisfies (7) for each $x \in \Delta_s$.

Employing Theorem 2, our differential equation technique [1], some compactness results [12], and several connectedness and smoothing techniques, we can prove:

Theorem 3: If we assume (8) is valid, then there exists a unique $\underline{x}^* \in \Delta_s$ for which

$$\|M(\cdot,\underline{x}^*)\| = \min_{\underline{x}\in\Delta_s} \|M(\cdot,\underline{x})\|.$$

$M(\cdot,\underline{x}^*)$ is uniquely determined by a set of s+1 points

$$0 = t_0 < t_1 < \dots < t_s = 1$$

such that

$$\|M(\cdot,\underline{x}^*)\| = (-1)^{N-\Sigma_{j=1}^{i} n_j} \; M(t_i,\underline{x}^*) \qquad (i = 0,1,\dots,s).$$

Here $0 = t_0 < x_1^* < t_1 < \dots < x_s^* < t_s = 1$.

Theorems 1 and 3 extend the classical results of Johnson [7]; Karlin [9, 10]; Barrar, Loeb [1]; and Barrar, Loeb, Werner [2]. In the special case where all the $\{n_j\}$ are even integers, our theorems yield some new results about one-sided approximations.

## REFERENCES

[1] R.B. Barrar and H.L. Loeb, On Monosplines with Odd Multiplicity of Least Norm, *J. Anal. Math. 33* (1978), 12-38.

[2] R.B. Barrar, H.L. Loeb and H. Werner, On the Uniqueness of the Best Uniform Extended Totally Positive Monospline, *J. Approx. Theory 28* (1980), 20-29.

[3] R.B. Barrar and H.L. Loeb, Fundamental Theorem of Algebra for Monosplines and Related Results, to appear *SIAM J. Num. Analysis*.

[4] R.B. Barrar and H.L. Loeb, Oscillating Tchebycheff Systems, Preprint.

[5] D. Barrow, On Multiple Node Gaussian Quadrature Formulae, *Math. Comp. 32* (1978), 431-439.

[6] B.D. Bojanov, A Generalization of Chebyshev Polynomials, *J. Approx. Theory 26* (1979), 293-300.

[7] R.S. Johnson, On Monosplines of Least Deviation, *Trans. Amer. Math. Soc. 96* (1960), 458-477.

[8] S. Karlin and W.J. Studden, *Tchebycheff Systems: With Applications in Analysis and Statistics;* Interscience, New York, 1966.

[9] S. Karlin, On a Class of Best Nonlinear Approximation Problems and Extended Monosplines, in: *Studies in Spline Functions and Approximation Theory,* (S. Karlin, C. Micchelli, A. Pinkus, and I.J. Schoenberg, eds.), Academic Press, New York, 1976, 19-59.

[10] S. Karlin, A. Global Improvement Theorem for Polynomial Monosplines, in: *Studies in Spline Functions and Approximation Theory* (S. Karlin, C. Micchelli, A. Pinkus and I.J. Schoenberg, eds.), Academic Press, New York, 1976, 67-74.

[11] S. Karlin and A. Pinkus, Gaussian Quadrature Formulae with Multiple Nodes, in: *Studies in Spline Functions and Approximation Theory,* (S. Karlin, C. Micchelli, A. Pinkus, and I.J. Schoenberg, eds.), Academic Press, New York, 1976, 113-137.

[12] C. Micchelli, The Fundamental Theorem of Algebra for Monosplines with Multiplicities, in: *Linear Operators and Approximation,* (P.L. Butzer, J.P. Kahane and B. Sz. Nagy, eds.), Birkhäuser Verlag, Basel, 1972, 419-430.

# SOME APPLICATIONS AND DRAWBACKS OF PADÉ APPROXIMANTS

J.C. Mason

Department of Mathematics and Ballistics,

Royal Military College of Science,

Shrivenham, Swindon, England

ABSTRACT

The Padé approximant method is discussed critically with regard to its area of applications and the reliability of its results, particular emphasis being given to approximation on an infinite range. A number of successful applications are described, but examples are also given in which numerical error or lack of convergence leads to poor or misleading results.

## 1. INTRODUCTION

### 1.1 The One-point Padé Approximant

If a function f(x) has a formal power series expansion

$$f(x) \sim f_0(x) = c_0 + c_1x + c_2x^2 + \dots + c_kx^k + \dots, \qquad (1)$$

then the [r,s] one-point Padé approximant $P_{rs}(x)$ to f(x) of the form





ISBN 0-12-780650-4

$$P_{rs}(x) = \frac{a_0 + a_1 x + \ldots + a_r x^r}{b_0 + b_1 x + \ldots + b_s x^s} \tag{2}$$

is defined by the relation

$$f_0(x) - P_{rs}(x) = \rho(x^{r+s+1}), \tag{3}$$

where $\rho(x^{r+s+1})$ denotes a formal power series starting with a term in $x^p$ for $p \geq r+s+1$.

From (1), (2), (3) it follows that

$$(b_0 + \ldots + b_s x^s)(c_0 + \ldots + c_{r+s} x^{r+s}) - (a_0 + \ldots + a_r x^r) = \rho(x^{r+s+1}).$$

Setting $b_0 = 1$ and equating coefficients of powers of x leads to a linear algebraic system for the determination of $b_1, \ldots, b_s$:

$$c_{i-1} b_1 + c_{i-2} b_2 + \ldots + c_{i-s} b_s = - c_i \quad (i = r+1, \ldots, r+s) \tag{4}$$

and formulae for the subsequent determination of $a_0, \ldots, a_r$:

$$a_i = c_i + c_{i-1} b_1 + \ldots + c_{i-s} b_s. \quad (i = 0, \ldots, r) \tag{5}$$

By convention $c_k$ is taken to be zero for $k < 0$ in (4) and (5). The solution is unique provided the matrix (a Hankel matrix) of the system (4) is non-singular. If the matrix happens to be singular, then $b_0$ is set equal to zero and $b_1, \ldots, b_s$ are determined from the homogeneous equations

$$c_{i-1} b_1 + c_{i-2} b_2 + \ldots + c_{i-s} b_s = 0 \quad (i = r+1, \ldots, r+s).$$

In either case the Padé approximant may be uniquely determined. In our algorithm, Gauss elimination is used to solve the equations (4).

A number of more sophisticated algorithms, such as the $\epsilon$-algorithm and the quotient difference algorithm have been proposed for the determination of Padé approximants, and these are discussed by Gragg [1] who also gives an excellent exposition of the properties of the "Padé table" - the array of all [r,s] approximants for r = 0, 1, 2, ...; s = 0, 1, 2, ... . However, the linear algebraic equation approach is convenient for our purposes here, and it offers the possibility of numerical error analysis.

The present discussion cannot attempt to be all-embracing, and in particular we shall have to neglect various theoretical aspects. For a broader treatment, the reader is referred to Baker [2], and recent developments are discussed in Wuytack [3].

### 1.2 The Two-point Padé Approximant

A number of extensions of Padé approximants have been proposed (such as multivariate approximants, approximants with branch cuts, etc), and in particular two-point Padé approximants exist for a function f(x) having in addition to a power series expansion (1), a formal asymptotic expansion

$$f(x) \sim f_\infty(x) = d_0 + d_1 x^{-1} + d_2 x^{-2} + \ldots + d_k x^{-k} + \ldots \tag{6}$$

in powers of $x^{-1}$. The two-point [r, s; m] Padé approximant $P_{rs}^{(m)}(x)$ of form (2) is defined by the relations

$$f_0(x) - P_{rs}^{(m)}(x) = \rho(x^{m+1}) \tag{7}$$

and $$f_\infty(x) - P_{rs}^{(m)}(x) = \rho((x^{-1})^{r+s-m}) . \tag{8}$$

On multiplying through by the denominator of $P_{rs}^{(m)}(x)$ and equating coefficients of powers of x in (7) and powers of $x^{-1}$ in (8), a system of r+s+1 linear equations is obtained for the determination of $a_0, \ldots, a_r, b_1, \ldots, b_s$, where $b_0$ is normally set equal to unity. (See [4] for full details.) With a suitable choice of r, s, and m, one can hope to obtain an approximation to f valid throughout $[0, \infty)$.

Two-point Padé approximants are discussed from the point of view of continued fractions by McCabe and Murphy [5].

### 1.3 Choice of Approximants

It is generally agreed that, for accurate approximation on a finite range by a one-point Padé approximant, it is best to choose the degrees r and s of numerator and denominator to be equal or nearly equal. However, in the case of a two-point Padé approximant the relation between r and s is dictated by the asymptotic expansion (6); specifically if $d_0 \neq 0$ then r = s. If a one-point Padé approximant is required on an infinite range, and f(x) does not possess an asymptotic expansion in powers of $x^{-1}$, then it may still be possible to choose r and s so that $P_{rs}(x)$ and f(x) have the same dominant asymptotic behaviour, as we shall see in §2.4 below.

## 2. SOME APPLICATIONS OF PADE APPROXIMANTS

### 2.1 Extrapolation

The power series expansion of a function f(x) frequently converges slowly, and indeed it diverges if the analytic extension f(z) has a pole closer to 0 than x. However, a Padé approximant, and specifically one on the main diagonal (r = s) of the Padé table, is typically accurate for a much greater range of x than the truncated power series with the same number of coefficients. We give no examples here, since this application is very well known and appears to be widely applicable. Considerable theoretical work has been done on the convergence of these approximants (see for example Pommerenke [6]), but the method still seems to work better in practice than the theory can rigorously predict.

### 2.2 Inclusion of Poles

If the function f(x) has poles on (or near) the real axis, then Padé approximants can give good results over ranges which include poles, and indeed they can be used to approximately determine the positions of poles. This possibility has been well known for a very long time and de Montessus de Ballore [7] gave some early theoretical results.

As an example, consider the function f(x) = tan x, which has a power series expansion

$$f_0(x) = x(c_0 + c_1 t + c_2 t^2 + \dots + c_k t^k + \dots), \quad t = x^2, \tag{9}$$

where $c_0, c_1, \dots, c_k, \dots$ may be determined (for example) by applying the Frobenius method to the differential equation

$$f' = 1 + f^2, \quad f(0) = 0.$$

Clearly (9) converges only on an interval interior to $(-\pi/2, \pi/2)$, since $f(x)$ has poles at $\pm(k - \tfrac{1}{2})\pi$ $(k = 1, 2, \dots)$. However, an approximation of the form

$$f^* = x \, . \, P_{rr}(t),$$

where $P_{rr}(t)$ is the $[r, r]$ Padé approximant in t to $x^{-1}f(x)$, would be expected to be valid in a range well beyond $(-\pi/2, \pi/2)$ for r sufficiently large. Moreover the poles and zeros of $f^*$ would be expected to converge to the poles and zeros of f.

In Table 1, we give computed values of the first four poles in t of $P_{rr}(t)$, and these can clearly be seen to be converging to the corresponding poles $(k - \tfrac{1}{2})^2\pi^2$ of f.

TABLE 1. Poles of $P_{rr}(t)$

| r | Pole 1 | Pole 2 | Pole 3 | Pole 4 |
|---|---|---|---|---|
| 1 | 2.5 | | | |
| 2 | 2.46744 | 30. | | |
| 3 | 2.4674011 | 22.29 | 90. | |
| 4 | 2.4674011 | 22.2074 | 64. | 200. |
| 5 | 2.4674011 | 22.20660 | 61.61 | 125. |
| $(k - \tfrac{1}{2})^2\pi^2$ | 2.4674011 | 22.20661 | 61.69 | 121. |

### 2.3 Infinite Range: Two-point Padé Approximants

If f has a power series expansion in powers of $x^\alpha$, say, for some $\alpha > 0$, then for a two-point Padé approximant to exist it is necessary that f should have an asymptotic expansion in powers of $x^{-\alpha}$. McCabe and Murphy [5] give a number of examples where this occurs.

One interesting example (which is discussed by McCabe and Murphy [5]) is Dawson's integral

$$f(x) = e^{-x^2} \int_0^x e^{t^2}\, dt, \quad x \text{ in } [0, \infty),$$

which has the power series expansion

$$f_0 = x \left[1 - \tfrac{2}{3} t + \ldots + \frac{(-2)^k\, t^k}{1.3.5.\ \ldots\ (2k+1)} + \ldots\right], \quad t = x^2,$$

and the asymptotic expansion

$$f_\infty = \tfrac{1}{2}x^{-1}\left[1 + \tfrac{1}{2}t^{-1} + \ldots + 1.3.5.\ \ldots\ (2k-1)2^{-k}t^{-k} + \ldots\right].$$

Clearly it is possible to obtain approximations to f(x) of the form

$$f^* = x\, P^{(m)}_{r,\, r+1}(t),$$

where $P^{(m)}_{r,\, r+1}(t)$ is the two-point Padé approximant in t corresponding to the expansions $x^{-1}f_0$ and $2x.f_\infty$.

For the choice m = r+1, f* is observed in practice to converge fairly well, and indeed for r = 7, f* has an accuracy of about 3 significant decimal places throughout $[0, \infty)$.

## 2.4 Infinite Range: One-point Padé Approximants

We now give examples of three functions for which two-point Padé approximants cannot be defined, but for which one-point Padé approximants may be chosen to match the leading term in the asymptotic behaviour. In these examples we exploit the idea of raising a power series to a power $\alpha$. Given an expansion of form (1), we define a new expansion

$$e_0 + e_1 x + \ldots + e_k x^k + \ldots = (f_0)^\alpha = (c_1 + c\, x + \ldots + c_k x^k + \ldots)^\alpha,$$

for some specified real value of $\alpha$. By taking logs and differentiating, we may obtain the following relations between $\{e_k\}$ and $\{c_k\}$:

$$\left.\begin{aligned} e_0 &= (c_0)^{\alpha}, \\ e_k &= (kc_0)^{-1} \sum_{i=0}^{k-1} \{(k-i)\alpha - i\} e_i\, c_{k-i}. \end{aligned}\right\} \tag{10}$$

<u>Example 1</u> <u>Thomas Fermi Initial Value Problem</u> Define $f(x)$ as the solution on $[0, \infty)$ of

$$x(f'')^2 = f^3,\ f(0) = 1,\ f(\infty) = 0. \tag{11}$$

Kobayashi et al [8] have computed a value -1.5880710 for $f'(0)$, and hence we may solve this problem as an initial value problem. By the Frobenius method an expansion may be obtained for $f$ in powers of $t = x^{\frac{1}{2}}$ of the form

$$f \sim f_0 = c_0 + c_1 t + c_2 t^2 + c_3 t^3 + \dots$$

where $c_0 = 1$, $c_1 = 0$, $c_2 = f'(0) = -1.5880710$, etc.

Since

$$f \sim 144\, x^{-3} \text{ as } x \to \infty \tag{12}$$

and $f$ is positive, a squared approximation of the form

$$f_0^* = [P_{r,\ r+3}(t)]^2 \tag{13}$$

was proposed by Mason [9], where $P_{r,\ r+3}$ is the $[r, r+3]$ Padé approximant in $t$ to $(f)^{\frac{1}{2}}$. The power series of $f^{\frac{1}{2}}$ may be determined from (10) with $x$ replaced by $t$ and $\alpha = \frac{1}{2}$. Clearly $f_0^*$ is positive and has the same dominant asymptotic behaviour (12) as $f$.

For $r = 4$, computed values of $f_0^*$ (see [10]) agree with values tabulated in [8] with a maximum absolute error of about .00001 throughout $[0, \infty)$.

<u>Example 2</u> <u>Thomas-Fermi Boundary Value Problem</u> The problem (11) can be tackled without assuming a value for $f'(0)$. The function $f$ has an asymptotic expansion of form

$$f \sim f_\infty = 144\, x^{-3}(d_0 + d_1 w + d_2 w^2 + \dots + d_k w^k + \dots)$$

where $d_0 = d_1 = 1$, $w = -F\, x^{-\lambda}$, $\lambda = \frac{1}{2}(-7 + \sqrt{73})$.

The coefficients $d_2$, $d_3$, ... may be determined by the Frobenius method, and F is an unknown asymptotic parameter. Consider approximations of the form

$$f^*_\infty = 144\ x^{-3}\ [P_{rs}(w)]^{1/\alpha} \tag{14}$$

where $P_{rs}(w)$ is the [r, s] Padé approximant in w to $(144^{-1}x^3f)^\alpha$. This involves the determination of the power series

$$(144^{-1}x^3f)^\alpha \sim (d_0 + d_1w + \ldots + d_kw^k + \ldots)^\alpha,$$

which may be obtained by algorithm (10).

By choosing

$$\alpha = \lambda(s-r)/3 \text{ and } r \neq s, \tag{15}$$

we ensure that

$$f^* \sim x^0 \text{ at } x = 0 \quad (w = \infty).$$

Equating f* to f at x = 0 gives the equation

$$(-F)^{r-s} = 144^{-\alpha}\ b_s/a_r \tag{16}$$

for the determination of F from the Padé coefficients.

In practice this method works rather effectively in achieving a modest uniform accuracy with very little computation. For example for r = 1, s = 3 (based on just 5 terms of the asymptotic series) a uniform relative error of about 5 per cent is achieved throughout [0, ∞). And for r = 3, s = 5, an accuracy of about 2 significant decimal places is obtained.

Example 3 Blasius Function Define f(x) by the equation

$$f''' + f\ .\ f'' = 0,\quad f(0) = f'(0) = 0,\ f' \to B \text{ as } x \to \infty, \tag{17}$$

where B is some specified constant. If $f''(0) = \alpha$, say, is specified, then a power series expansion may be obtained (see Davis [11]) of the form

$$f \sim f_0 = \alpha x^2\ [c_0 + c_1t + c_2t^2 + \ldots + c_kt^k + \ldots],\quad t = \alpha x^3. \tag{18}$$

It may also be shown that for large x

$$f \sim A + Bx - \psi(x), \tag{19}$$

where $\psi(x) \sim x^{-2}e^{-x^2/2}$ as $x \to \infty$.

In this case Mason [10] chose the approximation

$$f^* = \alpha x^2 \left[P_{r,\ r+1}(t)\right]^{1/3}, \qquad (20)$$

where $P_{r,\ r+1}(t)$ is the $[r, r+1]$ Padé approximant in t to $\alpha^{-3}x^{-6}f^3$. This involves cubing the power series $c_0 + \ldots + c_k t^k + \ldots$ in (18), and algorithm (10) may again be used.

For the case B = 2, L. Howarth computed solutions [11] from which he determined the value $f''(0) = \alpha = 1.32824$. With this choice of $\alpha$, the values of the approximations (20) were compared in [10] with Howarth's values and some results are given in Table 2. It can be seen that, apart from rounding errors, f* converges on the range [0, 4] at least. Indeed, for r = 7, f* is correct to about 5 decimals on [0, 4], and beyond this range f is dominated by its linear asymptotic behaviour. At $x = \infty$, f* has a gradient of 1.930 for r = 7, which is reasonably close to the true value of 2.

TABLE 2 Comparison between f* and Howarth's Results

| x | Howarth's solution f | f* r = 4 | f* r = 7 | f* r = 9 |
|---|---|---|---|---|
| 1.2 | 0.92230 | 0.92230 | 0.92230 | 0.92230 |
| 2.8 | 3.88031 | 3.88032 | 3.88032 | 3.88032 |
| 3.6 | 5.47925 | 5.47952 | 5.47926 | 5.47926 |
| 4.0 | 6.27923 | 6.28050 | 6.27925 | 6.27925 |
| 4.4 | 7.07923 | 7.08371 | 7.07927 | 7.07925 |

## 2.5 Infinite Range: Pairs of Padé Approximants

In some cases uniform relative accuracy may be achieved on an infinite range by combining a pair of one-point Padé approximants, one defined from the power series expansion and the other from the asymptotic expansion. For example, for the ordinary Thomas-Fermi function, defined by (11), the approximation $f_0^*$ given by (13) achieves 5 significant decimal places of accuracy for r = 6 throughout [0, 15] of x. And, if F is given the correct value of 13.270974 (see [8]), then the approximation f* given by (14), for $\alpha = 1$, r = 6, and s = 9, achieves 5 significant decimal places of accuracy throughout $[2, \infty)$ of x. (In the latter case there is no need to define $\alpha$ by (15), since we do not require the approximation to extend to x = 0).

### 2.6 Infinite Range: Exponential Decay

Functions which decay exponentially can sometimes be well approximated using inverse powers of polynomials, and examples such as $e^{-x}$ and erf x are considered by Hastings [12]. Another example is the function $\psi(x)$ defined by (19) in connection with the asymptotic behaviour of the Blasius function (17), and an appropriate approximation in this case (see [10]) is of the form

$$f(x) \simeq f^*(x) = A + Bx - \psi^*(x),$$

where (21)

$$\psi^*(x) = A(e_0 + e_1 x + \ldots + e_r x^r)^{-4}$$

Here A = - 1.72077 and B = 2, based on Howarth's solution [11].

The polynomial $e_0 + \ldots + e_r x^r$ in (21) is chosen as the partial sum of degree r of the power series expansion of the function

$$\begin{aligned}[\psi(x)/A]^{-1/4} &= [1 + A^{-1}Bx - A^{-1}f(x)]^{-1/4} \\ &= [1 + A^{-1}Bx - \alpha A^{-1}x^2(c_0 + c_1 t + c_2 t^2 + \ldots)]^{-1/4},\end{aligned}$$

where $t = \alpha x^3$. Once more a power (-1/4) of a power series is required, and algorithm (10) can be used.

This simple application of a power series expansion leads to remarkably good results. For r = 11, f* agrees with all values of f tabulated by Howarth (to 5 decimals) and is in fact valid throughout $[0, \infty)$. Full details are given in [10].

More compact approximations of the form (21) have been obtained by more sophisticated methods by Osborne and Watson [13], who developed an iterative algorithm for obtaining minimax rational approximations to the solutions of certain differential equations.

## 3. DRAWBACKS OF PADÉ APPROXIMANTS

Some of the examples used above also serve to point out certain difficulties which may arise with Padé approximants.

### 3.1 Inaccurate Power Series

Padé approximants are defined from power series expansions, and any inaccuracies in the power series coefficients lead to induced inaccuracies in the Padé coefficients. Unfortunately, since power series coefficients are usually determined successively, there is a fair chance of rounding errors occurring and there is no easy way to eliminate them.

Consider the example $f(x) = \tan x$ of §2.2. In this case rounding errors build up rapidly when the power series coefficients are calculated by the Frobenius method, and a precision of about 6 decimals in this calculation was found to lead to a series for which the corresponding (accurately calculated) approximations $x\,P_{rr}(t)$ were incorrect for $r > 3$. Indeed all Padé approximants $P_{rr}$ for $r > 3$ were found to reduce approximately or exactly to [3, 3] approximants. For example $P_{55}(t)$ was found to have two common zeros in its numerator and denominator. And $P_{66}(t)$ had a spurious pole at 3.81595 and a spurious zero at 3.81597, as well as two common zeros in its numerator and denominator. To cope with these problems, the power series coefficients in §2.2 were calculated in double precision.

Clearly great care must be taken to ensure adequate accuracy in power series coefficients. However, in some applications it may simply be impossible to perform the calculation with sufficient precision on the available computer. For this reason K.O. Geddes [14] has developed procedures using rational arithmetic for generating exact power series expansions from non-linear differential equations with polynomial coefficients. These procedures are expensive in terms of computer time, but they do give correct results!

### 3.2 Ill-conditioned Linear Algebraic Equations

Unfortunately the linear equations (4) and (5) defining Padé coefficients are often ill-conditioned, and this means that small changes in the power series coefficients lead to large changes in the Padé coefficients. A physical interpretation of this occurrence is that Padé approximants extrapolate a power series and so are sensitive to errors in the series.

As an illustration, the asymptotic coefficients $d_0$, $d_1$, $d_2$, ... of Example 2 of §2.4 (Thomas-Fermi boundary value problem) were calculated in double precision (about 14 decimals) by two different methods. Although all the resulting coefficients up to $c_{19}$ agreed to about 6 significant decimals, the Padé coefficients based on them (for the approximation (14) with r = 9, s = 10) did not in any case agree to a single figure. On the other hand all the actual values of the Padé approximants agreed uniformly to about 2 significant figures and the computed values (16) of the asymptotic parameter F agreed to 3 figures.

Our observations in this example are consistent with some recent analysis of Luke [15] who has shown that, although ill-conditioning may affect Padé coefficients, it has a much less adverse effect on the actual values taken by Padé approximants, provided Padé coefficients are calculated from the linear equations (4) and (5).

To deal with severe ill-conditioning it appears to be necessary either to use a computer with a longer word length or to find a better Padé algorithm. The latter approach has been adopted by Geddes [16], whose fraction-free Padé algorithm uses modular arithmetic to generate an exact solution of the linear equations (4) and (5). Again his algorithm is expensive, but again it is reliable.

### 3.3 Lack of Convergence - Asymptotic Constants

Although good approximations were obtained to functions on infinite ranges in all three examples of §2.4,less success was experienced in determining some of the asymptotic constants in these problems.

Example 1 Although the approximation (13) achieved absolute accuracy on $[0, \infty)$, high relative accuracy only occurred on a smaller range. For example for r = 6, f* was accurate to 5 significant decimals only on $[0, 15]$. More specifically the constant $(a_r/b_{r+3})^2$, which presumably should converge to the asymptotic constant 144 as $r \to \infty$, was equal to 148.0 for r = 6. So convergence to 144 (if there was convergence!) was rather gradual.

Example 2 Although approximation (14) achieved about 2 significant figures throughout $[0, \infty)$ for r = 3, s = 5, very little improvement was obtained by increasing r. More specifically we give in Table 3 computed values (16) of the asymptotic parameter F, and we see that a value more accurate that 13.3 could not be determined. Rounding errors were undoubtedly affecting the results for r+s > 12, but it was not clear that much more accurate results would have been obtained even with exact calculations.

If the entry for r = 6 was omitted from the first column of Table 3, we might be tempted to deduce from the consistency of the results that

F = 13.30, but we should be incorrect in doing so!

TABLE 3 Thomas-Fermi Asymptotic Parameters

| r | s | F | r | s | F | r | s | F |
|---|---|---|---|---|---|---|---|---|
| 3 | 5 | 13.34 | 1 | 5 | 13.42 | 4 | 0 | 13.65 |
| 4 | 6 | 13.32 | 2 | 6 | 13.35 | 8 | 0 | 13.51 |
| 5 | 7 | 13.30 | 3 | 7 | 13.32 | 12 | 0 | 13.45 |
| 6 | 8 | 13.46 | 4 | 8 | 13.31 | 16 | 0 | 13.42 |
| 7 | 9 | 13.30 | 5 | 9 | 13.28 | 20 | 0 | 13.39 |
| 8 | 10 | 13.30 | 6 | 10 | 13.30 | 24 | 0 | 13.37 |

Example 3 Although the approximation (20) achieved 5 significant decimals on $[0, 4]$ for $r = 7$, it was not so effective on the complete range $[0, \infty)$. Specifically the asymptotic gradient B, computed as $\alpha(a_r/b_{r+1})^{1/3}$, did not appear to be converging convincingly to the correct value of 2, as can be seen from Table 4. Moreover this lack of convergence was not attributable to rounding error; for Geddes [17] has computed values by his exact methods, and these are just as irregular - see Table 4.

TABLE 4 Blasius Asymptotic Gradients

| r | B (Mason: Inexact) | B (Geddes: Exact) |
|---|---|---|
| 5 | 1.852 | 1.852 |
| 6 | 1.852 | 1.852 |
| 7 | 1.930 | 1.915 |
| 8 | 1.930 | 1.883 |
| 9 | 1.881 | - |
| 10 | 1.890 | - |

Probably the main cause for the slightly disappointing results in the three examples above is that in each case the asymptotic expansion of the Padé approximant only agrees with that of the true solution to one term, thus presumably limiting the accuracy which can be achieved. However, the results are still slightly disturbing, and suggest that considerable caution should be taken in deducing values for asymptotic constants by Padé approximant methods. Consistent values may be incorrect, especially if rounding errors have not been properly controlled or if convergence is irregular.

ACKNOWLEDGEMENTS

We wish to thank Professor J.L. Fields of the University of Alberta for helpful discussions about this topic, and Professor K.O. Geddes and Professor Y.L. Luke for some invaluable contributions.

REFERENCES


1 W.B. Gragg, "The Padé table and its relation to certain algorithms of numerical analysis", SIAM Review 14 (1972), 1-62.

2 G.A. Baker, Jr., "Essentials of Padé Approximants", Academic Press, London, 1975.

3 L. Wuytack (Editor), "Padé Approximation and its Applications", Proceedings of a Conference in Antwerp, Springer-Verlag, Berlin, 1979.

4 P.A. Frost and E.Y. Harper, "Extended Padé procedure for constructing global approximations from asymptotic expansions; an explication with examples", SIAM Rev. 18 (1976), 62-91.

5 J.H. McCabe and J.A. Murphy, "Continued fractions which correspond to power series expansions at two points", JIMA 17 (1976), 233-247.

6 C. Pommerenke, "Padé approximants and convergence in capacity", J. Math. Anal. Appl. 41 (1973), 775-780.

7 R. de Montessus de Ballore, "Sur les fractions continues algebriques", Bull. Soc. Math. France 30 (1902), 28-36.

8 S. Kobayashi, T. Matsukuma, S. Nagai, and K. Umeda, "Accurate value of the initial slope of the ordinary T.-F. function", J. Phys. Soc. of Japan 10, (1955), 759.

9 J.C. Mason, "Rational approximations to the ordinary T.-F. function and its derivative", Proc. Phys. Soc. 84 (1964), 357-359.

10 J.C. Mason, "Some New Approximations for the Solution of Differential Equations", D. Phil. Thesis, Oxford, 1965.

11 H.T. Davis, "Introduction to Nonlinear Differential and Integral Equations", U.S. Atomic Energy Commission, 1960.

12 C. Hastings, Jr., "Approximations for Digital Computers", Princeton University Press, 1955.

13 M.R. Osborne and G.A. Watson, "An algorithm for minimax approximation in the nonlinear case", Computer J. 12 (1969), 64-69.

14 K.O. Geddes, "Algorithms for Analytic Approximation", Ph.D. Thesis, Department of Computer Science, University of Toronto, 1973.

15 Y.L. Luke, "On computing the polynomials in Padé approximations by solving linear equations", Proceedings of a Symposium in honour of George Lorentz, University of Texas, 1980, Academic Press (to appear).

16 K.O. Geddes, "Symbolic computation of Padé approximants", ACM Trans. on Math. Software 5 (1979), 218-233.

17 K.O. Geddes (Private communication).

# FROM DIRAC DISTRIBUTIONS TO MULTIVARIATE REPRESENTATION FORMULAS

Jean Meinguet

Institut Mathématique
Université de Louvain
Louvain-la-Neuve, Belgium

It is shown how easily convolutions (or Fourier transforms) can be used, together with basic integral inequalities, to find in a unified way "appropriate" expressions of distributions (or functions) in terms of a prescribed subset of partial derivatives. Three types of significant situations are analyzed in succession: the "compact support" case, the "whole space" case, the "restriction to bounded sets" case. Various applications are considered, in connection with the multivariate extension problem of the Peano kernel theorem and with the optimization of the famous inequalities due to Friedrichs and to Poincaré.

## I. INTRODUCTION

Throughout this paper, all functions and vector spaces we shall consider are complex, m and n are fixed integers, $m,n \geq 1$ $\mathbb{R}^n$ is the Euclidean n-space of all n-tuples of real numbers,



ISBN 0-12-780650-4

$\Omega$ is an open subset of $\mathbb{R}^n$, $P_{m-1}$ is the vector space of dimension $\binom{m+n-1}{n}$ of all polynomials in n variables of (total) degree $\leq$ m-1,

$$M = \binom{n+m-1}{m} \tag{1}$$

is the dimension of the space of polynomials in n variables of degree m, $|\cdot|_{k,\Omega}$ is the (Sobolev-like) seminorm generated by the semi-inner product

$$(v,w)_{k,\Omega} := \sum_{i_1,\ldots,i_k=1}^{n} \int_\Omega \partial_{i_1\ldots i_k} v(x)\, \partial_{i_1\ldots i_k} \overline{w(x)}\, dx \tag{2}$$

where $\partial_{i_1\ldots i_k} := \partial^k/\partial x_{i_1}\ldots\partial x_{i_k}$ is to be interpreted in the distributional sense; for $k \geq 1$, the kernel of the seminorm $|\cdot|_{k,\Omega}$ (which is <u>rotation invariant</u> when $\Omega \equiv \mathbb{R}^n$) is known to be the vector space of all functions in $\Omega$ whose restriction to each connected component of $\Omega$ belongs to $P_{k-1}$. All integrals in $\mathbb{R}^n$ will be taken with respect to the Lebesgue measure. As usual, supp u (where u stands for a function or a distribution in $\Omega$) denotes the <u>support</u> of u (i.e., the smallest closed subset of $\Omega$ outside which $u \equiv 0$), $\mathcal{D}(\Omega)$ is the vector space of <u>test functions</u> in $\Omega$ (i.e., infinitely differentiable functions with compact support in $\Omega$), provided with the canonical Schwartz topology, $\mathcal{D}'(\Omega)$ (i.e., the topological dual of $D(\Omega)$ is the vector space of distributions in $\Omega$; $L_p(\Omega)$ is the (Banach) space of (Lebesgue measurable) functions in $\Omega$ which are integrable to the p-th power if $1 \leq p < \infty$ (resp. essentially bounded if $p = \infty$), equipped with the norm $\| v \|_{p,\Omega} := (\int_\Omega |v(x)|^p\, dx)^{1/p}$

if $1 \leq p < \infty$ (resp. $\|v\|_{\infty,\Omega} := \text{ess sup}|v(x)|$); $L_{p,c}(\Omega)$ is the space of $L_p(\Omega)$ functions with compact support in $\Omega$. It should be noted that, when $\Omega \equiv \mathbb{R}^n$, $(\mathbb{R}^n)$ will always be omitted.

By representation formula (in a specified vector subspace X of $\mathcal{D}'$), we mean a solution of the following problem: Find an "appropriate" way of expressing any distribution $v \in X$ in terms of a prescribed subset of its partial derivatives. Needless to say, this problem is of a basic importance in Approximation Theory and in Numerical Analysis. In particular, these representation formulas can provide appropriate substitutes for the truncated Taylor series and for its integral remainder term, such as requested for extending the Peano kernel theorem to the representation of errors in intrinsically multivariate situations. It turns out that a reasonably simple mathematical theory, underlying in fact a wide range of relevant topics (both of a theoretical and practical nature), can be developed in a unified way by resorting systematically to convolutions or to Fourier transforms of distributions.

As regards the specification of X, we will consider here in succession three types of significant situations: the "compact support" case in Section II, the "whole space" case in Section III, the "restriction to bounded sets" case in Section IV. Let us recall in passing a most interesting connection of the "whole space" case with the concrete method of surface spline interpolation (which is based on the minimization of a Sobolev seminorm under interpolatory constraints):

as analyzed in detail in recent papers (see e.g. [3], [4], [5]), the intrinsic structure of such surface splines is in fact that of a multivariate extension of the classical natural splines.

Error estimates (as opposed to truly quantitative bounds), involving generic constants, are known to pervade the modern literature on error estimation (typically in spline analysis, in connection with the rate of convergence of the finite element method). In recent years, a method of some practical value was devised for obtaining, from representation formulas, realistic upper bounds of approximation errors in a wide variety of situations (see e.g. [2]). We will emphasize here the additional and somewhat surprising fact that optimal error coefficients can sometimes be found explicitly, not only for pointwise approximation problems but also for mean-square approximation problems; this possibility will be illustrated here in connection with two famous integral inequalities in Sobolev spaces, due to Friedrichs and to Poincaré.

## II. THE "COMPACT SUPPORT" CASE

Solving the representation problem proves particularly simple if $X := \mathcal{E}'$, i.e., for every distribution with compact support in $\mathbb{R}^n$. Indeed we have the following result.

Theorem 1. Every $v \in \mathcal{E}'$ can be expressed in the convolution form

$$v = \sum_{i_1,\ldots,i_m = 1}^{n} \partial_{i_1\ldots i_m} e * \partial_{i_1\ldots i_m} v, \tag{3}$$

where e is the locally integrable function in $\mathbb{R}^n$:

$$e(x) := \begin{cases} c|x|^{2m-n}\ell n|x|, & \text{if } 2m \geq n \text{ and } n \text{ is even} \\ d|x|^{2m-n} & \text{otherwise} \end{cases} \tag{4a}$$

with

$$c := \frac{(-1)^{n/2+1}}{2^{2m-1}\,\pi^{n/2}(m-1)!(m-n/2)!}, \tag{4b}$$

$$d := \frac{(-1)^m\,\Gamma(n/2-m)}{2^{2m}\,\pi^{n/2}(m-1)!}, \tag{4c}$$

$|x|$ denoting as usual the radial coordinate (or Euclidean norm) of $x \in \mathbb{R}^n$.

It must be emphasized that each distributional partial derivative of e in (3) is actually defined by the corresponding classical derivative, which indeed exists everywhere in the complement of the origin and is locally integrable in $\mathbb{R}^n$ (see e.g. [9], p.57, Theor.V.1°, except for n=1, in which case it trivially follows from (4a) that the only property to verify is the absolute continuity in $\mathbb{R}$ of the function: $x \to |x|^m$, $m \geq 1$).

As stated in [3], Theorem 1 originates quite simply from the identities

$$v = \delta * v = \Delta^m e * v, \quad \forall v \in \mathcal{D}', \tag{5}$$

in view of the following facts (borrowed from the theory of distributions): — the (n-dimensional) Dirac distribution (or measure) $\delta \equiv \delta_{(0)}$ is the unit of convolution. More generally,

$$\tau_a v = \delta_{(a)} * v, \ \forall v \in \mathcal{D}', \tag{6a}$$

where $\delta_{(a)}$, the shifted Dirac function relative to the point $a \in \mathbb{R}^n$, is the distribution assigning to each test function $\varphi$ the value $\varphi(a)$, while $\tau_a v$, the transform of v under the translation $\tau_a : x \to x+a$ of $\mathbb{R}^n$, is the distribution $\varphi \to \langle v, \tau_{-a} \varphi \rangle$ (the symbol $\langle \cdot, \cdot \rangle$ denotes the duality pairing between dual topological vector spaces such as, for example, $\mathcal{D}'$ and $\mathcal{D}$); it follows that, according to an incorrect but usual notation, we have:

$$(\tau_a v)(x) = v(x-a), \ \forall x \in \mathbb{R}^n, \tag{6b}$$

as in the case of functions.

— there exists a (tempered) distribution e satisfying the partial differential equation

$$\Delta^m e = \sum_{i_1, \ldots, i_m = 1}^{n} (\partial_{i_1 \ldots i_m})^2 e = \delta \ ; \tag{7}$$

such a fundamental solution of the iterated Laplacian $\Delta^m$ in $\mathbb{R}^n$ is precisely the rotation invariant function defined by (4a,b,c) in the complement of the origin.

— the convolution v*w of the distributions v and w, which is the distribution uniquely defined (if existent at all!) by the equation

$$\langle v*w, \varphi \rangle := \langle v_\xi, \langle w_\eta, \varphi(\xi+\eta) \rangle \rangle, \ \forall \ \varphi \in \mathcal{D}, \tag{8}$$

is commutative, associative and commutes with differentiation (at least conditionally); as a matter of fact, all these remarkable properties hold valid if the addition $(\xi,\eta) \to \xi+\eta$ in $\mathbb{R}^n$ is a proper mapping from supp v × supp w into $\mathbb{R}^n$ (that is, the inverse image under this mapping of any compact subset of

$\mathbb{R}^n$ is compact in supp v × supp w). Hence the representation formula (3) in $E'$, and the corresponding generalization of the usual <u>Poisson formula for potentials</u>, viz.,

$$v = e*\Delta^m v, \ \forall v \in E'. \tag{9}$$

It must be emphasized that the compact support assumption is here an extreme precaution; as a matter of fact, any convolution v*w can be given a quite satisfactory meaning provided only that v "decreases sufficiently fast at infinity", so as to match the "growth at infinity" of w; unlike differentiation and multiplication (which are local operations), convolution is indeed a global operation. Furthermore there are clearly various alternatives to formulas (3), (4) and (9), for example, the following representation formula:

$$v = \sum_{k=0}^{m} \binom{m}{k} (-1)^k \sum_{i_1,\ldots,i_k=1}^{n} \partial_{i_1\ldots i_k} f*\partial_{i_1\ldots i_k} v, \forall v \in E', \tag{10}$$

where f denotes any fundamental solution of the partial differential operator $(1-\Delta)^m$, typically the rotation invariant function defined in $\mathbb{R}^n \setminus \{0\}$ by

$$f(x) := \frac{1}{2^{m-1}(2\pi)^{n/2}(m-1)!} \ \frac{K_{n/2-m}(|x|)}{|x|^{n/2-m}} \ , \tag{11}$$

where $K_\nu(\cdot)$ denotes the modified Bessel function of the third kind of order $\nu$ (called also Macdonald's function, or Basset's function, of order $\nu$); of course, (10) can be rewritten in the Poisson type form

$$v = f*(1-\Delta)^m v \ , \ \forall v \in E'. \tag{12}$$

As defined explicitly above, the distribution f (resp. $(-1)^m e$) is the so-called <u>Bessel-Macdonald kernel</u> (resp. <u>Riesz kernel</u>)

of order 2m, while the convolution of f (resp. $(-1)^m e$) and any $v \in \mathcal{E}'$ is known as the Bessel potential (resp. Riesz potential) of order 2m of v (see e.g. [8], Chap. 8, and [12], Chap. V). In view of the classical asymptotic estimate

$$K_\nu(z) = \left(\frac{\pi}{2z}\right)^{\frac{1}{2}} e^{-z}\left\{1+O(z^{-1})\right\}, \text{ as } z \to \infty \text{ in } |\arg z| < \frac{3\pi}{2} \tag{13}$$

the global behavior ($|x| \to \infty$) of f is much more favorable than the one of $(-1)^m e$ (in the sense that $K_\nu(|x|)$ is always rapidly recessive at infinity), whereas the local behavior ($|x| \to 0$) of the two kernels is asymptotically the same whenever $m \le n/2$ (in the opposite case, the above definitions imply that the origin is no longer a discontinuity of either kernel). It follows that the representation formula (10) can be extended by continuity to much larger classes of distributions than the representation formula (3); on the other hand, (3) is much simpler and often more appropriate to the needs in the sense that the only partial derivatives of v it involves are those of order m (for examples of application, see e.g. [6]). As a matter of fact, it can be proved that Bessel-Macdonald kernels of positive order are always in $L_1$ and, which is particularly significant, that the vector space of Bessel potentials of order $m > 0$ of all functions in $L_p (1 < p < \infty)$ is isomorphic to the well known Sobolev space $W_p^m$.

Another interesting representation formula, of a tensor product type, is the following:

$$v = \left\{ \frac{(x_1)_+^{\alpha_1 - 1}}{(\alpha_1 - 1)!} \otimes \ldots \otimes \frac{(x_n)_+^{\alpha_n - 1}}{(\alpha_n - 1)!} \right\} * \frac{\partial^{\alpha_1 + \ldots + \alpha_n}}{\partial x_1^{\alpha_1} \ldots \partial x_n^{\alpha_n}} v, \forall v \in \mathcal{D}'_+ , \tag{14}$$

where $\mathcal{D}'_+$ is the space of distributions in $\mathbb{R}^n$ with support in the positive orthant, it being understood that any normalized truncated power function of exponent 0 (resp. -1) is nothing else than the univariate Heaviside function (resp. univariate Dirac distribution).

## III. THE "WHOLE SPACE" CASE

As stated above, the compact support assumption on $v \in \mathcal{D}'$ can often be removed (at the expense of convenience!); this is clearly required to reach global conclusions. For the particular vector space

$$X: = \{v \in \mathcal{D}' : \partial_{i_1 \ldots i_m} v \in L_2 \text{ for } i_1, \ldots, i_m \in [1,n]\} \tag{15}$$

equipped with the semi-inner product $(\cdot,\cdot)_m$ defined by (2), whose theoretical and practical significance can be taken for granted, we will summarize here some of the main global results obtained earlier (see e.g. [3], [4], [5], [6]).

A. The potential-based subcase: $m < n/2$. One of the fundamental results that shall be exploited here is the following:

$$\partial_{i_1 \ldots i_m} e \in \mathcal{D}'_{L_2} \quad \text{for} \quad i_1, \ldots, i_m \in [1,n]. \tag{16}$$

where $\mathcal{D}'_{L_2}$ denotes the vector space of all finite sums of distributional derivatives of $L_2$ functions. From the remarkable identities

$$\partial_{i_1 \ldots i_m} e(x) = |x|^{m-n} \partial_{i_1 \ldots i_m} e(x/|x|), \quad \forall x \in \mathbb{R}^n \setminus \{0\}, \qquad (17a)$$

which are readily proved by making use of elementary homogeneity arguments in connection with the definition (4a,c) of $e(x)$, it follows that, for all $i_1,\ldots,i_m \in [1,n]$,

$$|\partial_{i_1 \ldots i_m} e(x)| \leq c_1 |x|^{m-n}, \quad \forall x \in \mathbb{R}^n \setminus \{0\}, \qquad (17b)$$

where $c_1$ denotes any upper bound (depending only on m and n) on the unit sphere in $\mathbb{R}^n$ for the absolute values of all the m-th partial derivatives of e. Owing to (17b), the proof of (16) essentially amounts to noting that the product of each m-th partial derivative of e by the characteristic function of a relatively compact neighborhood of the origin (resp. of the complement in $\mathbb{R}^n$ of such a neighborhood) belongs to $L_{1,c}$(resp. $L_2$).

A further basic preliminary result is the so-called $L_p$ <u>inequality for potentials</u> (see e.g. [12], p.119), more precisely its specialization for p=2, to wit: if $1/q := 1/2 - m/n > 0$, then the (Riesz-like) potential of order m:

$$(r^{-n+m} * f)(x) := \int_{\mathbb{R}^n} |x-y|^{-n+m} f(y)\,dy, \quad \forall f \in L_2, \qquad (18)$$

exists for almost every $x \in \mathbb{R}^n$, and

$$\| r^{-n+m} * f \|_q \leq c_2 \| f \|_2, \quad \forall f \in L_2, \qquad (19)$$

where the constant $c_2$ depends only on m and n. Due essentially to Sobolev (1938), this most important theorem can also be regarded as deriving from the specially deep <u>Marcinkiewicz interpolation theorem</u> (1939), which indeed succeeds in turning "weak-type" integrability assumptions into "strong" integrability results (see e.g. [12], pp. 120-121). It should be realized that the only value of q for which (19) may hold is merely a reflection of the homogeneity of the radial kernel.

These arguments produce the following important result.

Theorem 2. Suppose $m < n/2$. Then for every $v$ in the (semi-inner product) space $X$ defined by (15) and the corresponding distribution

$$w := \sum_{i_1,\ldots,i_m=1}^{n} \partial_{i_1\ldots i_m} e * \partial_{i_1\ldots i_m} v \ , \tag{20a}$$

we have the global estimate

$$\|w\|_q \leq c_3 |v|_m \text{ with } 1/q := 1/2 - m/n \ , \tag{21}$$

where $c_3$ is a constant depending only on $m$ and $n$. Moreover, the mapping $Q : X \to X$ such that

$$v = Qv + w \ , \ \forall v \in X \ , \tag{20b}$$

is a linear projector of $X$ onto $\mathcal{P}_{m-1}$.

By using in conjunction the definition of distributional differentiation and the Fubini theorem (for functions) it is easily seen that, just as throughout Section 2 , the convolution in $\mathcal{D}'_{L_2}$ is always defined (and therefore commutative), is associative and commutes with differentiation. It follows that each m-th partial derivative (in the distributional sense) of the expression (20a) for w coincides in $\mathcal{D}'$ with the corresponding derivative of $v \in X$, which leads eventually to the representation formula (20a,b) in $X$ for $m < n/2$. It should be noted that various more sophisticated theorems can be exploited here to good purpose (see e.g. [9], p.203 and p. 270); for example, the one asserting that the Fourier transformation can be defined in $\mathcal{D}'_{L_2}$ (by restriction of the Fourier transformation in the space $S'$ of tempered distributions in $\mathbb{R}^n$, which itself is the transpose of the classical Fourier automorphism for the space $S$ of infinitely differentiable functions with each derivative rapidly decreasing at infinity) and enjoys the essential property of exchanging convolution (of unit $\delta$) and multiplication (of unit 1); for a significant application, see ([6], p. 208).

B. The surface spline-based subcase: $m > n/2$. As explained in detail in [3], the crux of the constructive proof of the important theorems recalled hereafter is simply that

$$\mu * e \in X, \tag{22}$$

for every measure $\mu$ with compact support in $\mathbb{R}^n$ which is summable and annihilates $P_{m-1}$. This can be readily verified on the Fourier level, owing to the classical results: the Fourier transform $\hat{\mu}$ of the (compactly supported) distribution $\mu$ is a $C^\infty$ function in $\mathbb{R}^n$, which is bounded and vanishes at the origin together with all its partial derivatives of order $\leq m-1$, while the Fourier transform $\hat{e}$ of the (tempered) distribution e defined by (4a,b,c) is the pseudofunction

$$\hat{e}(\xi) = (-4\pi^2)^m \text{ Pf } |\xi|^{-2m} . \tag{23}$$

In consequence, the Fourier transform of every partial derivative of order m of the (tempered) distribution $\mu * e$ is a $C^\infty$ function in $\mathbb{R}^n \setminus \{0\}$, which is bounded and $0(|\xi|^{-m})$ as $|\xi| \to \infty$, hence certainly in $L_2$ whenever $m > n/2$, which was to be proved.

Now the convolution process in $L_2$, which of course enjoys the remarkable properties of the convolution process in $\mathcal{D}'_{L_2}$, is known to be continuous and to yield uniformly continuous functions vanishing at infinity in $\mathbb{R}^n$ (by virtue of a noteworthy special case of Young's inequality for convolutions in Lebesgue spaces). Hence the following general result.

Theorem 3. Suppose $m > n/2$. Then for any measure $\mu$ with compact support in $\mathbb{R}^n$ which is summable and annihilates $P_{m-1}$, we have in the space $X$ defined by (15) the representation formula:

$$\mu * v = \sum_{i_1,\ldots,i_m=1}^{n} \partial_{i_1 \ldots i_m}(\mu * e) * \partial_{i_1 \ldots i_m} v, \forall v \in X, \tag{24}$$

and the global estimate:

$$\| \mu * v \|_\infty \leq |\mu * e|_m |v|_m, \ \forall v \in X, \tag{25}$$

which is sharp.

Provided only it annihilates $P_{m-1}$, any finite linear combination of Dirac distributions can be substituted for $\mu$ in the concrete application of Theorem 3; as discussed in great detail in several recent papers (see e.g. [3], [4], [5]), this most natural specialization of the measure $\mu$ leads eventually to the following fundamental result.

Theorem 4. Let $\ell$ denote the largest integer such that

$$0 \leq \ell < m-n/2. \tag{26}$$

Then every distributional derivative of order $\leq \ell$ of $v$ in the semi-inner product space $X$ defined by (15), say $Dv$, can be expressed in the form

$$(Dv)(x) = (D(Pv))(x)+(v,D_x k_x)_m, \ \forall x \in \mathbb{R}^n; \tag{27}$$

here

$$Pv := \sum_{j \in J} v(a_j)p_j \tag{28}$$

is the (uniquely defined) $P_{m-1}$-interpolant of v on any prescribed $P_{m-1}$-unisolvent subset $(a_j)_{j \in J} \subset \mathbb{R}^n$ written in the Lagrange form, and

$$k_x(y) := (-1)^m \Big\{ e(x-y) - \sum_{i \in J} p_i(x)e(a_i-y) - \sum_{j \in J} p_j(y)e(x-a_j) + \sum_{i \in J} \sum_{j \in J} p_i(x)p_j(y)e(a_i-a_j) \Big\}, \tag{29}$$

with e defined by (4a,b,c), is the reproducing kernel of the Hilbert function space

$$X_0 := \{v \in X : v(a_j) = 0, \forall j \in J\}. \tag{30}$$

All the derivatives D(v-Pv) of order $\leq \ell$ are continuous functions which vanish at infinity.

An essential argument underlying Theorem 4 is the following consequence of Theorem 3: the measure-valued function $\mu_x$ defined by

$$\mu_x := \delta_{(-x)} - \sum_{j \in J} p_j(x)\, \delta_{(-a_j)}, \quad \forall x \in \mathbb{R}^n, \tag{31a}$$

is such that, for every v in X,

$$(\mu_x * v)(0) = \Big[\tau_{-x} v - \sum_{j \in J} p_j(x)\, \tau_{-a_j} v\Big](0) \equiv (v-Pv)(x) \tag{31b}$$

is a continuous function of x in $\mathbb{R}^n$ which vanishes at infinity; the proof amounts to exploiting in conjunction the apprasial (25) where $\mu := \mu_x$ and the property of all partial derivatives of order m of $\mu_x * e$ to depend continuously on x (qua $L_2$ functions!).

Theorem 4 is closely connected with the concrete method of surface spline interpolation, which was recently devised for solving (in a numerically stable way!) the classical minimization problem of a Sobolev seminorm under interpolatory constraints. Another far-reaching result we will recall here is the following multivariate extension (to the Sobolev-like space X) of the classical Peano kernel theorem (relative to $C^m[a,b]$).

Theorem 5. Let L denote any bounded linear functional on X equipped with the seminorm $|\cdot|_m$. Then, provided that $m > n/2$, the representation formula

$$< L,v > = (v,K)_m, \forall v \in X, \tag{32a}$$

where

$$\overline{K(x)} \equiv (-1)^m < L_y, e(x-y) > (\text{mod } P_{m-1}), \ \forall x \in \mathbb{R}^n, \tag{32b}$$

is valid, the associated Peano-like kernel being accordingly the (distributional) m-th Fréchet derivative $D^m K \in (L_2)^M$ of the (generalized) Rodrigues function $x \to K(x)$. Moreover, the error coefficient $c_L$ in the sharp appraisal

$$|<L,v>| \leq c_L |v|_m, \ \forall v \in X, \tag{33a}$$

is given explicitly by

$$c_L \equiv \| L \| = [(-1)^m < L_x, \overline{<L_y, e(x-y)>}>]^{1/2}. \tag{33b}$$

## IV. THE "RESTRICTION TO BOUNDED SETS" CASE

For simplicity, we shall limit ourselves to the particular (but significant!) case where the distribution v to be represented belongs to the Sobolev space of order 1:

$$H^1(\Omega) := \{v \in \mathcal{D}'(\Omega) : v, \partial_1 v, \ldots, \partial_n v \in L_2(\Omega)\} \tag{34a}$$

or to its closed subspace:

$$H_0^1(\Omega) := \text{ the closure in } H^1(\Omega) \text{ of } \mathcal{D}(\Omega). \tag{34b}$$

We assume that the open subset $\Omega$ of $\mathbb{R}^n$ is bounded and connected, has a piecewise $C^1$ boundary $\Gamma$, and lies on one side of it. Many important properties are then known to hold, among which the illuminating ones:

— the restrictions to $\Omega$ of the functions belonging to $C^\infty$ form a dense subspace of $H^1(\Omega)$, which is nothing else but $C^\infty(\bar{\Omega})$ (i.e., the space of $C^\infty$ functions in $\Omega$ whose derivatives of all order extend as continuous functions to the closure $\bar{\Omega}$ of $\Omega$).

— the trace mapping $\gamma : C^\infty(\bar{\Omega}) \to L_2(\Gamma)$, $v \to \gamma(v)$, where $\gamma(v)$ denotes the restriction of v to $\Gamma$ (this boundary is naturally equipped with the Leray form, or hypersurface measure $d\gamma$, induced by the Euclidean measure dx on $\mathbb{R}^n$), is linear and continuous, and therefore can be extended by continuity to $H^1(\Omega)$; moreover, the kernel of $\gamma$ in $H^1(\Omega)$ is exactly $H_0^1(\Omega)$.

— the linear mapping "restriction to $\Omega$ of distributions in $\mathbb{R}^n$", R, maps $H^1$ continuously onto $H^1(\Omega)$ and has a continuous right inverse (called extension operator for $\Omega$, from $H^1(\Omega)$ into $H^1$); $H_0^1(\Omega)$ is the image under R of the space of $H^1$ functions with support in $\bar{\Omega}$.

In the following, the $n_i (1 \leq i \leq n)$ are the components of the unit vector $\nu$ along the outer normal to $\Gamma$ (which is defined almost everywhere), $\delta_\Gamma$ stands for the surface Dirac measure associated with $\Gamma$, i.e.,

$$\langle \delta_\Gamma, \varphi \rangle := \int_\Gamma \varphi \, d\gamma, \ \forall \varphi \in \mathcal{D}, \tag{35}$$

while $\tilde{v}$ denotes the extension of any $v \in \mathcal{D}'(\Omega)$ by setting it equal to zero outside of $\Omega$; it should be noted that $\tilde{v} \in \mathcal{D}'$ if v does not "grow" at the boundary $\Gamma$ faster than some power of the inverse of the distance to $\Gamma$ (which extendability assumption is always satisfied here).

A. The $H_0^1(\Omega)$ subcase. Since $\tilde{v}$ is compactly supported in $\mathbb{R}^n$, (3) and (4) yield directly a candidate for representation formula in $H^1(\Omega)$, viz.,

$$\tilde{v} = \sum_{i=1}^{n} \partial_i (e+h) * \partial_i \tilde{v}, \quad \forall v \in H^1(\Omega). \tag{36}$$

where e is the locally integrable function in $\mathbb{R}^n$:

$$e(x) := \begin{cases} (2\pi)^{-1} \ell n |x|, & \text{if } n = 2 \\ -\Gamma(n/2-1)/(4\pi^{n/2}|x|^{n-2}), & \text{if } n \neq 2, \end{cases} \tag{37}$$

and h stands for an arbitrary harmonic function in $\mathbb{R}^n$. Now by using in conjunction the very definition of distributional differentiation and the Green-Gauss formula (of integration by parts), it can be shown easily that

$$(\partial_i v)^{\sim} - \partial_i v = \gamma(v) n_i \delta_\Gamma, \quad \forall v \in H^1(\Omega). \tag{38}$$

Hence the required representation formula in $H_0^1(\Omega)$:

$$\tilde{v} = \sum_{i=1}^{n} \partial_i (e+h) * (\partial_i v)^{\sim}, \quad \forall v \in H_0^1(\Omega), \tag{39a}$$

which indeed can be rewritten in the functional form:

$$\tilde{v}(x) = \sum_{i=1}^{n} \int_\Omega \partial_i (e+h)(x-y) \partial_i v(y) dy, \quad \forall v \in H_0^1(\Omega), \tag{39b}$$

at least almost everywhere in $\mathbb{R}^n$ (in accordance with Fubini's theorem).

Representation formulas can be exploited to good purpose in various situations, in particular for the analysis and the quantitative estimation of the so-called error coefficients (see e.g. [2]). By way of illustration, we will obtain here, in connection with the classical Friedrichs inequality

$$|v|_{0,\Omega} \leq c|v|_{1,\Omega}, \quad \forall v \in H_0^1(\Omega) \tag{40}$$

the optimal error coefficient:

$$c_F := \sup_{\substack{v \in H_0^1(\Omega) \\ |v|_{1,\Omega}=1}} |v|_{0,\Omega} \equiv \min\{c \in \mathbb{R} \text{ verifying } (40)\} \tag{41}$$

First of all, since $H_0^1(\Omega)$ is a Hilbert space, it follows from the Rellich compactness lemma that the set of admissible v's in (41) is a compact subset of $L_2(\Omega)$ (see [10], p.420), so that the supremum must be attained. Now this stationary point formulation (of Ritz type) is classically equivalent (see e.g. [13], Chap. 6) to the weak formulation (of Galerkin type): Find the greatest possible scalar $c_F^2$ such that there exists a function $v \in H_0^1(\Omega)$ for which

$$c_F^2(v,w)_{1,\Omega} - (v,w)_{0,\Omega} = 0, \ \forall w \in H_0^1(\Omega). \tag{42}$$

But $\mathcal{D}(\Omega)$ is dense in $H_0^1(\Omega)$, so that finally this variational problem amounts to finding the smallest possible eigenvalue $c_F^{-2}$ and the associated eigenfunction(s) $v_F$ for the eigenvalue problem:

$$\begin{cases} \Delta v + \lambda v = 0 \text{ in } \mathcal{D}'(\Omega), \\ v \in H_0^1(\Omega), \end{cases} \tag{43}$$

that is, the fixed membrane problem in $\Omega$.

This typical result could have been derived directly from the representation formula (39). Indeed, the relation

$$|v|^2_{0,\Omega} \equiv \left(\tilde{v}, \sum_{i=1}^{n} \partial_i (e+h) * (\partial_i v)^{\sim}\right)_0 = \qquad (44)$$

$$= - (\tilde{v}, (e+h) * \tilde{v})_1, \ \forall v \in H_0^1(\Omega),$$

which is readily obtained by integrating by parts, strongly suggests, in view of the definition (41) and Schwarz's inequality in $H_0^1(\Omega)$, that $c_F^2$ must be the reciprocal of the smallest eigenvalue of the (integral equation) eigenvalue problem:

$$v + \lambda R(G*\tilde{v}) = 0 \text{ in } H_0^1(\Omega); \qquad (45)$$

here G denotes the well known <u>Green's function</u> for the Laplacian $\Delta$ in $\Omega$, and the convolution $G*\tilde{v}$ is accordingly to be interpreted in the sense of Volterra (see e.g. [1], p. 423); the mapping $v \to G*v$ (of $H_0^1(\Omega)$ into itself) is <u>compact</u> (by Rellich's lemma) and <u>self-adjoint</u> (by the Hermitian symmetry of Green's function), so that Riesz's spectral theory can be exploited to good purpose. As a matter of fact, owing to the distributional version of the second Green formula, viz.,

$$(\Delta v)^{\sim} - \Delta\tilde{v} = \gamma(\partial_\nu v)\delta_\Gamma + \partial_\nu[\gamma(v)\delta_\Gamma], \ \forall v \in H^2(\Omega),$$

the complete equivalence of the eigenvalue problems (43) and (45) can be readily verified.

In practical applications, it usually proves impossible to determine the precise values of optimal error coefficients whereas <u>realistic bounds</u> can often be found economically. This is clearly true for <u>lower bounds</u>, in view of the classical Rayleigh-Ritz method. On the other hand, there is apparently no unified method for finding satisfactory <u>upper</u>

<u>bounds</u> for $c_F$, but only a few specialized techniques (for an interesting survey of useful results along these lines, see [11], pp. 15-18); their applications to the optimized Friedrichs inequality yields, for example, the following realistic results. Let $2a > 0$ denote the length of the edge of a (n-dimensional) cube containing $\Omega$; by virtue of a classical <u>monotony principle</u>, the lowest eigenvalue of the fixed membrane problem for $\Omega$ is bounded below by that of the hypercube; from the explicit value of the latter, which is readily obtained by separating variables, it follows that

$$c_F^2(\Omega) \leq 4a^2/(n\pi^2); \tag{47a}$$

as expected, this appraisal is much sharper than those obtained by more conventional methods (see e.g. [7], p. 14). Another interesting result immediately follows from the <u>Faber-Krahn inequality</u>, which states that the lowest eigenvalue of the fixed membrane problem for $\Omega$ is not smaller than that for the (n-dimensional) ball whose volume is the same as that of $\Omega$; this appraisal is

$$c_F(\Omega) \leq \rho^n/j^2_{(n-2)/2}, \tag{47b}$$

where $\rho$ stands for the radius of that equivalent ball and $j_{(n-2)/2}$ denotes the first zero of the Bessel function $J_{(n-2)/2}$.

B. <u>The</u> $H^1(\Omega)$ <u>subcase</u>. As it readily follows from (36) and (38), a possibly appropriate candidate for representation formula in $H^1(\Omega)$ is

(48a)

$$\tilde{v} = \sum_{i=1}^{n} \partial_i(e+h)*(\partial_i v)^{\sim} - \partial_\nu(e+h)*[\gamma(v)\delta_\Gamma], \quad \forall v \in H^1(\Omega),$$

are equivalently, almost everywhere in $\mathbb{R}^n$,

(48b)

$$\tilde{v}(x) = \sum_{i=1}^{n} \int_\Omega \partial_i(e+h)\,\partial_i v(y)\,dy - \int_\Gamma \partial_\nu(e+h)(x-y)[\gamma(v)](y)\,d\gamma,$$

where again h denotes an arbitrary harmonic function in $\mathbb{R}^n$.

By proceeding essentially as explained above in connection with Friedrichs' inequality, the famous generalization known as the Poincaré inequality can also be optimized; the final result is

$$|v|^2_{0,\Omega} \le c_p^2 |v|^2_{1,\Omega} + \left|\int_\Omega v(x)\,dx\right|^2 \Big/ \int_\Omega dx, \quad \forall v \in H^1(\Omega), \tag{49}$$

where $c_p^{-2}$ is the first non-zero eigenvalue of the free membrane problem in $\Omega$, to wit:

$$\begin{cases} \Delta v + \lambda v = 0 \text{ in } \mathcal{D}'(\Omega) \\ v \in H^1(\Omega). \end{cases} \tag{50}$$

In the particularly simple case where $\Omega$ is a (n-dimensional) cube of side 2a, it is easily found, by separating variables, that

$$c_p^2(\text{hypercube}) = 4a^2/\pi^2, \tag{51}$$

which again is significantly smaller than that given in [7] (see p. 16), to wit: $2na^2$. More generally, it is known that

$$c_p^2(\Omega) < (\text{Euclidean diameter of } \Omega)^2/\pi^2, \tag{52}$$

provided only that $\Omega$ is convex (see [11], p. 17); the coefficient $1/\pi^2$ in this very specialized result is of course much smaller than that deduced from the appraisal (3.9d bis) in [2], which was obtained by resorting to a quite general method of wide applicability.

As a matter of fact, (49) is equivalent (by Pythagoras' theorem) to the sharp appraisal

$$|v - Pv|^2_{0,\Omega} \le c_p^2 |v|^2_{1,\Omega}, \quad \forall v \in H^1(\Omega), \tag{53}$$

where P denotes the orthogonal projector of $L_2(\Omega)$ with range $P_0$, so that

$$Pv := \int_\Omega v(x)dx \Big/ \int_\Omega dx, \quad \forall v \in L_2(\Omega) \tag{54}$$

Let us introduce now, for every $x \in \Omega$, a Neumann function $N_x$ (for $\Delta$ in $\Omega$), that is, a $C^2$ function in $\bar{\Omega}\setminus\{x\}$ such that

$$\begin{cases} \Delta N_x = \delta_{(x)} - 1 \Big/ \int_\Omega dx, \text{ in } \mathcal{D}'(\bar{\Omega}), \\ \partial_\nu N_x = 0, \text{ a.e. on } \Gamma; \end{cases} \tag{55a}$$

this boundary value problem is compatible, and has a unique solution verifying the "natural" condition

$$\int_\Omega N_x(y)dy = 0. \tag{55b}$$

In terms of such a Neumann function, we have, almost everywhere in $\Omega$,

$$v(x) = Pv(x) - \sum_{i=1}^n \int_\Omega \partial_i N_x(y) \partial_i v(y)dy, \quad \forall v \in H^1(\Omega), \tag{56a}$$

which is, written in functional form, the representation formula in $H^1(\Omega)$ to which (48) reduces for $h(y) := N_x(y) - e(x-y)$ at every point $x \in \Omega$, that is finally,

$$\tilde{v} = P\tilde{v} + \sum_{i=1}^n \partial_i N^* (\partial_i v)^\sim, \quad \forall v \in H^1(\Omega),$$

where $N(x,y) := N_x(y)$ for $x \in \Omega$ and $y \in \bar{\Omega}$ is the so-called Neumann kernel for the Laplacian $\Delta$ in $\Omega$.

REFERENCES

1. Choquet-Bruhat, Y., C. De Witt-Morette and M. Dillard-Bleick, Analysis, manifolds and physics, North-Holland Publishing Company, Amsterdam-New York-Oxford, 1977.

2. Meinguet, J., A practical method for estimating approximation errors in Sobolev spaces, in Multivariat Approximation, D.C.Handscomb,ed.,Academic Press Inc., London, 1978, 169-187.

3. Meinguet, J., An intrinsic approach to multivariate spline interpolation at arbitrary points, in Polynomial and Spline Approximation, B.N.Sahney,ed., D. Reidel Publishing Company, Dordrecht, 1979, 163-190.

4. Meinguet, J., Multivariate interpolation at arbitrary points made simple, *Z.Angew.Math.Phys.* 30(1979),292-304.

5. Meinguet, J., Basic mathematical aspects of surface spline interpolation, in Numerische Integration, G. Hämmerlin,ed., Birkhäuser Verlag, Basel, 1979, 211-220.

6. Meinguet, J., A convolution approach to multivariate representation formulas, in Multivariate Approximation Theory, W. Schempp and K. Zeller,eds., Birkhäuser Verlag, Basel, 1979, 198-210.

7. Nečas, J., Les méthodes directes en théorie des équations elliptiques, Masson, Paris, 1967.

8. Nikol'skii,S.M., Approximation of Functions of Several Variables and Imbedding Theorems, Springer-Verlag, Berlin-Heidelberg, 1975.

9. Schwartz, L., Théorie des distributions, Herman, Paris, 1966.

10. Schwartz, L., Analyse: Topologie générale et analyse fonctionnelle, Hermann, Paris, 1970.

11. Sigillito, V. G., Explicit "a priori" inequalities with applications to boundary value problems, Pitman Publishing Ltd., London, 1977.

12. Stein, E. M., Singular Integrals and Differentiability Properties of Functions, Princeton University Press, Princeton, 1970.

13. Strang, G., and G. J. Fix, An analysis of the finite element method, Prentice-Hall, Inc., Englewood Cliffs, 1973.

# A NEW ITERATIVE METHOD FOR THE SOLUTION OF SYSTEMS OF NONLINEAR EQUATIONS

Beny Neta

Department of Mathematical Sciences
Northern Illinois University
DeKalb, Illinois

A new quasi-Newton method for the solution of systems of nonlinear algebraic equations is introduced. This method is a generalization of a sixth-order one developed by the author for approximating the solution of one nonlinear equation. The R-order of the method is four. Numerical experiments comparing the method to Newton's show that one can save over 20% of the cost of solving a system of algebraic equations. The saving is greater when the dimension is higher or the number of iterations needed is larger.

## I. INTRODUCTION

Let $F : D \subset \mathbb{R}^n \longrightarrow \mathbb{R}^n$ be a nonlinear mapping with both its domain and its range in the n-dimensional real linear space $\mathbb{R}^n$. In this paper we consider the numerical solution of the system of n equations in n variables

$$\underline{F}(\underline{x}) = \underline{0}. \tag{1}$$

Two special cases of (1), in particular, are much better understood than most others - namely, the n-dimensional linear




ISBN 0-12-780650-4

systems, and the one-dimensional nonlinear equations. See Varga [10], Householder [2] and Young [14] for the first case, and Ostrowski [7], Traub [9] and Householder [3] for the second one. In recent years there has been much interest in n-dimensional variations of Newton's method, the secant method, and other classical one-dimensional iterative methods. See for example Rheinboldt [8], Ortega and Rheinboldt [6], Werner [13], Voight [11],[12], Dennis and Moré [1], and references there.

The algorithm in most common usage for problem (1) can be written as follows:

Given $\underline{x}_0$

$$\underline{x}_{k+1} = \underline{x}_k - J^{-1}(\underline{x}_k)\underline{F}(\underline{x}_k), \quad k = 0,1,2,\ldots , \tag{2}$$

where

$$J_{i\ell}(\underline{x}) = \left(\frac{\partial F_i(\underline{x})}{\partial x^\ell}\right) \quad \text{is the Jacobian matrix.}$$

The iteration formula (2) is certainly not the way Newton's method should be implemented on a computer. The following form is much more like an actual implementation.

(i) Given $\underline{x}_k$, $\underline{F}(\underline{x}_k)$ and $J(\underline{x}_k)$

(ii) Solve the n x n linear system

$$J(\underline{x}_k)\underline{\sigma}_k = -\underline{F}(\underline{x}_k) \tag{3}$$

for the Newton step $\underline{\sigma}_k$ .

(iii) Using $\underline{\sigma}_k$ and perhaps some other values of $\underline{F}(\underline{x})$, choose $\underline{x}_{k+1}$.

(iv) Evaluate $\underline{F}(\underline{x}_{k+1})$ and test for convergence. Either terminate the computation or proceed to (v).

(v) Evaluate (or approximate) $J(\underline{x}_{k+1})$, set the counter to k+1 and return to (ii).

The traditional area of research on quasi-Newton methods are steps (iii) and (v). The reason is that for real problems, evaluations of $\underline{F}$ and J dominate the cost of solution and so it is in these steps that the potential saving is greatest.

In the next section the new method will be introduced. In section 3 the order of convergence will be established. In the last section we present some of the numerical experiments performed and compare the performance of the method to Newton's.

## II. DESCRIPTION OF ALGORITHM

Let us consider the one-dimensional case for a moment. In [5] the author developed a sixth-order method to approximate the solution x* of f(x) = 0. An iteration consists of a Newton substep followed by two substeps, of "modified" Newton (i.e., using the derivative of f at the first substep instead of the current one). Given $x_k$ solve the three equations:

$$\begin{cases} w_k = x_k - \dfrac{f(x_k)}{f'(x_k)} \\[2ex] z_k = w_k - \dfrac{f(w_k)}{f'(x_k)} \cdot \dfrac{f(x_k) + Af(w_k)}{f(x_k) + (A-2)f(w_k)} \\[2ex] x_{k+1} = z_k - \dfrac{f(z_k)}{f'(x_k)} \cdot \dfrac{f(x_k) - f(w_k)}{f(x_k) - 3f(w_k)} \end{cases} \tag{1}$$

where A is a parameter. If we choose A = -1 then the correcting term in the last two substeps is the same.

For a system of n equations the algorithm will be as follows:

(i) Given $\underline{x}_k$, $\underline{F}(\underline{x}_k)$ and $J(\underline{x}_k)$

(ii) Solve

$$J(\underline{x}_k)(\underline{w}_k - \underline{x}_k) = -\underline{F}(\underline{x}_k) \tag{2}$$

for $\underline{w}_k$.

(iii) Evaluate $\underline{F}(\underline{w}_k)$ and test for convergence. Either terminate the computation or proceed to (iv).

(iv) Evaluate the entries of the diagonal matrix D

$$D_{ii}(\underline{x}_k, \underline{w}_k) = \begin{cases} \dfrac{F_i(\underline{x}_k) - F_i(\underline{w}_k)}{F_i(\underline{x}_k) - 3F_i(\underline{w}_k)}, & \text{if denominator} \neq 0 \\ 1 & \text{otherwise} \end{cases} \tag{3}$$

(v) Solve

$$J(\underline{x}_k)(\underline{z}_k - \underline{w}_k) = - D(\underline{x}_k, \underline{w}_k)F(\underline{w}_k) \tag{4}$$

for $\underline{z}_k$ .

(vi) Evaluate $\underline{F}(\underline{z}_k)$ and test for convergence. Either terminate the process or proceed to (vii).

(vii) Solve

$$J(\underline{x}_k)(\underline{x}_{k+1} - \underline{z}_k) = - D(\underline{x}_k, \underline{w}_k)F(\underline{z}_k) \tag{5}$$

for $\underline{x}_{k+1}$ .

(viii) Evaluate $\underline{F}(\underline{x}_{k+1})$ and test for convergence. Either terminate the computation or proceed to (ix).

(ix) Evaluate $J(\underline{x}_{k+1})$, set the computer to k+1 and return to (ii).

Remark: Since the evaluation and factorization of the Jacobian J is costly, one can save by keeping the Jacobian fixed. This idea is not new. Our claim is that if one modifies the righthand side as described in steps (v) and (vii) there will be no serious reduction in the rate of convergence.

Let us now compare the number of multiplications and divisions needed in one step of this algorithm with two consecutive steps of Newton's method (both are of order four). The number $N_C$ of multiplications needed to calculate the entries of the Jacobian is given by:

$$N_C \sim nb \tag{6}$$

where b is half the bandwidth. The number $N_F$ of multiplications needed to factor the Jacobian is

$$N_F \sim \frac{1}{2}nb^2 \tag{7}$$

and the number $N_S$ of multiplications needed to back solve the two systems is given by:

$$N_S \sim 2nb \ . \tag{8}$$

If J is a full matrix instead of banded, one has to replace b by n in (6) - (8). The number $N_D$ of multiplications needed to evaluate the entries of D and multiply by F in both steps is:

$$N_D \sim 4n \ . \tag{9}$$

Note that if J is not symmetric $N_C$ and $N_F$ should be doubled. Thus, the total number of multiplications needed for one step of our algorithm is:

$$T_0 = nb + \frac{1}{2}nb^2 + 6nb + 4n = n(7b + \frac{1}{2}b^2 + 4). \tag{10}$$

The total number of multiplications needed for two steps of Newton's method is:

$$T_N = 2(nb + \frac{1}{2}nb^2 + 2nb) = n(8b + b^2) \ . \tag{11}$$

Clearly, $T_0$ is smaller than $T_N$ if $b > 2$ .

Remark: It is known that the natural extension of a procedure to higher dimensions does not preserve the order of convergence exhibited by the one-dimensional procedure (see Voigt [12]). Therefore, one cannot expect the method to have a sixth-order in either $O_Q$ or $O_R$ measures.

In the next section we recall some definitions of measures and prove that our algorithm has at least fourth order.

## III. ORDER OF CONVERGENCE

In this section we are interested in measuring how fast the sequence $\{\underline{x}_k\}$ converges to $\underline{x}^*$ (the solution of (1.1)). We shall use two measures of the order of convergence denoted by $O_Q$ and $O_R$ which depend on the asymptotic convergence factors $Q_p$ and $R_p$, respectively. A complete discussion of these measures may be found in Ortega and Rheinboldt [6].

<u>Definition</u>: Let $F$ be the iterative procedure

$$F: \underline{x}_{k+1} = G(\underline{x}_k, \dots, \underline{x}_{k-m+1}), \quad k = m-1, m, \dots \quad (1)$$

Let $S(F, x^*)$ denote the set of all sequences generated by an iterative procedure $F$ with limit point $\underline{x}^*$. Then

$$Q_p(F, \underline{x}^*) = \sup\left\{Q_p\{\underline{x}_k\} \,\middle|\, \{\underline{x}_k\} \in S(F, \underline{x}^*)\right\}, \quad (2)$$

where

$$Q_p\{\underline{x}_k\} = \begin{cases} 0 & \text{if } \underline{x}_k = \underline{x}^* \text{ for all } k \geq k_0 \\ \limsup\limits_{k\to\infty} \dfrac{\|\underline{x}_{k+1} - \underline{x}^*\|}{\|\underline{x}_k - \underline{x}^*\|^p}, & 1 \leq p < \infty, \text{if } \underline{x}_k \neq \underline{x}^* \text{ for all but finitely many } k \\ \infty & \text{otherwise,} \end{cases} \quad (3)$$

are the <u>Q-convergence factors</u> of $F$ at $\underline{x}^*$.

<u>Definition</u>: Let $Q_p(F, \underline{x}^*)$ be the Q-convergence factors of an iterative procedure $F$ at $\underline{x}^*$. Then

$$O_Q = \inf\left\{p \in [1, \infty) \,\middle|\, Q_p(F, \underline{x}^*) = \infty\right\} \quad (4)$$

is the <u>Q-order</u> of $F$ at $\underline{x}^*$.

Definition: Let $S(F,\underline{x}^*)$ denote the set of all sequences $\{\underline{x}_k\}$ generated by an iterative procedure $F$ with limit point $\underline{x}^*$.

Then

$$R_p(F,\underline{x}^*) = \operatorname{Sup}\left\{R_p\{\underline{x}_k\} \mid \{\underline{x}_k\} \in S(F,\underline{x}^*)\right\} , \tag{5}$$

where

$$R_p\{\underline{x}_k\} = \begin{cases} \limsup\limits_{k\to\infty} \|\underline{x}_k - \underline{x}^*\|^{1/k} & \text{if } p = 1 \\ \limsup\limits_{k\to\infty} \|\underline{x}_k - \underline{x}^*\|^{1/p^k} & \text{if } p \in (1,\infty), \end{cases} \tag{6}$$

are the R-convergence factors of $F$ at $\underline{x}^*$.

Definition: Let $R_p(F,\underline{x}^*)$ be the R-convergence factors of an iterative procedure $F$ at $\underline{x}^*$. Then

$$O_R(F,\underline{x}^*) = \inf\left\{p \in [1,\infty) \mid R_p(F,\underline{x}^*) = 1\right\} \tag{7}$$

is the R-order of $F$ at $\underline{x}^*$.

Definition: The mapping $F: D \subset \mathbb{R}^n \to \mathbb{R}^n$ is Frechet- (or F-) differentiable at $x \in \operatorname{int}(D)$ if there is an $A \in L(\mathbb{R}^n, \mathbb{R}^n)$ such that

$$\lim_{h\to 0} \frac{\|F(x+h) - Fx - Ah\|}{\|h\|} = 0 . \tag{8}$$

The linear operator A is denoted by $F'(x)$, and is called the F-derivative of F at x.

Theorem: Let $F: D \subset \mathbb{R}^n \to \mathbb{R}^n$ be F-differentiable in an open ball $S = S(x^*,\delta) \subset D$ and satisfy

$$\|J(\underline{x}) - J(\underline{x}^*)\| \leqslant \gamma\|\underline{x} - \underline{x}^*\| , \quad , \text{ for all } \underline{x} \in S.$$

Assume, further, that $F(\underline{x}^*) = 0$ and $J(\underline{x}^*)$ is nonsingular. Then $\underline{x}^*$ is a point of attraction of the iteration $F$ defined by

$$\underline{x}_{k+1} = H\underline{x}_k \tag{9}$$

where

$$\begin{cases} N\underline{x} = \underline{x} - J^{-1}(\underline{x})\underline{F}(\underline{x}) \\ G\underline{x} = N\underline{x} - J^{-1}(\underline{x})D(\underline{x}, N\underline{x})\underline{F}(N\underline{x}) \\ H\underline{x} = G\underline{x} - J^{-1}(\underline{x})D(\underline{x}, N\underline{x})\underline{F}(G\underline{x}) \end{cases}, \tag{10}$$

and

$$O_R(F, \underline{x}^*) \geq O_Q(F, \underline{x}^*) \geq 4 \tag{11}$$

Proof: Since $N\underline{x} = \underline{x} - J^{-1}(\underline{x})F(\underline{x})$ it is clear that

$$\|N\underline{x} - \underline{x}^*\| \leq \eta\|\underline{x} - \underline{x}^*\|^2 \text{ on } S_1 \subset S . \tag{12}$$

$G\underline{x} = N\underline{x} - J^{-1}(\underline{x})D(\underline{x}, N\underline{x})F(N\underline{x})$ is well defined on $S_2 \subset S_1$.

Therefore, if $\|J^{-1}(\underline{x})\| \leq \beta$ for all $x \in S_2$, then

$$\begin{aligned} G\underline{x} - \underline{x}^* &= N\underline{x} - \underline{x}^* - J^{-1}(\underline{x})D(\underline{x}, N\underline{x})F(N\underline{x}) = \\ &= J^{-1}(\underline{x})\{J(\underline{x})(N\underline{x} - \underline{x}^*) - D(\underline{x}, N\underline{x})F(N\underline{x})\} = \\ &= J^{-1}(\underline{x})\{-D(\underline{x}, N\underline{x})F(N\underline{x}) + F(\underline{x}^*) + J(\underline{x}^*)(N\underline{x}-\underline{x}^*) - \\ &\quad -[J(\underline{x}^*) - J(\underline{x})](N\underline{x} - \underline{x}^*)\} \end{aligned} \tag{13}$$

$$\begin{aligned} \|G\underline{x}-\underline{x}^*\| &\leq \beta\|D(\underline{x}, N\underline{x})F(N\underline{x}) - F(\underline{x}^*) - J(\underline{x}^*)(N\underline{x}-\underline{x}^*)\| + \\ &+ \beta\gamma\|\underline{x}-\underline{x}^*\| \, \|Nx - \underline{x}^*\| \end{aligned} \tag{14}$$

In order to bound the first term on the right, let us examine the i-th component of the vector

$$A_i = D_{ii}(\underline{x}, N\underline{x})F_i(N\underline{x}) - F_i(\underline{x}^*) - \sum_{\ell=1}^{n} J_{i\ell}(\underline{x}^*)(N\underline{x} - \underline{x}^*)_\ell$$

$$\begin{aligned} D_{ii}(\underline{x}, N\underline{x})F_i(N\underline{x}) &= \frac{F_i(\underline{x}) - F_i(N\underline{x})}{F_i(\underline{x}) - 3F_i(N\underline{x})} F_i(N\underline{x}) = \\ &= \left(1 + \frac{2F_i(N\underline{x})}{F_i(\underline{x}) - 3F_i(N\underline{x})}\right)F_i(N\underline{x}) = \\ &= F_i(N\underline{x}) + \frac{2F_i^2(N\underline{x})}{F_i(\underline{x}) - 3F_i(N\underline{x})} \end{aligned} \tag{15}$$

Expanding $F_i(\underline{x})$ and $F_i(N\underline{x})$ in Taylor series one obtains

$$A_i = \frac{\sum_{j=1}^{n} \sum_{\ell=1}^{n} F''_{ij\ell} \frac{(N\underline{x} - \underline{x}^*)_j (N\underline{x} - x^*)_\ell}{2} + \sum_{j=1}^{n} J_{ij}(\underline{x}^*)(N\underline{x} - \underline{x}^*)_j + \text{H.O.T.}^2}{\sum_{j=1}^{n} J_{ij}(\underline{x}^*)(\underline{x} - \underline{x}^*)_j + \text{H.O.T.}} \tag{16}$$

Thus

$$\|G\underline{x} - \underline{x}^*\| \leqslant \beta\rho\|\underline{x} - \underline{x}^*\|^3 + \beta\gamma\eta\|\underline{x} - \underline{x}^*\|^3 = \beta(\rho + \gamma\eta)\|\underline{x} - \underline{x}^*\|^3 . \tag{17}$$

$H\underline{x} = G\underline{x} - J^{-1}(\underline{x})D(\underline{x},N\underline{x})F(G\underline{x})$ is well defined on $S_3 \subset S_2$.

$$\begin{aligned} H\underline{x} - \underline{x}^* &= G\underline{x} - \underline{x}^* - J^{-1}(\underline{x})D(\underline{x},N\underline{x})F(G\underline{x}) = \\ &= J^{-1}(\underline{x})\{J(\underline{x})[G\underline{x} - \underline{x}^*] - D(\underline{x},N\underline{x})F(G\underline{x})\} = \\ &= J^{-1}(\underline{x})\{[-D(\underline{x},N\underline{x})F(G\underline{x}) + F(\underline{x}^*) + \\ &\quad + J(\underline{x}^*)(G\underline{x} - \underline{x}^*)] + [J(\underline{x}^*) - J(\underline{x})](G\underline{x} - \underline{x}^*)\} \end{aligned} \tag{18}$$

$$\begin{aligned} \|H\underline{x} - \underline{x}^*\| \leqslant \beta\|D(\underline{x},N\underline{x})F(G\underline{x}) - F(\underline{x}^*) - J(\underline{x}^*)(G\underline{x}-\underline{x}^*\| + \\ + \beta\gamma\|G\underline{x} - \underline{x}^*\| \; \|\underline{x} - \underline{x}^*\| \end{aligned} \tag{19}$$

The first term on the right can be bounded in a similar way to yield

$$\|D(\underline{x},N\underline{x})F(G\underline{x}) - F(\underline{x}^*) - J(\underline{x}^*)(G\underline{x} - \underline{x}^*)\| \leqslant \lambda\|\underline{x}-\underline{x}^*\|^4 \tag{20}$$

Combining (19) - (20) with (17) one obtains

$$\|H\underline{x} - \underline{x}^*\| \leqslant \beta\lambda\|\underline{x} - \underline{x}^*\|^4 + \beta^2\gamma(\rho + \gamma\eta)\|\underline{x} - \underline{x}^*\|^4 =$$
$$= \beta[\lambda + \beta\gamma(\rho + \gamma\eta)]\|\underline{x} - \underline{x}^*\|^4 . \qquad (21)$$

This implies that

$$O_R \geqslant O_Q \geqslant 4 .$$

## IV. NUMERICAL EXPERIMENTS

In this section we present some of the numerical experiments and compare the performance of our algorithm to Newton's. It is clear that the saving is in calculating the entries of the Jacobian and in factoring it. Thus one cannot expect to see any difference in the performance of the two algorithms when solving systems of $2 \times 2$ or $3 \times 3$. We have compared the CPU time needed to solve 5 different systems of $2 \times 2$ and 3 different systems of $3 \times 3$. The results are summarized in Table 1.

TABLE 1

| Problem No. | CPU time in seconds NETA | NEWTON |
|---|---|---|
| 1 | .68 | .71 |
| 2.1 | .84 | .84 |
| 2.2 | .76 | .79 |
| 3 | .68 | .76 |
| 4 | .82 | .74 |
| 5 | .81 | .75 |
| 6.1 | 1.16 | 1.22 |
| 6.2 | Divergence | 1.05 |
| 7 | 1.04 | Divergence |
| 8 | .73 | .93 |

TABLE 2

| Problem No. | System of Equations | Initial Value |
|---|---|---|
| 1 | $x + 3\log x - y^2 = 0$<br>$2x^2 - xy - 5x + 1 = 0$ | (1, -2) |
| 2 | $x^2 + xy^3 - 9 = 0$<br>$3x^2y - y^3 - 4 = 0$ | 1. (1.2, 2.5)<br>2. (-1.2, -2.5) |
| 3 | $x + 2y - 3 = 0$<br>$2x^2 + y^2 - 5 = 0$ | (1.5, 1) |
| 4 | $3x^2 + 4y^2 - 1 = 0$<br>$y^3 - 8x^3 - 1 = 0$ | (-.5, .25) |
| 5 | $4x^2 + y^2 - 4 = 0$<br>$x + y - \sin(x-y) = 0$ | (1, 0) |
| 6 | $x^5 + y^3z^4 + 1 = 0$<br>$x^2yz = 0$<br>$z^4 - 1 = 0$ | 1. (-1000, -1000, -1000)<br>2. (-100, 0, 100) |
| 7 | $x^2 + y = 37$<br>$x - y^2 = 5$<br>$x + y + z = 3$ | (5, 0, -2) |
| 8 | $12x - 3y^2 - 4z = 7.17$<br>$x^2 + 10y - z = 11.54$<br>$y^3 + 7z = 7.631$ | (3, 0, 1) |

In Table 2 we list the systems solved and initial values used. We found one example of each where only one of the methods converged.

Note that the CPU time is for the execution step only. All numerical results were obtained on IBM 370/148 computer.

In our last experiment we consider a system of algebraic equations arising in the finite element approximation to the solution of:

$$-\nabla(|\nabla u(\underline{x})|^{p-2}\nabla u(\underline{x}) = f(\underline{x}) \quad \underline{x} \in \Omega \ , \ p \geq 2 \tag{1}$$

$$u(\underline{x}) = 0 \qquad \underline{x} \in \partial\Omega.$$

Fix and Neta [4] showed that the finite element approximation $u^h$ to the solution u can be written as follows

$$u^h(\underline{x}) = \sum_{i=1}^{n} u_i \ \phi_i(\underline{x}) \tag{2}$$

where $\phi_i(\underline{x})$ are the basis functions of the finite dimensional subspace S (dim S = n). The weights $u_i$ can be computed by solving the algebraic system of equations

$$K(\underline{u})\underline{u} = \underline{g} \ , \tag{3}$$

where

$$K_{ij} = \int_{\Omega} \sum_{\ell=1}^{n} u_\ell \ \nabla\phi_\ell|^{p-2} \nabla\phi_j \cdot \nabla\phi_i \ dx \ , \tag{4}$$

$$g_i = \int_{\Omega} f(\underline{x})\phi_i(\underline{x})d\underline{x}. \tag{5}$$

The results are summarized in Table 3.

Note that the system (3) is linear when p = 2 and the saving is a result of only one less computation and factorization of the Jacobian.

It is clear from Table 3 that the saving is larger when either the dimension is higher or the number of iterations is larger.

TABLE 3

| Exact Solution | P | Dimension of Matrix K | No. of Iterations NETA | No. of Iterations NEWTON | CPU time (sec) NETA | CPU time (sec) NEWTON | % Change |
|---|---|---|---|---|---|---|---|
| sin(x+y) | 2 | 25x25, 49x49, 81x81 | 1 | 2 | 28.89 | 35.33 | 22.3 |
| sin(x+y) | 2 | 121 x 121 | 1 | 2 | 32.74 | 40.23 | 23 |
| sin(x+y) | 2 | 169 x 169 | 1 | 2 | 46.75 | 60.43 | 29.3 |
| xy(1-x)(1-y) | 2 | 25x25, 49x49, 81x81 | 1 | 2 | 25.80 | 30.97 | 20 |
| xy(1-x)(1-y) | 2 | 121 x 121 | 1 | 2 | 32.48 | 39.47 | 21.5 |
| xy(1-x)(1-y) | 2 | 169 x 169 | 1 | 2 | 46.10 | 57.15 | 24 |
| x(1-x) | 2 | 25x25. 49x49, 81x81 | 1 | 2 | 26.23 | 32.26 | 23 |
| xy(1-x)(1-y) | 3 | 25x25, 49x49, 81x81 | 2 | 4 | 54.40 | 67.60 | 24 |
| xy(1-x)(1-y) | 3 | 121 x 121 | 2 | 4 | 70.30 | 90.70 | 29 |
| xy(1-x)(1-y) | 3 | 169 x 169 | 2 | 4 | 91.10 | 119.30 | 31 |
| xy(1-x)(1-y) | 4 | 121 x 121 | 3 | 6 | 82.50 | 107.20 | 30 |
| xy(1-x)(1-y) | 4 | 169 x 169 | 3 | 7 | 112.40 | 149.50 | 33 |
| sin(x+y) | 3 | 121 x 121 | 2 | 4 | 70.50 | 93.00 | 32 |
| sin(x+y) | 3 | 169 x 169 | 2 | 5 | 90.60 | 122.30 | 35 |
| sin(x+y) | 4 | 121 x 121 | 3 | 6 | 80.90 | 108.40 | 34 |
| sin(x+y) | 4 | 169 x 169 | 3 | 7 | 110.70 | 152.80 | 38 |

## REFERENCES


1. Dennis, J.E. and More, J.J., Quasi-Newton Methods, Motivation and Theory, *SIAM Rev., Vol. 19* (1977), 46-89.
2. Householder, A., *The Theory of Matrices in Numerical Analysis,* Ginn, Boston, 1964.
3. Householder, A., *The Numerical Treatment of a Single Nonlinear Equation,* McGraw-Hill, New York, 1970.
4. Fix, G.J. and Neta, B., Finite Element Approximation of a Nonlinear Diffusion Problem, *Computers and Math. with Applic. Vol. 3* (1977), 287-298.
5. Neta, B., A Sixth-Order Family of Methods for Nonlinear Equations, *Int. J. Computer Math., Vol. 7* (1979), 157-161.
6. Ortega, J. and Rheinboldt, W., *Iterative Solution of Nonlinear Equations in Several Variables,* Academic Press, New York, 1970.
7. Ostrowski, A., *Solution of Equations and Systems of Equations,* Academic Press, New York, 1966.
8. Rheinboldt, W., *Methods for Solving Systems of Nonlinear Equations,* SIAM, Phila. 1974.
9. Traub, J., *Iterative Methods for the Solution of Equations,* Prentice-Hall, Englewood Cliffs, N.J., 1964.
10. Varga, R., *Matrix Iterative Analysis,* Prentice-Hall, Englewood Cliffs, N.J., 1962.
11. Voigt, R., Rates of Convergence for a Class of Iterative Procedures, *SIAM J. Numer. Anal. Vol. 8* (1971), 127-134.
12. Voigt, R., Orders of Convergence for Iterative Procedures, *SIAM J. Numer. Anal. Vol. 8* (1971), 222-243.

13. Werner, W., Über ein Verfahren der Ordnung $1+\sqrt{2}$ zur Nullstellenbestimmung, *Numer. Math. Vol. 32* (1979), 333-342.

14. Young, D., *Iterative Solution of Large Linear Systems*, Academic Press, New York, 1971.

# POLYNOMIALS AND RATIONAL FUNCTIONS

D. J. Newman
Department of Mathematics
Temple University
Philadelphia, Pennsylvania

## INTRODUCTION

Our four lectures will alternate between considerations of polynomials and of rational functions. Lectures I and III will treat certain estimates of polynomials in the complex plane which arise from the theory of optimal recovery. Lectures II and IV, on the other hand, will treat the classical problems of rational approximation over real intervals

First, however, we take this opportunity to give thanks to Zvi Ziegler and the Technion for its kindness in inviting us to this meeting, and to Israel for its utter magnificence!

## LECTURE I. DERIVATIVE BOUNDS FOR POLYNOMIALS WITH A PRESCRIBED ZERO

Bernstein's theorem says that, over the space of nth degree polynomials normed by the maximum modulus in the unit disc, the norm of the derivative operator is exactly n. We investigate the norm of this derivative operator over the subspace of nth degree polynomials which vanish at 1. It seems



ISBN 0-12-780650-4

quite difficult to pin this value down with any great precision but we are able to show that it lies between $n - \frac{C_1}{n}$ and $n - \frac{C_2}{n}$ where the $C_i$ are positive constants.

It is curious that the restriction of vanishing at 1 has so little effect on the norm of the derivative. Indeed in the analogous situation for polynomials under the $H^2$ norm the correct answer is $n - \frac{C}{\log n}$ rather than $n - \frac{C}{n}$ and this indicates a much bigger effect. (This bigger effect is shared by $H^p$, $p < \infty$).

We turn to the proofs.

I. $\|D\| \le n - \frac{C_2}{n}$

So let $P(z)$ be of degree $n$, satisfy $P(1) = 0$, and have $\|P\| = 1$. We need to show that $|P'(\zeta)| \le n - \frac{C_2}{n}$ for all $\zeta$, $|\zeta| < 1$.

Begin with the formula

$$P'(\zeta) = \frac{1}{2\pi i} \int_{|z|=1} \frac{P(z)dz}{(\zeta-z)^2} = \frac{1}{2\pi i} \int_{|z|=1} \frac{P(z)}{(\zeta-z)^2} \left(1 - \frac{\zeta^n}{z^n}\right)^2 dz,$$

and also observe that, since $P(1) = 0$, Bernstein's theorem itself ($P' << n$) insures that $|P(z)| < \frac{1}{2}$ on the arc $|z| = 1$, $|\arg z| < \frac{1}{2n}$.

Thus we have

$|P'(\zeta)| \le$

$$\le \frac{1}{2\pi} \int_{|z|=1} \left|\frac{1-\zeta^n/z^n}{\zeta-z}\right|^2 |dz| - \frac{1}{2}\cdot\frac{1}{2\pi} \int_{\substack{|z|=1 \\ |\arg z| \le 1/2n}} \left|\frac{1-\zeta^n/z^n}{\zeta-z}\right|^2 |dz|$$

$$\leq n - \frac{1}{4\pi} \int\limits_{\substack{|z|=1 \\ |\arg z|\leqslant 1/2n}} \left|\frac{1-\zeta^n/z^n}{\zeta - z}\right|^2 |dz|$$

$$\leq n - \frac{1}{16\pi} \int\limits_{\substack{|z|=1 \\ |\arg z|\leqslant 1/2n}} |1-\zeta^n/z^n|^2 |dz|$$

But this integral can be explicitly evaluated. The result is $\dfrac{1-4 \sin \frac{1}{2} \cdot \text{Re } \zeta^n + |\zeta^n|^2}{n}$ and this is in turn equal to

$$\frac{1-4 \sin^2 \frac{1}{2}}{n} + \frac{(2 \sin \frac{1}{2} - \text{Re } \zeta^n)^2 + (\text{Im } \zeta^n)^2}{n} \geq \frac{1-4 \sin^2 \frac{1}{2}}{n}$$

which proves I with $C_2 = \dfrac{1-4 \sin^2 \frac{1}{2}}{16\pi}$

II. $\|D\| \geq n - \dfrac{c_1}{n}$

This time we are required to construct a P(z) of degree n for which P(=) = 0, $\|P\| = 1$, and $\|P'\| \geq n - \dfrac{c_1}{n}$. We shall restrict ourselves to odd n (as it will be clear that the polynomial zP(z) will then serve for the next higher degree) and we shall produce a P(z) with P(1) = 0, $\|P\| = 1$, $|P'(-1)| \geq n - \dfrac{c_1}{n}$.

We choose namely the P(z) which satisfies the identity $|P(z)|^2 + \dfrac{1}{(n+1)^2} \left|\dfrac{1-z^{n+1}}{1-z}\right|^2 \equiv 1$ all along $|z| = 1$ and which has all its zeros in $|z| \leq 1$.

By comparing leading terms it is clear that $P(z)$ is exactly of degree $n$. Also it satisfies $P(1) = 0$, $\|P\| = 1$ and so it only remains to get the proper estimate for $P'(-1)$. Equivalently since obviously $|P(-1)| = 1$ it suffices to estimate $\frac{P'(-1)}{P(-1)}$. So note that $\log |z^n P(\frac{1}{z})|$ is harmonic in $|z| < 1$ and has the boundary values $-\frac{1}{2}\phi(\theta)$,

$$\phi(\theta) = \log\left(1 - \frac{\sin^2 \frac{n+1}{2}\theta}{(n+1)^2 \sin^2 \frac{\theta}{2}}\right)^{-1}, \text{ at } z = e^{i\theta}.$$

Thus Poisson's integral formula gives

$$\log\left(z^n P(\tfrac{1}{z})\right) = \frac{1}{4\pi}\int_{-\pi}^{\pi} \frac{z+e^{i\theta}}{z-e^{i\theta}}\phi(\theta)\,d\theta \quad \text{for} \quad |z| < 1,$$

(the additive imaginary constant being chosen as 0).

Differentiating and setting $z = -1/w$ gives

$$-w\frac{P'(-w)}{P(-w)} = n - \frac{w}{2\pi}\int_{-\pi}^{\pi} \frac{e^{i\theta}}{(1+we^{i\theta})^2}\phi(\theta)\,d\theta \quad \text{for} \quad |w| > 1.$$

We must show this integral is $0(\frac{1}{n})$ as $w \to 1$ and we show in fact that this is true for all positive $w$. So assume $w > 1$ and proceed by splitting this integral into

$$I_1 = \int_{-\pi/2}^{\pi/2} \frac{e^{i\theta}}{(1+we^{i\theta})^2}\phi(\theta)\,d\theta \quad \text{and}$$

$$I_2 = \int_{\pi/2}^{3\pi/2} \frac{e^{i\theta}}{(1+we^{i\theta})^2}\phi(\theta)\,d\theta = -\int_{-\pi/2}^{\pi/2} \frac{e^{i\theta}}{(1-we^{i\theta})^2}\phi(\theta+\pi)\,d\theta$$

Now we have $\left|\frac{e^{i\theta}}{(1+e^{i\theta}w)^2}\right| \le 1$ on $(-\pi/2, \pi/2)$ and so

$$|I_1| < \int_{-\pi/2}^{\pi/2}\phi(\theta)\,d\theta = \int_{-\pi/2}^{\pi/2}\log\left(1 - \frac{\sin^2\frac{n+1}{2}\theta}{(n+1)^2\sin^2\frac{\theta}{2}}\right)^{-1} d\theta =$$

$$= \frac{4}{n+1}\int_0^{\pi(n+1)/4}\log\left(1 - \frac{\sin^2 t}{(n+1)^2\sin^2 t/(n+1)}\right)^{-1} dt\ .$$

To estimate this integral we split it at $\frac{\pi}{2}$. On $(0,\frac{\pi}{2})$ we have

$$\frac{\sin^2 t}{(n+1)^2 \sin^2 \frac{t}{n+1}} \le \frac{\sin^2 t}{2^2 \sin^2 \frac{t}{2}} = \cos^2 \frac{t}{2} = 1 - \sin^2 \frac{t}{2} \le 1 - \frac{t^2}{\pi^2}$$

so that

$$\int_0^{\pi/2} \log\left(1 - \frac{\sin^2 t}{(n+1)^2 \sin^2 \frac{t}{n+1}}\right)^{-1} dt \le \int_0^{\pi/2} \log \frac{\pi^2}{t^2}\, dt \; .$$

On $(\frac{\pi}{2}, \frac{\pi}{4}(n+1))$ we have $\sin^2 \frac{t}{n+1} \ge (\frac{2}{\pi})^2 \cdot \frac{t^2}{(n+1)^2}$ and $\sin^2 t \le 1$ so that

$$\int_{\pi/2}^{\frac{\pi}{4}(n+1)} \log\left(1 - \frac{\sin^2 t}{(n+1)^2 \sin^2 \frac{t}{n+1}}\right)^{-1} dt \le \int_{\pi/2}^{\infty} \log\left(1 - \frac{\pi^2}{4t^2}\right)^{-1} dt .$$

As for $I_2$ we have

$$\left|\frac{e^{i\theta}}{(1-e^{i\theta}w)^2}\right| \le \frac{1}{|1-e^{i\theta}|^2} = \frac{1}{4 \sin^2 \frac{\theta}{2}} \le \left(\frac{\pi}{2\theta}\right)^2 ,$$

and so since n is odd

$$|I_2| \le \frac{\pi^2}{4} \int_{-\pi/2}^{\pi/2} \frac{1}{\theta^2} \log\left(1 - \frac{\sin^2 \frac{n+1}{2}\theta}{(n+1)^2 \cos^2 \frac{\theta}{2}}\right)^{-1} d\theta$$

$$< \frac{\pi^2}{4} \int_{-\pi/2}^{\pi} \frac{1}{\theta^2} \log\left(1 - \frac{\sin^2 \frac{n+1}{2}\theta}{4 \cdot \frac{1}{2}}\right)^{-1} d\theta$$

$$= \frac{\pi^2}{2(n+1)} \int_{-\pi(n+1)/4}^{\pi(n+1)/4} \frac{1}{t^2} \log\left(1 - \frac{\sin^2 t}{2}\right)^{-1} dt$$

$$< \frac{\pi^2}{2(n+1)} \int_{-\infty}^{\infty} \frac{1}{t^2} \log\left(1 - \frac{\sin^2 t}{2}\right)^{-1} dt \; .$$

Combining these three estimates does indeed establish II. Our results can be generalized to the case of k zeros at the kth roots of unity and, amusingly enough, the answer then turns out to be $n - \frac{ck^2}{n}$. We briefly outline the proofs.

A. $\|D\| \geq n - \frac{c_3 k^2}{n}$.

Simply choose as polynomial $P(z^k)$ where $P$ is the previously defined polynomial, but of degree $\frac{n}{k}$. The resultant bound becomes

$$\left\|\frac{dP}{dz}(z^k)\right\| = k\left\|P'\right\| \geq k\left(\frac{n}{k} - C_1\frac{k}{n}\right) = n - C_1\frac{k^2}{n}.$$

B. $\|D\| \leq n - C_4\frac{k^2}{n}$.

Again use the previous formula for $P'(\zeta)$ but notice that this time there is a zero of $P$ within $\frac{A}{k}$ of $\zeta$. Hence there is a savings commensurate with $\int_I \frac{1}{|\zeta-z|^2}$, $I$ an interval of length $\frac{1}{n}$ and $|\zeta-z|$ of size $\frac{1}{k}$. A savings, therefore, of size $\frac{k^2}{n}$.

## LECTURE II

The remarkable behavior of $|x|$ with regard to rational approximation has led to the general belief that functions enjoy good rational function approximation when they have smoothness at "most points", but not necessarily at all points. Led by this belief and the, well known fact, that Lip 1 functions are differentiable almost everywhere, I made the conjecture that for every single $f(x) \subset \text{Lip } 1$, $R_n(f) = o(\frac{1}{n})$.

This conjecture was finally proved by the Bulgarian mathematician V. Popov. His proof was quite beautiful and very brilliant but, perhaps because of its brilliance, did not seem to lend itself easily to generalizations. Later on I was able to give a much simpler and more direct proof of Popov's theorem. Apparently it is this proof that does lead to important generalizations. In mathematical slang these say that: to almost everywhere differentiable functions, rationals converge infinitely faster than do polynomials. In other rough terms we may state this as the inequality $R_{o(n)}(f) \le E_n(f)$. Of course the o(n) depends very delicately and crucially on the particular f in question, but the result definitely asserts that in each particular case rational approximation is "infinitely cheaper" than polynomial approximation.

Even the requirement of being a.e. differentiable is usually more than is needed. What is needed, again in our usual slang, is that the almost everywhere smoothness be better than the everywhere smoothness. Since we plan to publish the details of this general result at some later date let us not now even bother with a precise formulation, but simply record a typical example, the Lip $\alpha$ case of course.

Theorem: Suppose that $f(x) \in \text{Lip}\,\alpha$, $0 < \alpha < 1$, while, for almost all x, $f(x+h)-f(x) = o(h^\alpha)$ as $h \to 0$, then $R_n(f) = o(\frac{1}{n^\alpha})$.

Proof: As in our aforementioned proof of the Lip 1 theorem our assault will consist of four steps.

A. A polygonal approximation $\Pi$, will be made of f.

B. A polynomial approximation, P, will be made of f.

C. Using the same knots as $\Pi$ the polygonal fit, $\Pi_1$ will be made to P.

D. A rational approximation R, will be made to $\Pi - \Pi_1$.

These done, the choice of rational approximator is taken as R + P.

We now outline these four steps.

A. The finitistic form of our hypothesis can be construed as saying that for each $\varepsilon > 0$ and all large N, for all x except a set, S, of measure $\varepsilon$, we have $|f(x+h) - f(x)| < \varepsilon |h|^{\alpha}$ as long as $|h| \leq \frac{1}{N}$.

Now break the interval $[-1,1]$ into the usual 2N subintervals of length $\frac{1}{N}$ and call such a subinterval bad if it is contained in S, otherwise call it good. Next break the bad subintervals (at most $N\varepsilon$ in number) into $[\frac{1}{\varepsilon}]$ equal subintervals which we then call the small subintervals.

Note that we have now obtained at most 2N good intervals and also $\leq \varepsilon N[\frac{1}{\varepsilon}] \leq N$ small intervals. The endpoints of all these intervals are now taken as the knots for the linear interpolant to f and this is what we choose as $\Pi$.

We record the properties of this $\Pi$, namely

$$|f - \Pi| \leq C(\frac{\varepsilon}{N})^{\alpha} \qquad (1.)$$

(This is true on the good invervals since the $\varepsilon$ factor there is even smaller than the $C \cdot \varepsilon^{\alpha}$ here. Also it is true on small intervals since their lengths are less than $\frac{2\varepsilon}{N}$).

The knots form $\leq$ 2N intervals of length $\frac{1}{N}$ and $\leq$ N intervals of length $\frac{1}{N[\frac{1}{\varepsilon}]}$ (2.)

B. Choose P as the Jackson polynomial of degree M to f (on a slightly extended interval, to avoid endpoint troubles). We record its properties as

$$\| f-P \| \leq \frac{C}{M^{\alpha}} \qquad (3.)$$

$$\| P'' \| \leq CM^{2-\alpha} \qquad (4.)$$

(4. is obtained directly from the estimates on the Jackson kernel).

C. Pick $\Pi_1$ as the linear interpolator to P using the very same knots as those of $\Pi$. Because of the bound on P" and the fact that the intervals have lengths bounded by $\frac{1}{N}$ we deduce that

$$\| \Pi_1 - P \| \leq \frac{CM^{2-\alpha}}{N^2} \tag{5.}$$

D. To produce the rational approximation to the 3N-gon $\Pi-\Pi_1$ we note that it decomposes into 3N triangles and a triangle is basically a graph of $|x|$ and so we inherit the construction from the old $|x|$ one! (The details are given on pages 38,39 of the Regional Conference Series in Mathematics, number 41) At any rate this gives us

$$\| \Pi-\Pi_1-R \| \leq \frac{C}{Mm^3}, \tag{6.}$$

$$\deg R \leq 5Nm \tag{7.}$$

We may now just sit back and reap the harvest. Namely we have

$$\| f-(P+R) \| \leq \|f - \Pi\| + \| \Pi_1-P \| + \| \Pi-\Pi_1-R \|$$

$$\leq C(\frac{\varepsilon}{N})^{\alpha} + C\frac{M^{2-\alpha}}{N^2} + \frac{C}{Mm^3} \text{ (by 3., 5., and 6.)}$$

while $\deg(P+R) \leq 5Nm + M$.

So if we simply choose $N = \sqrt{\varepsilon}$, $M = n\varepsilon$, $m = \frac{1}{\sqrt{\varepsilon}}$ (or the greatest integers thereof) we complete the proof.

## LECTURE III. EXACT INEQUALITIES RELATED TO CERTAIN OPTIMAL RECOVERY PROBLEMS

To contrast with the crude bounds and estimates given in the last two lectures we will now present some sharp and best possible inequalities. These have their origins in two problems arising from the theory of optimal recovery. In rough terms we may describe them as the following two tasks:

(1.). From bounds on the real part of a polynomial in the unit disc, derive bounds on the absolute value of that polynomial.

(2.) From bounds on the k-th derivative of an analytic function together with the location of certain of its zeros, derive bounds on that function.

Problem (1.) stems directly from some work of T.J.Rivlin while problem (2.) arises in the research of C. Micchelli and S. Fisher.

So let us begin with problem (1.) and state the following best possible inequality:

If P(z) is an n-th degree polynomial with P(0) = 0, then

$$\max_{|z| \leq 1} |P(z)| \leq n \max_{|z| = 1} \operatorname{Re} P(z).$$

Note that the restriction P(0) = 0 is vital since the addition of a large imaginary constant doesn't change Re P(z) while it enormously changes |P(z)|. Also note that we only require a one-sided bound on the real part. We will write the result to emphasize this fact, and also to indicate an improvement. Thus we estimate the $\ell_1$ norm rather than only the sup norm. This is the following:

<u>Theorem 1</u>: Suppose $\operatorname{Re}(a_1 z + a_2 z^2 + \ldots a_n z^n) \geq -1$ all along $|z| = 1$ then $|a_1|+|a_2|+\ldots|a_n| \leq n$.

To see that this is best possible we need only look at the Fejer kernel. Thus we have

$$0 \leq |1+z+\ldots z^n|^2 = n+1+n(z+\bar{z})+(n-1)(z^2+\bar{z}^2)+\ldots 1(z^n+\bar{z}^n)$$

and, on |z| = 1, this is equal to $n+1+2\operatorname{Re}(nz+(n-1)z^2+\ldots z^n)$.

In short $\mathrm{Re}\left(\frac{2n}{n+1}z+2\frac{(n-1)}{n+1}z^2 +.. \frac{2}{n+1}z^n\right) \geq -1$, while, of course $\frac{2n}{n+1} + \frac{2(n-1)}{n+1} +...+ \frac{2}{n+1} = n$. S. Fischer pointed out that this Fejer kernel and its rotates are, in fact, the only cases of equality and this will emerge also from our proof.

Before that, however, it may prove interesting to note that the weaker inequality $|a_1|+|a_2|+ \; |a_n| \leq 2n$ follows immediately from the well known fact about the Caratheodory class to the effect that each $|a_k| \leq 2$. Now we give our

<u>Proof</u>: Writing $z = e^{i\theta}$ we recognize $1+\mathrm{Re}(a_1z+..a_nz^n)$ as a non-negative n-th degree trigonometric polynomial in $\theta$. As such it has a representation as $|\alpha_0+\alpha_1z+..\alpha_nz^n|^2$. Equating coefficients, then, gives $|\alpha_0|^2+|\alpha_1|^2+..|\alpha_n|^2 = 1$ and also for $k = 1,2,...n$, $a_k = 2(\overline{\alpha}_0\alpha_k+ \overline{\alpha}_1\alpha_{k+1}+..)$. Hence for appropriate $\zeta_k$ with $|\zeta_k| = 1$ we may write

$$|a_k| = 2\zeta_k(\overline{\alpha}_0\alpha_k+\overline{\alpha}_1\alpha_{k+1}+...) = 2\mathrm{Re}\; \zeta_k(\overline{\alpha}_0\alpha_k+\overline{\alpha}_1\alpha_{k+1}+...).$$

Now consider the double sum $\Sigma_{i,j}|\alpha_i-\zeta_{j-i}\alpha_j|^2$ taken over all i,j with $0 \leq i < j \leq n$. This is non-negative and it expands to

$$\sum_{i,j}(|\alpha_i|^2+|\alpha_j|^2)-\sum_k 2\mathrm{Re}(\zeta_k(\overline{\alpha}_0\alpha_k+\overline{\alpha}_1\alpha_{k+1}+..))$$

$$= \sum_{i,j}(|\alpha_i|^2+|\alpha_j|^2 -\sum_k |a_k| = n\sum_v |\alpha_v|^2-\sum_k|a_k|=n-\sum_k |a_k|$$

and this proves our inequality. A closer look shows in fact that equality can only occur if each of the quantities $\alpha_i-\zeta_{i-j}\alpha_j$ are 0. But this means that $\frac{\alpha_i}{\alpha_j}$ is a function of i-j alone, and it is a simple exercise to deduce from this that

$\alpha_i$ is an exponential function. Also $\frac{\alpha_i}{\alpha_j} = \sigma_{i-j}$ is of modulus 1 and so it is an imaginary exponential and Fischer's assertion that we have a Fejer kernel is verified.

Problem 2: We begin with a "baby" example. Suppose namely that $\| f' \| \leq 1$ ($\| \cdot \|$ here means sup norm in the unit disc) and also assume that $f(0) = 0$. How large can $f(z)$ be? (for a given z in the unit disc).

This is answered post haste by the simple observation that $f(z) = \int_0^z f'(\zeta)d\zeta \ll |z| \, \| f'(\zeta) \| \leq |z|$, a best possible result since the identity function satisfies our restrictions. But however trivial this conclusion is it can be rephrased so as to appear very respectable. Namely it says that the operator $I + z \frac{d}{dz}$ is a dilation on $H^\infty$. In less fancy terms this can also be written as the inequality $\|f(z) + zf'(z)\| \geq \| f(z) \|$. Problem 2 is thus seen to be intimately related to the question of identifying those differential operators which are dilations.

To better understand our approach to this form of the problem let us look at another, this time less babyish, example. So suppose that $\|f'\| \leq 1$ but now suppose that $f(1) = f(-1) = 0$, and again ask for the correct bound for $f(z)$. An educated guess seems to be $f(z) << \frac{z^2-1}{2}$, but now there is no trivial integration proof. The translation into operator language this time is that $z + \frac{z^2-1}{2} \frac{d}{dz}$ is a dilation on $H^\infty$, and into ordinary language says $\| f(z) + \frac{z^2-1}{2z} f'(z) \| \geq \| f(z) \|$.

To prove this inequality we may clearly restrict ourselves to very smooth $f(z)$, say even to polynomials. So let $\zeta$, $|\zeta|=1$, be a maximizing point for $|f(z)|$. By calculus, then, we have $\frac{d}{d\theta}\log|f(e^{i\theta})| = 0$ at $e^{i\theta} = \zeta$, or

$i\zeta\frac{d}{d\zeta}$ Re $\log f(\zeta) = 0$, or Re $i\zeta\frac{f'(\zeta)}{f(\zeta)} = 0$, or $\zeta\frac{f'(\zeta)}{f(\zeta)} =$ real.

But by viewing the derivative in the inward normal direction we conclude, moreover, that this real number must be non-negative. (Indeed by an old chestnut of Erdös' $f'(\zeta) \neq 0$ unless f is a constant so it must then be positive). At any rate we have $\zeta\frac{f'(\zeta)}{f(\zeta)} \geq 0$ and since $\frac{\zeta^2-1}{2\zeta^2} = \frac{1-\bar{\zeta}^2}{2}$ has non-negative real part we deduce that $\mathrm{Re}\left(\frac{\zeta^2-1}{2\zeta^2}\zeta\frac{f'(\zeta)}{f(\zeta)}\right) \geq 0$.

Hence $\left|1+\frac{\zeta^2-1}{2\zeta}\frac{f'(\zeta)}{f(\zeta)}\right| \geq \mathrm{Re}\left(1+\frac{\zeta^2-1}{2\zeta}\frac{f'(\zeta)}{f(\zeta)}\right) > 1$ so that

$|f(\zeta) + \frac{\zeta^2-1}{2\zeta}f'(\zeta)| \geq |f(\zeta)| = \|f\|$ and á fortiori

$\|f(z) + \frac{z^2-1}{2z}f'(z)\| \geq \|f\|$ as required.

A closer look at this proof shows that in fact we can obtain more generally

<u>Theorem 2</u>: If Re $\frac{u(z)}{z} \geq 0$ all along $|z| = 1$ (u isn't necessarily analytic) then $\|f(z) + u(z)f'(z)\| \geq \|f(z)\|$.

If u happens to be analytic then $1+u(z)\frac{d}{dz}$ is actually a dilation and the result can be compounded, the product of dilations being a dilation. For example from the identity

$$\left(\frac{d}{dz}\right)^3(z-\alpha)(z-\beta)(z-\gamma)\cdot = \left(3+(z-\alpha)\frac{d}{dz}\right)\left(2+(z-\beta)\frac{d}{dz}\right)\left(1+(z-\gamma)\frac{d}{dz}\right)$$

we conclude that $\frac{1}{6}\left(\frac{d}{dz}\right)^3(z-\alpha)(z-\beta)(z-\gamma)\cdot$ is a dilation when $\alpha,\beta,\gamma$ lie in the unit disc. As a (best possible) inequality this reads:

If f(z) vanishes at $\alpha,\beta,\gamma$ inside the unit disc and if $\|f'''\| \leq 1$ then at every z in this disc we have

$f(z) << \frac{(z-\alpha)(z-\beta)(z-\gamma)}{6}$.

Similarly scores of other inequalities can be obtained, but we are still unable to settle the Miccelli-Fischer conjecture which is as follows:

Let $\phi(z)$ vanish at $\alpha_1, \alpha_2, \ldots \alpha_{r+m}$ which lie inside the unit disc and be such that $\phi^{(r)}(z)$ is an m-th degree Blaschke product. If $f(z)$ vanishes at $\alpha_1, \alpha_2, \ldots, \alpha_{r+m}$ and $\| f^{(r)} \| \leq 1$ then $f(z) << \phi(z)$.

(The case of $\phi(z)$ = polynomial is covered by our method however).

## LECTURE IV. RATIONAL APPROXIMATION AND THE $L^1$ MODULUS OF CONTINUITY

In our second lecture we dedicated ourselves towards verifying the thesis that smoothness almost everywhere implies good approximation by rational functions. It has long been known, namely, that proximity to rational functions does not entail smoothness at every point but only perhaps at most points. But 'almost everywhere' is not the only concept of "nearly". A function can be "nearly" smooth by being smooth on the average. Long ago Turán had the idea that while the ('uniform') modulus of continuity told the story for polynomial approximation, the analogous story for rational approximations should be told by the $L^1$ modulus of continuity. Indeed we had many conversations about such a possibility but we ended up talking each other out of it. After all, we reasoned, the function $|x|$ has no stupendous smoothness in the $L^1$ sense any more than in the $L^\infty$ sense and so there really couldn't be any such connection.

We were wrong!

The fact, as is shown by $|x|$, that there is no implication of higher smoothness from the hypothesis of "higher" approximibility doesn't mean that ordinary type smoothness couldn't be equivalent to ordinary approximibility. Indeed the relation between Lip $\alpha$ and $\frac{1}{n^\alpha}$ does remain intact! Roughly speaking our present result is that $R_n(f)$ is of the "correct" order of $\omega_{L^1}(f\ ;\ \frac{1}{n})$. We say "roughly speaking" because there are a few extra logarithms in our estimates. (But, as K. Roth once said, "What's a few logarithms between friends?")

Notice, however, that implicit in our assertion of the equivalence or even rough equivalence of $R_n(f)$ with $\omega_{L^1}(f\ ;\frac{1}{n})$ is the assertion of lower bounds for rational approximation. Something which always seemed to call for those marvelous ad-hoc arguments! Of course, by contrast, what made lower bounds so easy in the polynomial case was the Bernstein derivative bound for polynomials. Thus a polynomial cannot make "sharp turns" without itself becoming exposively large. The simple example $\frac{\varepsilon}{x+i\varepsilon}$ shows that rational functions can. This function is bounded by 1 but its derivative $-\frac{\varepsilon}{(x+i\varepsilon)^2}$ is $\frac{1}{\varepsilon}$ at 0. Nevertheless there is something that a rational function cannot do. It cannot have a large derivative in the $L^1$ norm without somewhere being very large. Thus we have the following Bernstein type result which will be the basis of our lower bounds.

Theorem 1: Let $r_n(x)$ be an n-th degree rational function, then

$$\|r_n'(x)\|_{L^1} \leq 4n\|r_n(x)\|_{L^\infty}.$$

Proof: Immediate upon noting that this $L^1$ norm is the total variation of $r_n$. But for $n > 0$ an n-th degree rational function can only have $2(n-1)$ turning points and so, even counting endpoints, can only make the trip from peak to trough $2n-1$ times. Each such trip counts for at most $2 \cdot \| r_n \|$ in variation and so the total variation is at most $2(2n-1)\| r_n \|$ and if we weaken this to $4n\| r_n \|$ it also covers the case of $n = 0$.

Herein we have the tool for lower bounds and we will illustrate its use in a moment. But note how worthless this is for the higher smoothness case - for unlike the Bernstein derivative theorem for polynomials this cannot be iterated. We this time goes from $L^\infty$ to $L^1$ and so this does not give anything for the second derivative. So much for the Newman-Turán objections!

To illustrate the lower bound technique then, let us treat the case of functions which can be fit within $\frac{1}{n}$.

Theorem 2: If $R_n(f) = 0(\frac{1}{n})$ then

$$\omega_{L^1}(f\ ;\ \delta) = 0(\delta \log \frac{1}{\delta})$$

Proof: The hypothesis tells us that $f(x)$ has a series expansion $\sum_{k=1}^{\infty} \frac{1}{2^k} r_{2^k}(x)$ where the $r_{2^k}(x)$ are rational functions of degree $2^k$ and with uniformly bounded sup norms. We may then normalize so that all $\| r_{2^k}(x) \| \le 1$.

Now we have

$$f(x+\delta)-f(x) = \sum_{k=1}^{\infty} \frac{1}{2^k} [r_{2^k}(x+\delta)-r_{2^k}(x)]$$

so that

$$\int_0^{1-\delta} |f(x+\delta)-f(x)|dx \le \sum_{k=1}^{\infty} \frac{1}{2^k} \int_0^{1-\delta} r_{2^k}(x+s)-r_{2^k}(x)|dx$$

$$\le \sum_{k=1}^{N} \frac{1}{2^k} \int_0^{1-\delta} |r_{2^k}(x+\delta)-r_{2^k}(x)|dx + \sum_{k=N+1}^{\infty} \frac{1}{2^k} \cdot 2$$

$$= \sum_{k=1}^{N} \frac{1}{2^k} \int_0^{1-\delta} |r_{2^k}(x+\delta)-r_{2^k}(x)|dx + \frac{2}{2^N} .$$

Next observe that $\int_0^{1-\delta} |r_{2^k}(x+\delta)-r_{2^k}(x)|dx$

$$\le \int_0^{1-\delta} \int_x^{x+\delta} |r'_{2^k}(t)|dt\ dx = \int_0^1 \int_{(t-\delta)_+}^{t} dx|r'_{2^k}(t)|dt$$

$\le \delta \int_0^1 |r'_{2^k}(t)|dt \le \delta \cdot 4 \cdot 2^k$ by Theorem 1. Therefore we obtain $\omega_{L^1}(f;\delta) \le \delta \cdot 4 \cdot N + \frac{2}{2^N}$ and the proof is completed by the choice $N = [\log_2 \frac{1}{\delta}]$.

We turn now to the upper bound and we illustrate with the pivotal case, namely that of $\omega_{L^1}(f;\delta) \le \delta$, f continuous. This class can be equivalently described as the continuous functions of total variation $\le 1$. Thus at the cost of a factor of 2 we may restrict ourselves to continuous monotone f(x) with $f(0) = 0$, $f(1) = 1$

We utilize the technique of "drawing pictures" which was portrayed in Lecture II. In essence we say that at a cost of a factor of $\log^3 n$ or so we can do n-th degree rational approximation whenever we can do polygonal approximation by n-gons. Thus our task will be complete once we obtain an n-gon within

$\frac{1}{r}$ of our function f(x). But this job is easy! Simply break up the graph of y = f(x) along the y-axis. In other words pick the knots $x_k$ by $f(x_k) = \frac{k}{n}$, k = 0,1,2,...n, and draw the linear interpolant. The verification is trivial.

QUADRATURE FORMULAE BASED ON SHAPE PRESERVING INTERPOLATION[1]

David Hill
Eli Passow[2]

Department of Mathematics
Temple University
Philadelphia, Pennsylvania

## I. INTRODUCTION

Suppose $\{(x_i,y_i)\}_{i=0}^{N}$ , $x_0 < x_1 <\ldots< x_N$, is a set of data, where we presume that $y_i = g(x_i)$ for some unknown function $g(x)$. We wish to estimate $\int_{x_0}^{x_N} g(x)dx$. Suppose that we have some information about the monotonicity and convexity of g. In this case, we will outline a method founded upon shape preserving interpolation, a subject of much recent interest (1-5). (See (5) for additional references.)

Since such interpolants adhere well to the data in the sense that they do not display undesirable oscillations or inflections, it is natural to consider quadrature formulae based upon them. In this paper such formulae are developed using one of these methods. The outcome is somewhat sketchy and needs, in particular, a good deal of numerical work, as well as additional error analysis. Nevertheless, the results seem promising.

## II. BACKGROUND MATERIAL

Let $S_i = (y_i-y_{i-1})/(x_i-x_{i-1})$ , $i = 1,2,\ldots, N$, and suppose that $0 < S_1 < S_2 <\ldots< S_N$. The data $\{(x_i,y_i)\}_{i=0}^{N}$ are thus increasing and convex. In (4) a technique was devised for finding an increasing,

[1]Dedicated to Professor Donald Newman on the occasion of his 50th birthday.

[2]On leave at Technion, Haifa, Israel




ISBN 0-12-780650-4

convex spline interpolant of these data. The discussion which follows is a special case of these results.

Let $\Delta = \{x_0, x_1, \ldots, x_N\}$ and denote by $S_2(\Delta)$ the set of all quadratic splines with knots at $x_i$, $i = 1,2,\ldots, N-1$. Let $\bar{x}_i = (x_{i-1} + x_i)/2$, $i = 1,2,\ldots, N$.

Definition: The numbers $\{t_i\}$, $i = 1,2,\ldots, N$ are said to be $\frac{1}{2}$-admissible for the data $\{(x_i,y_i)\}$ if the piecewise linear function $L(x)$ generated by the points

$$(x_0,y_0),\ (\bar{x}_1,t_1),\ (\bar{x}_2,t_2),\ldots,(\bar{x}_N,t_N),\ (x_N,y_N)$$

passes through the points $(x_i,y_i)$, $i = 1,2,\ldots, N-1$, and is increasing and convex.

Theorem A (4): There exist $\frac{1}{2}$-admissible numbers for the data $\{(x_i,y_i)\}$ if and only if there exists an increasing, convex $f \in S_2(\Delta)$ satisfying $f(x_i) = y_i$, $i = 0,1,\ldots, N$.

The function f is obtained in the following manner. Let $L_i(x)$ be the restriction of $L(x)$ to the interval $[x_{i-1},x_i]$, $i = 1,2,\ldots, N$, and let $q_i(x)$ be the Bernstein polynomial of degree two of $L_i(x)$. Specifically,

$$q_i(x)=\frac{1}{(\Delta x_i)^2}\left[L_i(x_{i-1})(x_i-x)^2+2L_i(\bar{x}_i)(x-x_{i-1})(x_i-x)+L_i(x_i)(x-x_{i-1})^2\right] \quad (1)$$

where $\Delta x_i = x_i-x_{i-1}$. We define $f(x)$ on $[x_0,x_N]$ by letting $f(x) = q_i(x)$ on $[x_{i-1},x_i]$, $i = 1,2,\ldots, N$. It is shown in (4) that $f \in S_2(\Delta)$.

An important question that arises is whether a set of data $\{(x_i,y_i)\}$ has a $\frac{1}{2}$-admissible set of numbers $\{t_i\}$. An algorithm has been developed in (2) which tests for the existence of such a set and, in the affirmative case, actually calculates admissible numbers $\{t_i\}$. The algorithm is as follows. Let $m_0 = 0$ and $M_0 = S_1$. For $i = 1,2,\ldots, N-1$ define

$$m_i = 2S_i - M_{i-1}$$

$$M_i = \min(S_{i+1}, 2S_i-m_{i-1}).$$

Theorem B (2): There exists a $\frac{1}{2}$ - admissible set of numbers $\{t_i\}$ for the data $\{(x_i,y_i)\}$ if and only if $m_i \le S_{i+1}$, $i=1,2,\ldots, N-1$.

If the condition of Theorem B holds, then $t_N$ may be chosen to be any number between

$$a_N = y_{N-1} + m_{N-1}(\bar{x}_N - x_{N-1}) \tag{2}$$

and

$$b_N = y_{N-1} + M_{n-1}(\bar{x}_N - x_{N-1}). \tag{3}$$

The numbers $t_{N-1}, t_{N-2},\ldots, t_1$ are then uniquely determined recursively by the choice of $t_N$.

## III. A QUADRATURE FORMULA

Let us assume that $\frac{1}{2}$ - admissible numbers exist for $\{(x_i,y_i)\}_{i=0}^{N}$. (For what happens otherwise, see Section VI.) A quadrature formula is obtained by integrating the interpolant (1). A calculation shows that

$\int_{x_0}^{x_N} g(x)dx \sim \sum_{i=1}^{N} \int_{x_{i-1}}^{x_i} q_i(x)dx = \sum_{i=1}^{N} \frac{\Delta x_i}{3} [g(x_{i-1}) + t_i + g(x_i)]$. Thus

$$\int_{x_0}^{x_N} g(x)dx \sim \sum_{i=1}^{N} \frac{\Delta x_i}{3} [y_{i-1} + t_i + y_i]. \tag{4}$$

If $\Delta x_i = h$, $i = 1,2,\ldots, N$, we have

$$\int_{x_0}^{x_N} g(x)dx \sim \frac{h}{3} [y_0 + 2 \sum_{i=1}^{N-1} y_i + \sum_{i=1}^{N} t_i + y_N].$$

An interesting thing happens in this case when N is even. Let $\ell(x)$ be the line joining $(\bar{x}_i,t_i)$ and $(x_i,y_i)$. Then

$$\ell(x) = \left(\frac{y_i-t_i}{x_i-\bar{x}_i}\right) (x-x_i) + y_i = \frac{2}{h} (y_i-t_i) (x-x_i) + y_i.$$

Now $t_{i+1} = \ell(\bar{x}_{i+1}) = \frac{2}{h}(y_i - t_i)(\bar{x}_{i+1} - x_i) + y_i = 2y_i - t_i$, so that $t_i + t_{i+1} = 2y_i$. We thus obtain $\int_{x_{i-1}}^{x_{i+1}} g(x)dx = \frac{h}{3}[y_{i-1} + 4y_i + y_{i+1}]$, which is Simpson's rule. This fact is striking: even though completely different quadratics are used as interpolants the two methods agree! In Simpson, of course, we interpolate at $x_{i-1}, x_i$, and $x_{i+1}$, whereas here we use two quadratics, one on $[x_{i-1}, x_i]$, the other on $[x_i, x_{i+1}]$. Moreover, the result is independent of $\{t_i\}$. Recall that, in general, there is some freedom in the choice of $t_N$, which usually affects the outcome of the quadrature. In the equally spaced case, however, $t_N$ can be completely arbitrary.

## IV. ADDITIONAL FORMULAE

The method in (4) can actually be used to produce a whole class of quadrature formulae based on shape preserving interpolants. Specifically, it has been shown in (4) that there exists a set of $\frac{1}{2}$ - admissible numbers $\{t_i\}$ if and only if there exists an increasing, convex $f \in S_{2j}^j(\Delta)$ for all $j \geq 1$, where $S_{2j}^j(\Delta) = \{f \in C^j[x_0, x_N] : f \in P_{2j}$ on $(x_{i-1}, x_i)$, $i = 1, 2, \ldots, N\}$. Here $P_{2j}$ is the set of all algebraic polynomials of degree $\leq 2j$. The approach is essentially the same as outlined in Sections II and III, with the Bernstein polynomial of degree 2j taking the place of that of degree 2. Quadrature formulae are then obtained by integrating these interpolants. For example, using the fourth degree Bernstein polynomial one arrives at the formula

$$\int_{x_0}^{x_N} g(x)dx \sim \sum_{i=1}^{N} \frac{\Delta x_i}{10}[3y_{i-1} + 4t_i + 3y_i]. \qquad (5)$$

More generally, for $n = 2j$, we obtain instead of (1),

$$q_i(x) = \frac{1}{(\Delta x_i)^n} \sum_{k=0}^{n} L_i[x_{i-1} + \frac{k}{n}\Delta x_i] \binom{n}{k} (x - x_{i-1})^k (x_i - x)^{n-k}.$$

Now $\int_{x_{i-1}}^{x_i} (x-x_{i-1})^k (x_i-x)^{n-k} dx = B(k+1,n-k+1)(\Delta x_i)^{n+1} = \frac{(\Delta x_i)^{n+1}}{n+1} \frac{1}{\binom{n}{k}}$,

where $B(p,q)$ is the beta function. Hence $\int_{x_{i-1}}^{x_i} \binom{n}{k}(x-x_{i-1})^k (x_i-x)^{n-k} dx =$

$= \frac{(\Delta x_i)^{n+1}}{n+1}$. Also, $\sum_{k=o}^{n} L_i[x_{i-1}+\frac{k}{n}\Delta x_i] = \frac{1}{4}[(n+2)y_{i-1}+2nt_i+(n+2)y_i]$. Thus

we obtain the quadrature formula

$$\int_{x_0}^{x_N} g(x)dx \sim \sum_{i=1}^{N} \frac{\Delta x_i}{4(n+1)} [(n+2)y_{i-1} + 2n\, t_i + (n+2)y_i]. \tag{6}$$

We can also let $n \to \infty$. Since the Bernstein polynomials tend to $L(x)$ as $n \to \infty$, we are then using $\int_{x_0}^{x_N} L(x)dx$ as the approximation to the integral. We obtain

$$\int_{x_0}^{x_N} g(x)dx \sim \sum_{i=1}^{N} \frac{\Delta x_i}{4} [y_{i-1} + 2t_i + y_i]. \tag{7}$$

## V. A NUMERICAL EXAMPLE

The following example was considered in (2). Let $g(x) = 1/x^2$ on $[-2,.2]$, and use the data obtained at $x = -2,-1,-.3$ and $-.2$. In Table 1 we list the true value of $\int_{-2}^{-.2} g$, as well as the approximate values given by various methods. For the interpolatory cubic splines we have used the boundary values $S_0' = .75$, $S_3' = 138.89$ for #1 (these are the slopes of the first and last line segments), and $S_0' = .25$, $S_3' = 250$ for #2 (the values of $g'(x)$ at $x = -2$ and $-.2$, which are mysteriously assumed to be known). For these data there exists a $\frac{1}{2}$ - admissible set of numbers, and we can choose $t_3$ to be any number between $a_3 = 12.48$ and $b_3 = 12.52$. We have used $t_3 = 12.50$, and $n = 2,4,6, \infty$ in (6).

TABLE I.

| Method | Value |
|---|---|
| Trapezoidal | 6.67 |
| Cubic Lagrange interpolant | 14.89 |
| Natural cubic spline | 3.33 |
| Interpolatory cubic splines | |
| #1 | -5.94 |
| #2 | 3.14 |
| Shape preserving splines | |
| n = 2 | 5.34 |
| n = 4 | 5.06 |
| n = 6 | 4.96 |
| n = ∞ | 4.67 |
| Actual | 4.50 |

The value 4.67 obtained when $n = \infty$ is the closest to the actual value. In general, however, the optimal choice of n in (6) is not clear, nor is it obvious how best to select $t_N$. In this example the range of acceptable values for $t_3$ was limited, but in other cases the outcome may depend more strongly on this choice.

## VI. EXTENSIONS

Everything we have done till here is predicated upon the assumption that $\frac{1}{2}$ - admissible numbers exist for the data $\{(x_i, y_i)\}$. What do we do in the contrary case? We can then turn to a "point insertion algorithm", due to McAllister and Roulier (3). Their algorithm constructs a new data set containing the original one, has at most one addition knot between each pair $x_{i-1}$ and $x_i$, and for which a set of $\frac{1}{2}$ - admissible numbers exists. In this way, the quadrature formulae can be applied to <u>any</u> set of increasing, convex data. If the data are not increasing and convex throughout, then they may be divided into subintervals on which they are either increasing and convex, increasing and concave, decreasing and convex, or decreasing and concave, and use the quadrature formula (with the McAllister-Roulier modification) separately on each subinterval. The technique is thus universally applicable.

## VII. SUMMARY

The formulae we have presented give reasonable results in the sense that, say, a negative value can never be obtained from positive data, a phenomenon that can occur in Lagrange and interpolatory cubic spline methods. We thus feel confident in recommending our approach, especially in cases of sparse data, where a rough estimate of the integral will suffice. Error bounds for these formulae, however, have proved difficult to derive. This is due to the dependence of the method on the choice of $t_N$, which in turn determines $t_{N-1}, t_{N-2}, \ldots, t_1$. While an estimate of the error in the last interval can be obtained, it is hard to trace how the error (which depends on the $t_i$'s) propagates through the earlier intervals. This problem warrants additional study. It would also be interesting to derive quadrature formulae based on other shape preserving techniques. In particular, an investigation of the method of deBoor (1), who obtains a shape preserving cubic spline, could be productive.

In closing, it should be mentioned that some of these results for the equally spaced case have been obtained independently by D. F. McAllister and J. A. Roulier.

## REFERENCES


1. de Boor, C., "A Practical Guide to Splines", Springer-Verlag, New York, 1978, Chapter 16.
2. McAllister, D.F., Passow, E., and Roulier, J.A., *Math. Comp. 31*, 717-725 (1977).
3. McAllister, D.F., and Roulier, J.A., *Math. Comp. 32*, 1154-1162 (1978).
4. Passow, E., and Roulier, J.A., *SIAM J. Numer. Anal. 14*, 904-909 (1977).
5. Pruess, S., *Math. Comp. 33*, 1273-1281 (1979).

# OPTIMAL RECOVERY AMONG THE POLYNOMIALS

D. J. Newman

Department of Mathematics
Temple University
Philadelphia, Pennsylvania

T. J. Rivlin

Mathematical Sciences Department
Thomas J. Watson Research Center
Yorktown Heights, N. Y.

## I. INTRODUCTION

We recall the formulation of the notion of optimal recovery as given in Micchelli and Rivlin [4] (referred to as M-R henceforth). Let X be a linear space and Y and Z be normed linear spaces. K is a subset of X and U is a given linear operator from X into Z. We wish to estimate Ux for $x \in K$ using limited information about x. The information is provided by a linear operator, I, which maps X into Y. We assume that Ix for $x \in K$ is known within some error of limited size. That is, in our attempt to recover Ux we actually know some $y \in Y$ satisfying $\| Ix - y \| \leq \varepsilon$, for some given $\varepsilon \geq 0$. An <u>algorithm</u>, A, is any function from $IK + \varepsilon S$, where $S = \{y \in Y : \|y\| \leq 1\}$ with range in Z. Schematically we have




ISBN 0-12-780650-4

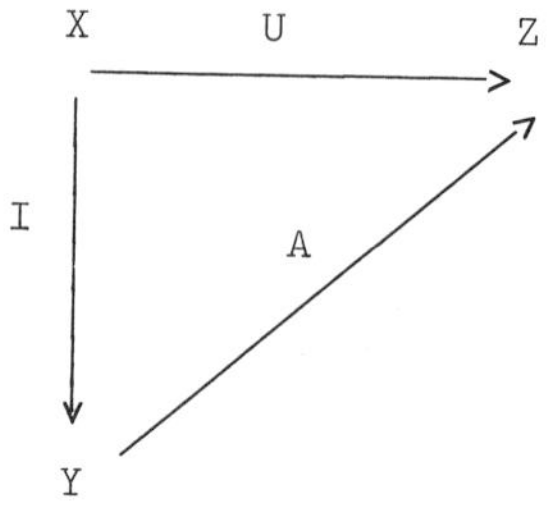


The algorithm A produces an error

$$E(A\ ;\ K\ ,\ \varepsilon) = \sup_{\left\{\substack{x \in K \\ \|Ix - y\| \le \varepsilon}\right.} \|Ux - Ay\| \tag{1}$$

$$E(K\ ,\ \varepsilon) = \inf_{A} E(A\ ;\ K\ ,\ \varepsilon) \tag{2}$$

is called the intrinsic error in the recovery problem and an A satisfying $E(A;K,\varepsilon) = E(K,\varepsilon)$ is called an optimal algorithm. Given X,Y,Z,U,I,K and ε we are interested in determining $E(K,\varepsilon)$ and finding optimal algorithms.

A useful tool in our work is the following. (Cf.M-R)

Proposition: If K is a balanced convex subset of X then

$$E = E(K\ ,\ \varepsilon) \ge \sup_{\left\{\substack{x \in K \\ \|Ix\| \le \varepsilon}\right.} \|Ux\| \ . \tag{3}$$

Let us illustrate these definitions with some examples.

Example 1. (Cf.M-R.) The Optimal Interpolation of Bounded Analytic Functions.

Let $X = H^{\infty}$, the set of bounded analytic functions in the unit disc $D: |z|<1$. Put $K = \{f \in X: \|f\| \le 1\}$ and $\varepsilon = 0$. Let $z_1,\ldots,z_n$ be given points of D and $Y = ¢^n$. Define I by $If = (f(z_1),\ldots,f(z_n))$ for $f \in X$, and suppose that $Z = ¢, \zeta \in D$ and $Uf = f(\zeta)$.

Put

$$B_n(z) = \prod_{j=1}^{n} \frac{z-z_j}{1-\bar{z}_j z},$$

then (3) implies that

$$E \geq |B_n(\zeta)|. \tag{4}$$

However, if $a_1,\ldots,a_n$ are defined by

$$Lf := f(\zeta) - \sum_{j=1}^{n} a_j f(z_j) := \frac{1}{2\pi i} \int_{|z|=1} \frac{B_n(\zeta)}{B_n(z)} \frac{1-|\zeta|^2}{1-z\bar{\zeta}} \frac{1}{z-\zeta} f(z)dz$$

then

$$|Lf| \leq |B_n(\zeta)| \frac{1}{2\pi} \int_0^{2\pi} \frac{1-|\zeta|^2}{|e^{i\theta}-\zeta|^2} d\theta = |B_n(\zeta)|.$$

Hence, if A denotes the algorithm

$$A : (f(z_1),\ldots f(z_n)) \rightarrow \sum_{j=1}^{n} a_j f(z_j) \tag{5}$$

we have $E(A;K,0) \leq |B_n(\zeta)| \leq E$, and in view of (2), A as defined in (5) is an optimal algorithm and $|B_n(\zeta)|$ is the intrinsic error in our recovery problem.

Example 2. The Optimal Interpolation of Subordinate Functions.

Now take X to be the set of all functions analytic in D. Y,Z,U,I and $\varepsilon(=0)$ are the same as in Example 1, however, g is a fixed element of X and $K=\{f\in X: |f(z)|\leq|g(z)|, z\in D\}$. Example 1 is the case $g(z) \equiv 1$.

If $g(\zeta)=0$ our problem is trivial. Suppose $g(\zeta) \neq 0$ and

$$g(z) = \prod_{j=1}^{m} (z-z_j)^{k_j} k(z), \quad k(z_j) \neq 0, \quad j=1,\ldots,n.$$

Suppose $m < n$. Put

$$b(z) = \prod_{j=m+1}^{n} \frac{z-z_j}{1-\bar{z}_j z},$$

then (3) yields

$$E(K,0) \geq |b(\zeta)\ g(\zeta)|. \tag{6}$$

Now $h(z) = f(z)/g(z)$ is analytic in D and satisfies $\|h\| \leq 1$. Using the result of Example 1, with sampling restricted to $z_{m+1},\dots,z_n$, we obtain $\alpha_{m+1},\dots,\alpha_n$ such that

$$|h(\zeta) - \sum_{j=m+1}^{n} \alpha_j h(z_j)| \leq |b(\zeta)|,$$

so that

$$|b(\zeta)| \geq \sup_{f \in K} \left| \frac{f(\zeta)}{g(\zeta)} - \sum_{j=m+1}^{n} \alpha_j \frac{f(z_j)}{g(z_j)} \right|$$

$$= \frac{1}{g(\zeta)} \sup_{f \in K} \left| f(\zeta) - \sum_{j=m+1}^{n} \left( \alpha_j \frac{g(\zeta)}{f(z_j)} \right) f(z_j) \right|.$$

Thus the algorithm

$$A_g : (f(z_1),\dots,f(z_n)) \rightarrow \sum_{j=m+1}^{n} \left( \alpha_j \frac{g(\zeta)}{g(z_j)} \right) f(z_j) \tag{7}$$

is optimal and $|b(\zeta)\ g(\zeta)|$ is the intrinsic error. $f \in K$ implies $f(z_j) = 0, j = 1,\dots,m$ and that is why this information is discarded in (7). If $m = n$ then $f(z_j) = 0$, $j = 1,\dots,n$ and the problem is easy, the optimal algorithm being $A : y \rightarrow 0$, and the intrinsic error $|g(\zeta)|$.

For example, suppose $g(z) = (1-z)^{-1}$. Then $m = 0$ and $E = |B_n(\zeta)/(1-\zeta)|$.

<u>Remark.</u> If

$$g(z) = \frac{\prod_{i=1}^{n} (1-\bar{z}_i z)}{1 - z\bar{\zeta}}$$

then polynomial interpolation is optimal.

The restriction to $x \in K$ in our general setting is needed, for if the set of allowable x is too large the intrinsic error becomes infinite and all algorithms are optimal. However, the worst case character of our formulation makes the analysis more tractable in cases when K is rich enough to contain distinctive "worst" elements (such as the Blaschke products above). Our aim here is to present some preliminary results for rather meager well-behaved sets, K, in particular, polynomials of fixed degree. These results could be called variations on some themes of I. Schur.

## 1. POLYNOMIALS ON AN INTERVAL

Let $P_n$ denote the set of polynomials of degree at most n with complex coefficients. For $p \in P_n$ put

$$\|p\| = \max_{0\leq x\leq 1} |p(x)|,$$

and $K_n = \{p \in P_n : \|p\| \leq 1\}$. I. Schur [6] in an improvement on A. Markov's inequality showed that if $p \in K_n$ satisfies $p(0) = 0$ then

$$\|p'\| \leq 2n^2 \cos^2 \frac{\pi}{4n}. \tag{8}$$

In view of (3), the optimal recovery of p' from the information $p \in K_n$ and $p(0)$ is suggested. That is: consider $X = P_n$, $Y = \mathbb{C}$, $Z = P_{n-1}$, $Up = p'$, $Ip = p(0)$, $K = K_n$ and $\varepsilon = 0$. According to (8) and (3) we have $E \geq 2n^2\cos^2(\pi/(4n))$. We want to determine an optimal algorithm for this problem. Let $T_n(x)$ denote the Chebyshev polynomial of degree n, and denote its extreme by $\eta_j = \cos(j\pi/n)$, $j = 0,\ldots,n$ and its zeros by $\xi_j = \cos((2j-1)\pi/(2n))$, $j = 1,\ldots,n$. Put

$$u_n(x) = T_n((1+\xi_1)x-\xi_1),$$

so that $u_n(x)$ is the Chebyshev polynomial with respect to the interval

$$\left[\frac{\xi_1 - 1}{\xi_1 + 1}, 1\right]$$

which contains $I : [0,1]$. Let $x_j$ be determined by $(1 + \xi_1)x_j - \xi_1 = \eta_j$, $j = 0,\ldots,n-1$. Hence $1 = x_0 > x_1 > \ldots > x_{n-1} > 0$.

Put $w(x) = x(x-x_0)\ldots(x-x_{n-1})$. Then if $p \in P_n$,

$$p(x) = \frac{w(x)}{x}\frac{p(0)}{w'(0)} + \sum_{j=0}^{n-1} \frac{w(x)}{x-x_j}\frac{p(x_j)}{w'(x_j)}$$

$$=: \frac{w(x)}{x}\frac{p(0)}{w'(0)} + q(x),$$

and

$$p'(x) - \left(\frac{w(x)}{x}\right)' \frac{p(0)}{w'(0)} = q'(x). \tag{9}$$

Suppose that $p \in K_n$, then for $j = 0,\ldots,n-1$, $|q(x_j)| = |p(x_j)| \leq 1$. But also, as Schur [6] observes

$$\left|q\left(\frac{\xi_1 - 1}{\xi_1 + 1}\right)\right| \leq \left|u_n\left(\frac{\xi_1 - 1}{\xi_1 + 1}\right)\right| = 1.$$

The hypotheses of Duffin and Schaeffer's [1] refinement of Markov's theorem are now satisfied and we conclude that for $x \in I$

$$|q'(x)| \leq u_n'(1) = 2n^2\cos^2\frac{\pi}{4n}.$$

Thus, in view of (9), $A : p(0) \to (w(x)/x)'(p(0)/w'(0))$ is an optimal algorithm and the intrinsic error is $2n^2\cos^2(\pi/(4n))$.

Schur [6] also treats the case that $p \in K_n$ satisfies $p(0) = p(1) = 0$ and shows that then

$$\|p'\| \leq 2n \operatorname{ctg}\frac{\pi}{2n}.$$

As in the case just concluded, we can show that $2n \operatorname{ctg}(\pi/(2n))$ is the intrinsic error in the optimal recovery of $p'$ from $p(0)$ and $p(1)$. Indeed, if we put $v_n(x) = T_n(\xi_1(2x-1)$

so that $v_n(x)$ is the Chebyshev polynomial with respect to

$$\left[\frac{\xi_1 - 1}{2\xi_1}, \frac{\xi_1 + 1}{2\xi_1}\right],$$

let $t_j$ be determined by $\xi_1(2t_j - 1) = \eta_j, j = 1,\ldots,n-1$ and put $w(x) = x(x-1)(x-t_1)\ldots(x-t_{n-1})$, then an optimal algorithm is provided by

$$A : (p(0),p(1)) \to \left(\frac{w(x)}{x}\right)' \frac{p(0)}{w'(0)} + \left(\frac{w(x)}{x-1}\right)' \frac{p(1)}{w'(1)} .$$

The details are above.

## 2. POLYNOMIALS ON THE UNIT DISC

We turn next to some problems in the complex plane. $P_n$ is as before, but for $p \in P_n$

$$\|p\| = \max_{|z|\leq 1} |p(z)|.$$

Lachance, Saff and Varga [2] recently showed that if $p \in P_n$, $n \geq 1$, then

$$\sup_{\{\substack{\|p\|\leq 1 \\ p(1)=0}} |p(0)| = \left[\cos\frac{\pi}{2(n+1)}\right]^{n+1} \qquad (10)$$

The quantity given by (10) is the intrinsic error in the optimal recovery problem specified by $X=P_n, Y=Z=\mathbb{C}, Up=p(0)$, $Ip=p(1)$, $K=K_n = \{p \in P_n : \|p\| \leq 1\}$ and $\varepsilon = 0$, since in this case, according to M-R, (3) is an equality. Let us determine the optimal linear algorithm whose existence is affirmed in M-R, as well.

Lachance, Saff and Varga [2] give the extremal polynomial in (10) explicitly. It has real coefficients. Call it

$P(z) = a_0 + a_1 z + \ldots + a_n z^n$. Let $A : p(1) \to \lambda p(1)$ be the optimal algorithm, then

$$\max_{p \in K_n} |p(0) - \lambda p(1)| = |P(0)| = |P(0) - \lambda P(1)|,$$

so that P is extremal for the linear functional on $P_n$, $p(0) - \lambda p(1)$. This linear functional has a canonical representation (Cf. Rivlin [5],)

$$p(0) - \lambda p(1) = \sum_{j=1}^{r} b_j p(\zeta_j), \quad r \leq n + 1 \tag{11}$$

and we must have $r = n$, since $r < n$ or $r = n+1$ leads to a contradiction. Thus $|P(\zeta_j)| = 1$, $j = 1, \ldots, n$ and hence $|\zeta_j| = 1$. Now on $|z| = 1$ we have for some $Q \in P_n$

$$|P(z)|^2 + |Q(z)|^2 \equiv 1$$

and so $Q(z) = c(z-\zeta_1)\ldots(z-\zeta_n)$. Therefore (11) implies that $Q(0) - \lambda Q(1) = 0$ and since $|Q(1)| = 1$ we obtain

$$\lambda = \frac{c \prod_{j=1}^{n} (-\zeta_j)}{c \prod_{j=1}^{n} (1-\zeta_j)} .$$

Since the coefficients of P are real $\Pi(-\zeta_j) = 1$. Also $(1-\zeta_j)(1-\bar{\zeta}_j) = 2 - 2\mathrm{Re}\,\xi_j$ hence $\lambda > 0$. If $c > 0$ then $Q(1) = 1$ and $\lambda = c$. Equating leading coefficients in (11) yields $2a_0 a_n + 2c^2 = 0$ or, finally $\lambda = \sqrt{-a_0 a_n}$ .

The general problem of optimal interpolation in $P_n$ is to recover $p(\zeta)$ from $p(z_1), \ldots, p(z_k)$, $k \leq n$ for $p \in K_n$. We want to comment on two special cases of this problem.

First suppose $k = n$. Put $w(z) = (z-z_1)\ldots(z-z_n)$. Then $E = |w(\zeta)|/\|w\|$. But if $|w(z')| = \|w\|$ then for $p \in K_n$ interpolation at $z_1, \ldots, z_n, \zeta$ yields

$$p(z') = \sum_{j=1}^{n} \ell_j(z')\, p(z_j) + \frac{w(z')}{w(\zeta)}\, p(\zeta)$$

from which we conclude that

$$A\colon (p(z_1),\ldots,p(z_n)) \to -\sum_{j=1}^{n} \frac{\ell_j(z')w(\zeta)}{w(z')}\, p(z_j) \tag{12}$$

is an optimal algorithm.

Next suppose $|z_j| = 1$, $j = 1,\ldots,k$. We claim that the intrinsic error is least when the k observation points coalesce (at 1, for example). (When this happens the information is, of course, $p(1),\ldots,p^{(k-1)}(1)$). That is, consider

$$E(z_1,\ldots,z_k) = \max_{\begin{cases} p \in K_n \\ p(z_j) = 0,\ j = 1,\ldots,k \end{cases}} |p(0)| = \frac{1}{\min_{\begin{cases} p(0) = 1 \\ p(z_j)=0, j=1,\ldots,k \end{cases}} \|p\|}$$

then we wish to minimize $E(z_1,\ldots,z_k)$ or, what is the same thing, maximize

$$\min_{c_j} \|P\|$$

where

$$P(z) = (1 - \frac{z}{z_1})\ldots(1 - \frac{z}{z_k})(1 + c_1 z + \ldots + c_{n-k} z^{n-k}).$$

We show that this happens where $z_1 = z_2 = \ldots = z_k$, by proving the following

<u>Lemma</u>. For every z on $|z| = 1$ and every choice of real $\theta_1,\ldots,\theta_k$ we have, with $\theta = (\theta_1 + \ldots + \theta_k)/k$

$$|1 - ze^{-i\theta}|^k \geq \prod_{j=1}^{k} |1 - ze^{-i\theta_j}|.$$

<u>Proof</u>. Note that $|1-e^{it}|=2|\sin(t/2)|$. So we need to show that

$$\left|\sin\left(\frac{t_1+\ldots+t_k}{k}\right)\right|^k \geq |\sin t_1 \sin t_2 \ldots \sin t_k|$$

for all $t_j$ in $[0,\pi]$, which follows from the convexity of $\log \csc x$ in $(0,\pi)$.

The Lemma immediately implies the

__Theorem__. For $|z_j| = 1$, $j = 1,\ldots k$

$$\min_{z_1,\ldots z_k} \quad \max_{\begin{cases} p \in K_n \\ p(z_j) = 0, \; j = 1,\ldots,k \end{cases}} |p(0)|$$

is attained when $z_1 = \ldots = z_k$. In particular, we may take all $z_j = 1$.

__Remark__. When $k = n$ the limiting form of (12) becomes

$$A : (p(1), p'(1), \ldots, p^{(n-1)}(1)) \to \sum_{k=0}^{n-1} (-1)^k \left(1 - \frac{1}{2^{n-k}}\right) p^{(k)}(1)$$

and the intrinsic error is $2^{-n}$.

Finally, let us recall the result of Lax [3], which says, in our notation, that if $|z_i| = 1$, $i = 1,\ldots,n$ and $p \in K_n$ satisfies $p(z_i) = 0$, $i = 1,\ldots,n$ then $\|p'\| = (n/2)$.

Consider the corresponding optimal recovery problem for which Lax's Theorem provides the intrinsic error. Choose $z_0$ such that $\|(z-z_1)\ldots(z-z_n)\| = |(z_0-z_1)\ldots(z_0-z_n)|$. Lagrange interpolation to a given $p \in K_n$ at $z_0, z_1, \ldots, z_n$ yields

$$p'(z) - \sum_{j=1}^{n} p(z_j)\ell_j'(z) = p(z_0)\ell_0'(z).$$

But, clearly, $\ell_0$ satisfies the hypotheses of Lax's Theorem, hence $\|\ell_0'\| = (n/2)$ and

$$A : (p(z_1), \ldots p(z_n)) \to \sum_{j=1}^{n} p(z_j)\ell_j'(z)$$

effects the optimal recovery of $p'$.

The same problem when we sample at $k < n$ points of the unit circle is much more difficult. D. J. Newman has obtained some results for the intrinsic error when $k = 1$, namely

$$n - \frac{c_1}{n} < E < n - \frac{c_2}{n},$$

and some fragmentary results for all k, which suggest that E is of the form

$$n - \frac{ck^2}{n}.$$

($k = n$, $c = \frac{1}{2}$ is Lax's Theorem).

REFERENCES

1. Duffin, R.J. and A.C. Schaeffer, "A refinement of an inequality of the brothers Markoff", *Trans. Amer. Math. Soc.* *50*, pp. 517-528 (1941).

2. Lachance,M., Saff,E.G. and R.S.Varga, "Inequalities for polynomials with a prescribed zero", *Math . Z.168*, pp.105-116 (1979).

3. Lax, P.D., "Proof of a conjecture of P. Erdös on the derivative of a polynomial", *Bull.Amer.Math.Soc.50*,pp.509-513 (1944).

4. Micchelli,C.A. and T.J.Rivlin, "A survey of optimal recovery",*Optimal Estimation in Approximation Theory,* Plenum Press, N. Y., pp. 1-54 (1977).

5. Rivlin,T.J., "The Chebyshev Polynomials", John Wiley & Sons, New York (1974).

6. Schur,I., Über das Maximum des absoluten Betrages eines Polynoms in einem gegebenen Intervall, *Math. Z. 4*, pp. 271-287 (1919).

# ON CARDINAL SPLINE INTERPOLANTS

Walter Schempp

Lehrstuhl für Mathematik I
University of Siegen
Siegen, Germany

## 1. INTRODUCTION

The theory of (univariate) spline functions has several very different aspects. The aspects of local bases (B-splines) and minimal properties (variational methods) are well-known. Recently some complex analytic features have been pointed out. It can be proved that the cardinal splines admit contour integral representations (with non-compact paths) that allow an investigation of their convergence behaviour for successively higher degree in the exponential case as well as in the logarithmic case. Moreover, a contour integral representation of the Euler-Frobenius polynomials can be established. The complex analytic method reveals also to be very useful for the treatment of cardinal exponential spline interpolants of higher order.



ISBN 0-12-780650-4

## 2. CARDINAL EXPONENTIAL SPLINES

For any integer $m \geq 1$ let $\mathfrak{S}_m(\mathbb{R};\mathbb{Z})$ denote the vector space over the field $\mathbb{C}$ of all complex cardinal spline functions of degree m on $\mathbb{R}$ having (equidistant) knots at the integer points. Moreover, let $U = \{z \in \mathbb{C} \mid |z| = 1\}$ denote the compact unit circle.

Theorem 1. The cardinal exponential splines $s_m \in \mathfrak{S}_m(\mathbb{R};\mathbb{Z})$ of degree $m \geq 1$ and weight $h \in \mathbb{C}^{\times} - U$ admit the contour integral representation with transcendental meromorphic integrand and non-compact path

$$s_m(x) = C_{m,h}\left(1-\frac{1}{h}\right)^{m+1} \frac{1}{2\pi i} \int_P \frac{e^{(x+1)z}}{(e^z-h)z^{m+1}}\,dz \qquad (x \in \mathbb{R}),$$

where $C_{m,h} \in \mathbb{C}$ denotes an arbitrary constant and P stands for the positively oriented boundary of any closed vertical strip in the open complex right resp. left half-plane according to the cases $|h|>1$ resp. $0<|h|<1$ that contains the line $\{w \in \mathbb{C} \mid \operatorname{Re} w = \log|h|\}$ in its interior.

The proof follows via the <u>inverse Laplace transform</u> by means of a line integral representation of the basis splines (cf. [5]). As a consequence, Theorem 1 implies

Theorem 2. Let $h \in \mathbb{C}^{\times} - U$ and $x \in [-1,0]$ be given. For all $z \in \mathbb{C}$ so that $|z|<|\log|h||$, the cardinal exponential splines $s_m \neq 0$ $(m \geq 1)$ give rise to the power series expansion

$$\frac{e^{(x+1)z}}{h-e^z} = \sum_{m \geq 0} \frac{h^{m+1}}{C_{m,h}(h-1)^{m+1}} s_m(x) z^m,$$

where $C_{0,h} = 1$ and $s_0 = \frac{1}{h}$.

Another consequence of Theorem 1 and the Cauchy residue

theorem is the following result (cf. Schoenberg [9], [10]):

Theorem 3. If $(z_k(h))_{k \in \mathbb{Z}}$ denotes the bi-infinite sequence of zeros of the function $z \rightsquigarrow e^z - h$ $(h \in \mathbb{C}^\times - U)$ then

$$s_m(x) = C_{m,h}\left(1-\frac{1}{h}\right)^{m+1} \sum_{k \in \mathbb{Z}} \frac{e^{xz_k(h)}}{z_k^{m+1}(h)} \qquad (m \geq 1)$$

holds for all $x \in \mathbb{R}$.

## 3. EULER-FROBENIUS POLYNOMIALS

In the case when the weight $h \in \mathbb{C}^\times - U$ is not a root of the m-th Euler-Frobenius polynomial $p_m$ $(m \geq 1)$ there exists one and only one cardinal exponential spline interpolant $s_m$ of degree $m \geq 1$ with respect to the bilateral geometric sequence $(h^n)_{n \in \mathbb{Z}}$. For these remarkable polynomials we can prove (cf. [7], [8]):

Theorem 4. For any $h \in \mathbb{C}^\times - U$ the Euler-Frobenius polynomials $(p_m)_{m \geq 1}$ admit the contour integral representation

$$p_m(h) = \frac{(h-1)^{m+1}}{h} \frac{m!}{2\pi i} \int_P \frac{e^z}{(e^z - h)z^{m+1}} \, dz \qquad (m \geq 1),$$

where the non-compact path $P$ is defined as in Theorem 1.

Corollary 1. For all $h \in \mathbb{C}^\times$ the reciprocal identity

$$h^{m-1} p_m\left(\frac{1}{h}\right) = p_m(h) \qquad (m \geq 1)$$

holds.

Corollary 2. For any $h \in \mathbb{C}^\times - U$ the derivatives $(p_m')_{m \geq 1}$ admit the contour integral representation

$$p'_m(h) = \frac{mh+1}{h(h-1)} p_m(h) + \frac{(h-1)^{m+1}}{h} \cdot \frac{m!}{2\pi i} \int_P \frac{e^z}{(e^z-h)^2 z^{m+1}} dz \quad (m \geq 1),$$

where the path P is defined as in Theorem 1.

An application of the calculus of residues to the preceding identity yields (cf. [8])

Theorem 5. The Euler-Frobenius polynomials $(p_m)_{m\geq 1}$ satisfy the three-term recurrence relation

$$p_m(h) = (mh+1)p_m(h)-h(h-1)p'_m(h) \qquad (m\geq 1).$$

It follows that for any integer $m\geq 2$ all the roots of the Euler-Frobenius polynomial $p_m$ are located on the open negative real half-line $\mathbb{R}^{\times}_{-}$. Therefore, Theorem 3 implies (cf. Schoenberg [9], [10])

Theorem 6. Let the weight $h \in \mathbb{C}-(U \cup \mathbb{R}_{-})$ be given. Then the cardinal exponential spline interpolants $(s_m)_{m\geq 1}$ of degree m with respect to the bilateral geometric sequence $(h^n)_{n \in \mathbb{Z}}$ satisfy the pointwise convergence property

$$\lim_{m\to\infty} s_m(x) = h^x$$

for all $x \in \mathbb{R}$.

In the logarithmic case the convergence behaviour of the interpolating cardinal spline functions is quite different. See Section 5 infra.

## 4. CARDINAL EXPONENTIAL SPLINE INTERPOLANTS OF HIGHER ORDER

Let the weight $h \in \mathbb{C}^\times - U$ be distinct from all the roots of the Euler-Frobenius polynomials $(p_m)_{m\geq 1}$. Construct the cardinal exponential spline interpolants $s_m \in \mathfrak{S}_m(\mathbb{R};\mathbb{Z})$ of degree $m\geq 1$ with respect to the bilateral geometric sequence $(h^n)_{n\in\mathbb{Z}}$. For any $r \in \mathbb{N}$, the function

$$s_{m,r} = \frac{1}{r!}\frac{\partial^r}{\partial h^r} s_m \qquad (m\geq 1)$$

on $\mathbb{R}$ is called the cardinal exponential spline interpolant of degree m and order r. For the case r=1 we obtain (cf. [8])

Theorem 7. Let the weight $h \in \mathbb{C}^\times - U$ satisfy $p_m(h) \neq 0$ for all $m \in \mathbb{N}^\times$. The cardinal exponential spline interpolant of degree $m\geq 1$ and of the first order with respect to the bi-infinite geometric sequence $(h^n)_{n\in\mathbb{Z}}$ admits the contour integral representation

$$s_{m,1}(x) = \frac{s_m(x)p_{m+1}(h)}{h(h-1)p_m(h)} + \frac{(h-1)^{m+1}}{hp_m(h)} \cdot \frac{m!}{2\pi i}\int_P \frac{e^{(x+1)z}}{(e^z-h)^2 z^{m+1}}\,dz$$

for all $x \in \mathbb{R}$. The path P is defined as in Theorem 1

Another application of the calculus of residues proves (cf. [8])

Theorem 8. For $x \in \mathbb{R}$ and $h \in \mathbb{C}^\times - U$ so that $p_m(h) \neq 0$ $(m\geq 1)$ the identity

$$s_{m,1}(x) = x.s_m(x-1) - \frac{p_{m+1}(h)}{(h-1)p_m(h)}\,(s_{m+1}(x-1) - s_m(x-1))$$

holds.

An induction argument establishes the following extension of Theorem 6 (cf. Greville-Schoenberg-Sharma [1]):

Theorem 9. Let the weight $h \in \mathbb{C}-(U \cup \mathbb{R}_-)$ be fixed. The convergence

$$\lim_{m\to\infty} s_{m,r}(x) = \binom{x}{r} h^{x-r} \qquad (r \geq 0)$$

holds pointwise for all $x \in \mathbb{R}$.

## 5. CARDINAL LOGARITHMIC SPLINE INTERPOLANTS

Let $h_0 > 1$ denote a fixed step width and $\mathfrak{k}_0$ the knot sequence $\mathfrak{k}_0 = (h_0^n)_{n \in \mathbb{Z}}$ in the open positive real half-line $\mathbb{R}_+^\times$. Consider the function

$$f_0 : \mathbb{R}_+^\times \ni x \rightsquigarrow \frac{\log x}{\log h_0} \in \mathbb{R}.$$

If $(\Gamma_m)_{m\geq 1}$ denotes the sequence of partial products in Euler's limit formula for the gamma function, we can prove

Theorem 10. The cardinal logarithmic spline interpolants $S_m \in \mathfrak{S}_m(\mathbb{R}_+^\times; \mathfrak{k}_0)$ of degree $m \geq 1$ with step width $h_0 > 1$ admit the contour integral representation with transcendental meromorphic integrand

$$S_m(x) = \frac{1}{2\pi i} \int_L \Gamma_m(z) h_0^{-zf_0(m)} \frac{1-x^{-z}}{1-h_0^{-z}} dz \quad (x \in \mathbb{R}_+^\times),$$

where $L = L_1 \vee L_2$ denotes the positively oriented boundary of a closed vertical strip in the complex plane $\mathbb{C}$ delimited by the lines $L_1 = \{z \in \mathbb{C} \mid \operatorname{Re} z = c\}$ with $c > 0$ and $L_2 = \{z \in \mathbb{C} \mid \operatorname{Re} z = d\}$ with $d \in ]-1,0[$. The contour integral is independent of the particular choices of the real constants c,d.

The proof follows via the inverse Mellin transform by means of a line integral representation of the elements of the truncated power basis (cf. [6]). In this connection also see [4].

An application of Cauchy's residue theorem entails the following striking fact ("Newman-Schoenberg phenomenon" [2], [4]).

Theorem 11. The condition $\lim_{m\to\infty} S_m(x) = f_o(x)$ is satisfied at the point $x \in \mathbb{R}_+^\times$ if and only if $x \in \mathcal{k}_o$.

In other words: The cardinal logarithmic spline interpolants $(S_m)_{m\geq 1}$ are pointwise convergent only at those points $x \in \mathbb{R}_+^\times$ where the convergence holds trivially. At all the other points of $\mathbb{R}_+^\times$ we have divergence.

For an investigation of the Newman-Schoenberg phenomenon by real transformation methods the reader is referred to the paper [3].

ACKNOWLEDGMENTS

The author wishes to express his gratitude to Professors A. Sard (Binningen/Switzerland) and I.J. Schoenberg (Madison/Wisconsin) for their encouragements and constant support. Sincere thanks are also due to Professor M.Z. Nashed (Newark/Delaware) for giving the opportunity to lecture on the subject at the University of Delaware.

## REFERENCES


1. Greville, T.N.E., Schoenberg, I.J., Sharma, A.: The spline interpolation of sequences satisfying a linear recurrence relation. J. Approx. Theory 17 (1976), 200-221
2. Newman, D.J., Schoenberg, I.J.: Splines and the logarithmic function. Pacific J. Math. 61 (1975), 241-258
3. Schempp, W.: On the convergence of cardinal logarithmic splines. J. Approx. Theory 23 (1978), 108-112
4. Schempp, W. A note on the Newman-Schoenberg phenomenon Math. Z. 167 (1979), 1-6
5. Schempp, W.: Cardinal exponential splines and Laplace transform. J. Approx. Theory (to appear)
6. Schempp, W.: Cardinal logarithmic splines and Mellin transform. J. Approx. Theory (to appear)
7. Schempp, W.: A contour integral representation of Euler-Frobenius polynomials. J. Approx. Theory (to appear)
8. Schempp, W.: Contour integral representation of cardinal spline functions, to appear
9. Schoenberg, I.J.: Cardinal interpolation and spline functions IV. The exponential Euler splines. In: Linear Operators and Approximation I, P.L. Butzer, J.P. Kahane, B. Sz.-Nagy editors. Basel-Boston-Stuttgart: Birkhäuser Verlag 1972
10. Schoenberg, I.J.: Cardinal spline interpolation. Regional Conference Series in Applied Mathematics, Vol. 12. Philadelphia, Pennsylvania: Society for Industrial and Applied Mathematics 1973

# APPROXIMATION BY LACUNARY POLYNOMIALS: A CONVERSE THEOREM

Maurice Hasson

Department of Mathematics
University of Alabama
University, Alabama 35486

Oved Shisha

Department of Mathematics
University of Rhode Island
Kingston, Rhode Island 02881

1. Let $-\infty < a < 0 < b < \infty$ and let $f$ be a real function, continuous on $[a, b]$ but coinciding there with no polynomial. Given integers $k,n$ $(0\le k\le n)$ we denote

(1) $$E_n(f) = \min \max_{a\le x\le b} |f(x) - p_n(x)|$$

where the minimum is taken over all polynomials $p_n$ of degree $\le n$, and we let $E_n{}^k(f)$ denote the right hand side of (1) with the minimum taken over all polynomials $p_n$ of degree $\le n$ whose coefficient of $x^k$ is 0.

If $f$ satisfies on $[a, b]$ a Lipschitz condition of order $\alpha$ $(0<\alpha\le 1)$, then [3, p.125, Corollary 1] $E_n(f) = 0(n^{-\alpha})$. If, in addition, $f(0) \neq 0$, then since $E_n{}^0(f) \ge |f(0)|$, $n = 1,2,\cdots$, we have, for some positive constant $A_0$,

$$E_n{}^0(f)/E_n(f) \ge A_0 n^{\alpha}, \quad n = 0,1,2,\cdots .$$

In [2] the ratio $E_n{}^k(f)/E_n(f)$ has been studied for $k = 1,2,\cdots$. Theorem 3.3 of [2] implies

Theorem 1. Let $k$ be an integer $\ge 0$, let $f^{(k)}$ exist and




ISBN 0-12-780650-4

*satisfy a Lipschitz condition of order* $\alpha\,(0<\alpha\leq 1)$ *throughout* $[a, b]$, *and suppose* $f^{(k)}(0) \neq 0$. *Then there exists a positive constant* $A_k$ *such that*

(1) $$E_n^{\,k}(f)/E_n(f) \geq A_k n^{\alpha}, \quad n = k, k+1, \cdots.$$

Our purpose is to invert Theorem 1 by proving

*Theorem 2. Let* $k$ *be an integer* $\geq 0$, *let* $0<\alpha<1$ *and suppose, for some positive constant* $A_k$, (1) *holds. Then* $f^{(k)}$ *exists throughout* $(a, b)$ *and, on each* $[a',b']$, $a < a' < b' < b$, *it satisfies a Lipschitz condition of order* $\alpha$.

2. In proving Theorem 2 we shall use the following two results.

*Theorem 3.* (Cf. [1, Theorem 3.4].) *Suppose, for some constants* $\lambda \geq 0$ *and* $L$,

$$E_n(f) \leq Ln^{-\lambda}, \quad n = 1,2,\cdots.$$

*For* $n = 0,1,2,\cdots$, *let* $P_n$ *be the polynomial of degree* $\leq n$ *of best approximation to* $f$ *on* $[a, b]$. *Let* $k$ *be a fixed integer* $> \lambda$ *and let* $a < a' < b' < b$. *Then*

$$\max_{a'\leq x\leq b'} |P_n^{(k)}(x)| = O(n^{k-\lambda}).$$

*Theorem 4. Suppose* $E_n(f) = O(n^{-k-\delta})$, $k$ *an integer* $\geq 0$, $\delta>0$. *Let* $a < a' < b' < b$, $0\leq r\leq k$. *Then there exists a constant* $A$ *such that, for* $n = 1,2,\cdots$, *the inequality*

$$\max_{a'\leq x\leq b'} |P_n^{(r)}(x) - f^{(r)}(x)| < An^{r-k-\delta}$$

*is meaningful and true, where* $P_1, P_2, \cdots$ *are as in Theorem 3.*

3. For every integer $q\,(\geq 0)$, let $C^q$ denote the set of all $2\pi$-periodic real functions having, throughout $(-\infty, \infty)$, a continuous $q$th derivative. For every noninteger $q > 0$, $C^q$

will denote the set of all $2\pi$-periodic real functions $F$ for which $F^{([q])}$ exists throughout $(-\infty,\infty)$ and satisfies there a Lipschitz condition of order $q-[q]$. As usual, $[q]$ is the integral part of $[q]$.

Set

$$t(x) \equiv f\left(\frac{(b-a)\cos x + a + b}{2}\right), \quad -\infty < x < \infty .$$

Then [3, p.120, Lemma 1]

$$E_n(f) = \min \max_{-\infty<x<\infty} |t(x)-t_n(x)|, \quad n = 0,1,2,\cdots,$$

where the minimum is taken over all trigonometric polynomials $t_n$ of degree $\leq n$. Also (i) if $q$ is an integer $\geq 0$, then $E_n(f) = o(n^{-q})$ if $t \in C^q$; (ii) if $q$ is a noninteger $>0$, then $E_n(f) = O(n^{-q})$ iff $t \in C^q$ [3, p.84, Theorem 1; p. 88, Theorem 2; p.86, Corollary 1; p.89, Corollary 1; p.97, Theorem 1; p.104, Theorem 2].

Theorem 2 follows readily from

Theorem 5. Assume the hypotheses of Theorem 2. Then $t \in C^{k+\alpha}$.

Proof of Theorem 5. Suppose the conclusion is false and let $k_0$ be the smallest integer $m \geq 0$ for which $t \notin C^{m+\alpha}$. For $n = k, k+1, k+2,\cdots$, let $P_n$ be as in Theorem 3 and let $a_n^k$ be the coefficient of $x^k$ in $P_n$ so that

$$E_n^k(f) \leq E_n^k(f(x)-a_n^k x^k) + E_n^k(a_n^k x^k) = E_n(f)+|a_n^k|E_n^k(x^k) .$$

By Theorem 2.7 of [2], for some constant $M_k$ and $n = k+1, k+2,\cdots$, $E_n^k(x^k) \leq M_k n^{-k}$ and so

$$(2) \quad A_k \leq E_n^k(f)n^{-\alpha}/E_n(f) \leq n^{-\alpha} + [M_k n^{-k-\alpha}|a_n^k|/E_n(f)].$$

I. Suppose $t \in C^{k_0}$ and $k_0 < k$. Then $E_n(f) = o(n^{-k_0}) = O(n^{-k_0})$. Theorem 3 clearly implies $|a_n^k| \leq B_k n^{k-k_0}$ for some constant $B_k$ and all $n \geq k$. Let

$$b_n = E_n(f)n^{k_0+\alpha}, \quad n = 1,2,\cdots.$$

Since $t \notin C^{k_0+\alpha}$, $b_n$ is unbounded. By (2), for $n = k+1, k+2, \cdots,$

$$A_k \leq n^{-\alpha} + M_k B_k (b_n)^{-1}$$

which contradicts the positivity of $A_k$.

II. $k_0$ must be $\geq 1$. For suppose $k_0 = 0$. Theorem 6 below clearly implies $|a_n^k| \leq \Gamma_k n^k$ for some constant $\Gamma_k$ and all $n \geq k$. Let

$$\gamma_n = E_n(f) n^{\alpha}, \quad n = 1,2,\cdots.$$

Since $t \notin C^{\alpha}$, $\gamma_n$ is unbounded. By (2), for $n = k+1, k+2, \cdots,$

$$A_k \leq n^{-\alpha} + M_k \Gamma_k \gamma_n^{-1}$$

which contradicts the positivity of $A_k$.

III. We show that $t \in C^{k_0-1+\beta}$ for every $\beta \in (0, 1)$. Choose such $\beta$ and let $N$ be a positive integer with $\gamma = \beta/N \leq \alpha$. Then $t \in C^{k_0-1+\gamma}$. Suppose $t \notin C^{k_0-1+\beta}$. Let $\mu$ be the smallest positive integer $m \leq N$ for which $t \notin C^{k_0-1+m\gamma}$. Then $\mu > 1$, $t \in C^{k_0-1+(\mu-1)\gamma}$ and $E_n(f) = O(n^{-k_0+1-(\mu-1)\gamma})$. By Theorem 3, $|a_n^k| \leq D_k n^{k-k_0+1-(\mu-1)\gamma}$ for some constant $D_k$ and all $n \geq k$. Let

$$d_n = E_n(f) n^{k_0-1+\mu\gamma}, \quad n = 1,2,\cdots.$$

Since $t \notin C^{k_0-1+\mu\gamma}$, $d_n$ is unbounded. Therefore, by (2), for $n = k+1, k+2, \cdots,$

$$A_k \leq n^{-\alpha} + M_k D_k n^{\gamma-\alpha} d_n^{-1}$$

which contradicts the positivity of $A_k$.

IV. We prove now that $t \in C^{k_0+(\alpha/2)}$ which, by I, shows that $k_0 = k$. Let

(3) $$1 - (\alpha/2) < \beta < 1 .$$

Then, by III, $E_n(f) = O(n^{-k_0+1-\beta})$. By Theorem 3, $|a_n^k| \le G_k n^{k-k_0+1-\beta}$ for some constant $G_k$ and all $n \ge k$. If $t \notin C^{k_0+(\alpha/2)}$, then, for infinitely many $n > k$, we have $E_n(f) n^{k_0+(\alpha/2)} \ge 1$. For these $n$, by (2),

$$A_k \le n^{-\alpha} + M_k G_k n^{1-(\alpha/2)-\beta}$$

which, by (3), contradicts the positivity of $A_k$.

V. Thus $t \in C^{k+(\alpha/2)}$ and, hence, $E_n(f) = O(n^{-k-(\alpha/2)})$ which, by Theorem 4 (with $r = k$), shows that $(a_n^k)_{n=k}^{\infty}$ is bounded. Since $t \notin C^{k+\alpha}$, $E_n(f) n^{k+\alpha}$ is unbounded which, by (2), contradicts the positivity of $A_k$.

4. The following is well known.

<u>Theorem 6</u>. [3, p.134, Corollary 2]. Let $a < a' < b' < b$, let $n, k$ be integers $\ge 0$ and $P(x)$ a real polynomial of degree $\le n$. Then

$$\max_{a' \le x \le b'} |P^{(k)}(x)| \le (Kkn)^k \max_{a \le x \le b} |P(x)|$$

where $K = \max((a'-a)^{-1}, (b-b')^{-1})$.

<u>Proof of Theorem 3</u>. Let $n$ be an integer $\ge 2$, say $2^m \le n < 2^{m+1}$, $m$ an integer $\ge 1$. Observe that if $0 \le u \le v$, then $\max_{a \le x \le b} |P_v - P_u| \le 2E_u(f)$. Let $\|\ \|$ denote $\max_{a' \le x \le b'} |\ \ |$.

By Theorem 6, $\|P_n^{(k)}\| \le \|(P_n - P_{2^m})^{(k)}\| + [\sum_{j=1}^m \|(P_{2^j} - P_{2^{j-1}})^{(k)}\|]$

$+\|P_1^{(k)}\| \le 2(Kkn)^k E_{2^m}(f) + [\sum_{j=1}^m 2(Kk2^j)^k E_{2^{j-1}}(f)] + \|P_1^{(k)}\|$

$< 2L(2Kk)^k 2^{m(k-\lambda)} + [\sum_{j=1}^m 2L(2Kk)^k 2^{j(k-\lambda)}] + \|P_1^{(k)}\|$

$< [2L(2Kk)^k \{1 + 2^{k-\lambda}(2^{k-\lambda}-1)^{-1}\} + \|P_1^{(k)}\|] n^{k-\lambda}$.

5. <u>Proof of Theorem 4</u>. By (ii) preceding Theorem 5, $f^{(k)}$ exists and is continuous in $(a, b)$. Set

$$Q_m(x) \equiv P_{2^{m-1}}(x) - P_{2^m}(x), \quad m = 1,2,\cdots .$$

Then, for $m = 0,1,2,\cdots$ and every $x \in [a,b]$,

$$P_{2^m}(x) - f(x) = \sum_{j=m+1}^{\infty} Q_j(x).$$

Let

$$E_n(f) \leq Bn^{-k-\delta}, \quad n = 1,2,\cdots .$$

Let $m$ and $s$ be integers, $m \geq 0$, $0 \leq s \leq k$. Then, by Theorem 6, for $j = m+1, m+2,\cdots$,

$$\max_{a' \leq x \leq b'} |Q_j^{(s)}(x)| \leq (Ks2^j)^s 2E_{2^{j-1}}(f) \leq 2B(Ks)^s 2^{k+\delta}(2^{s-k-\delta})^j$$

and hence $\sum_{j=m+1}^{\infty} Q_j^{(s)}(x)$ converges uniformly in $[a',b']$. Therefore, throughout $[a',b']$,

$$(4) \qquad |P_{2^m}^{(r)}(x) - f^{(r)}(x)| \leq 2B(Kr)^r 2^{k+\delta} \sum_{j=m+1}^{\infty} (2^{r-k-\delta})^j = D(2^m)^{r-k-\delta}$$

where $D = 2B(2Kr)^r(1-2^{r-k-\delta})^{-1}$.

Let $n$ be an integer $\geq 1$, say, $2^m \leq n < 2^{m+1}$, $m$ an integer $\geq 0$. Then

$$\max_{a \leq x \leq b} |P_n(x) - P_{2^m}(x)| \leq 2E_{2^m}(f) \leq 2B2^{-m(k+\delta)}$$

and by Theorem 6, throughout $[a',b']$,

$$(5) \qquad |P_n^{(r)}(x) - P_{2^m}^{(r)}(x)| \leq 2B(2Kr)^r(2^m)^{r-k-\delta} .$$

By (4) and (5),

$$\max_{a' \leq x \leq b'} |P_n^{(r)}(x) - f^{(r)}(x)| < 2^{-r+k+\delta}[D+2B(2Kr)^r]n^{r-k-\delta} .$$


## ACKNOWLEDGMENT

The research of the second author was sponsored by the Air Force Office of Scientific Research, Air Force Systems Command, USAF, under Grant No. AFOSR 77-3174. The United States Government is authorized to reproduce and distribute reprints for Governmental purposes notwithstanding any copyright notation hereon.

REFERENCES

1. Hasson, M., Derivatives of the algebraic polynomials of best approximation, J. Approximation Theory (to appear).

2. Hasson, M., Comparison between the degrees of approximation by lacunary and ordinary algebraic polynomials, J. Approximation Theory (to appear).

3. Natanson, I. P., "Constructive Function Theory", Vol. I. Ungar, New York, 1964.

# AN INTERPOLATORY RATIONAL APPROXIMATION

A. K. Varma

Department of Mathematics
University of Florida
Gainesville, Florida

J. Prasad
Department of Mathematics
California State University
Los Angeles, California

Let us denote by

$$-1 = x_{nn} < x_{n-1,n} < \ldots < x_{2n} < x_{1n} = 1 \tag{1.1}$$

the n distinct zeros of

$$\pi_n(x) = (1-x^2)P'_{n-1}(x) \tag{1.2}$$

where $P_n(x)$ is the Legendre polynomial of degree n with normalization

$$P_n(1) = 1. \tag{1.3}$$

Let $f(x)$ be a given continuous function defined on $[-1,+1]$. The Hermite-Fejer interpolation polynomial based on the zeros of (1.2) is given by

$$H_n[f,x] = \sum_{k=1}^{n} f(x_{kn})h_{kn}(x) \tag{1.4}$$

where

$$\ell_{kn}(x) = -\frac{(1-x^2)P'_{n-1}(x)}{n(n-1)\,(x-x_{kn})\,P_{n-1}(x_{kn})}, \quad k=1,2,\ldots,n, \tag{1.5}$$




ISBN 0-12-780650-4

$$h_{kn}(x) = \ell_{kn}^2(x), \quad k = 2,3,\ldots, n-1, \tag{1.6}$$

$$h_{1n}(x) = \left[1 + \frac{n(n-1)(1-x)}{2}\right]\ell_{1n}^2(x) \tag{1.7}$$

and

$$h_{nn}(x) = \left[1 + \frac{n(n-1)(1+x)}{2}\right]\ell_{nn}^2(x). \tag{1.8}$$

According to a well known theorem of L. Fejer [5]

$$\lim_{n\to\infty} H_n[f,x] = f(x) \tag{1.9}$$

uniformly on $[-1,+1]$, for every f continuous there. A quantitative version of Fejer's result was given by A. K. Varma and J. Prasad [9]. They proved that

$$|H_n[f,x]-f(x)| \le \frac{c}{n} \sum_{k=1}^{n} \omega\left(f, \frac{\sqrt{1-x^2}}{k}\right) \tag{1.10}$$

where $\omega_f$ is the modulus of continuity of f on $[-1,1]$.

(1.10) is closely related to the work of Teljakovskii [8] who proved the following

<u>Theorem A</u>. Let $f \in C[-1,1]$ then there exist a sequence of algebraic polynomials $p_n(f,x)$ of degree $\le cn$ such that

$$|p_n(f,x) - f(x)| \le c\, \omega\left(f, \frac{\sqrt{1-x^2}}{n}\right). \tag{1.11}$$

For $f \in \mathrm{Lip}1$ (1.11) is much better than (1.10), but $p_n(f,x)$ may not be interpolatory. Therefore, we may ask the following question: Does there exist a sequence of linear positive operators $L_n(f)$ interpolating f at the zeros of (1.2) and for which

$$|L_n[f,x]-f(x)| < c\, \omega\left(f, \frac{\sqrt{1-x^2}}{n}\right) \tag{1.12}$$

holds?

Following the interesting theorem of A. Meir [6] we consider a sequence of positive, linear, interpolatory rational functions of degree $\leq 4n-2$ defined by

$$L_n[f,x] = \frac{\sum_{k=1}^{n} f(x_{kn})\, h_{kn}^2(x)}{\sum_{k=1}^{n} h_{kn}^2(x)} \tag{1.13}$$

It is easy to see that

$$L_n[f,x_{in}] = f(x_{in}),\quad L_n'[f,x_{in}] = 0 \qquad i = 1,2,\ldots,n, \tag{1.14}$$

and

$$L_n[1,x] \equiv 1. \tag{1.15}$$

We now state the main theorem.

<u>Theorem 1</u>. Let $f \in C[-1,+1]$. Then for $L_n[f,x]$ given by (1.13) we have

$$|f(x) - L_n[f,x]| \leq c\,\omega\left(f\ ,\ \frac{\sqrt{1-x^2}}{n}\right) \tag{1.16}$$

where c is an absolute constant independent of n and x.

## 2. Preliminaries

L. Fejer [5] proved that

$$\sum_{k=1}^{n} \ell_{kn}^2(x) \leq 1\ ,\ |\ell_{kn}(x)| \leq 1\ ,\ k = 1,2,\ldots,n. \tag{2.1}$$

The following results are well known. A good reference is J. Balazs and P. Turán [1] or G. Szego [7].

$$(1-x^2)^{\frac{3}{4}} P_{n-1}'^{\,2}(x) \leq \sqrt{2n} \qquad n \geq 4, \tag{2.2}$$

$$(1-x^2)^{\frac{1}{2}}\ |P_{n-1}'(x)| \leq n-1, \tag{2.3}$$

$$|P'_{n-1}(x)| \le \frac{n(n-1)}{2}, \tag{2.4}$$

$$P^2_{n-1}(x_{kn}) \ge (8\pi k)^{-1}, \quad k = 2,3,\ldots,[\tfrac{n}{2}], \tag{2.5}$$

$$P^2_{n-1}(x_{kn}) \ge (8\pi(n+1-k))^{-1}, \quad k = [\tfrac{n}{2}]+1,\ldots,n-1, \tag{2.6}$$

$$\sin\theta_{kn} \ge \frac{k}{2(n-1)}, \quad k = 2,3,\ldots,[\tfrac{n}{2}], \tag{2.7}$$

$$\sin\theta_{kn} \ge \frac{(n+1-k)}{2(n-1)}, \quad k = [\tfrac{n}{2}]+1,\ldots,n-1. \tag{2.8}$$

From (2.5) — (2.8) it follows that

$$(1-x^2_{kn})^{\frac{1}{2}} P^2_{n-1}(x_{kn}) \ge \frac{1}{16\pi(n-1)}, \quad k = 2,3,\ldots,n-1 \tag{2.9}$$

Since $x_{kn} = \cos\theta_{kn}$, $x = \cos\theta$

$$|\sin\theta_{kn} - \sin\theta| = |\sqrt{1-x^2_{kn}} - \sqrt{1-x^2}|.$$

Hence

$$|\sin\theta_{kn} - \sin\theta| = \frac{|1-x^2_{kn} - (1-x^2)|}{\sqrt{1-x^2_{kn}} + \sqrt{1-x^2}} \le \frac{2|x-x_{kn}|}{\sqrt{1-x^2_{kn}}}, \tag{2.10}$$

$k = 2,3,\ldots,n-1.$

## 3. Main estimates

Here we shall prove the following

<u>Lemma 3.1</u>. For $-1 \le x \le 1$, $x = \cos\theta$, the following estimates are valid:

$$(1-x)h^2_{1n}(x) \le 3\frac{\sqrt{1-x^2}}{n}, \quad (1+x)h^2_{n,n}(x) \le 3\frac{\sqrt{1-x^2}}{n}, \tag{3.1}$$

and

$$\sum_{k=2}^{n-1} |x-x_{kn}|h^2_{kn}(x) \le c\frac{\sqrt{1-x^2}}{n}, \tag{3.2}$$

where c is an absolute constant independent of n and x.

Proof. It is known that $\sum_{k=1}^{n} h_{kn}(x) = 1$ and $h_{kn}(x) \geq 0$. It follows that

$$0 \leq h_{kn}(x) \leq 1 \ , \ k = 1,2,\ldots, n \tag{3.3}$$

From (1.7), (1.5), (2.1) and (2.4) we obtain

$$(1-x)h_{1n}^2(x) \leq (1-x)h_{1n}(x) = 1 + \left[\frac{n(n-1)(1-x)}{2}\right](1-x)\ell_{1n}^2(x) =$$

$$= \frac{(1-x)^2(1+x)P_{n-1}'^2(x)}{n^2(n-1)^2} + \frac{(1-x^2)^2P_{n-1}'^2(x)}{2n(n-1)} \leq$$

$$\leq \frac{\sqrt{1-x^2}}{n}\left[2\,\frac{\sqrt{1-x^2}\,|P_{n-1}'(x)|\,|P_{n-1}'(x)|}{n(n-1)^2} + \right.$$

$$\left. + \frac{(1-x^2)^{\frac{3}{2}}P_{n-1}'^2(x)}{2(n-1)}\right] \leq$$

$$\leq \frac{\sqrt{1-x^2}}{n}\left[\frac{2(n-1)n(n-1)}{n(n-1)^2 2} + \frac{2n}{2(n-1)}\right] \leq$$

$$\leq 3\,\frac{\sqrt{1-x^2}}{n} .$$

Similarly

$$(1+x)h_{n,m}^2(x) \leq 3\,\frac{\sqrt{1-x^2}}{n} .$$

This proves (3.1). Next, we turn to prove (3.2). On using (2.1), (2.3), (2.9) and (2.10) we obtain $(k=2,3,\ldots,n-1)$

$$(1-x_{kn}^2)^{\frac{1}{2}}\ell_{kn}^2(x) \leq |\ell_{kn}(x)|\,[\,|\sin\theta_k - \sin\theta| + \sin\theta] \leq$$

$$\leq |\ell_{kn}(x)|\,\frac{2|x-x_{kn}|}{\sqrt{1-x_{kn}^2}} + \sin\theta$$

$$\leq \frac{2(1-x^2)|P'_{n-1}(x)|}{n(n-1)\sqrt{1-x_{kn}^2}\,|P_{n-1}(x_{kn})|} + \sin\theta$$

$$\leq \frac{2(n-1)(1-x^2)^{\frac{1}{2}}}{n(n-1)\sqrt{1-x_{kn}^2}\;P_{n-1}^2(x_{kn})} + \sin\theta$$

Therefore

$$(1-x_{kn}^2)^{\frac{1}{2}}\ell_{kn}^2(x) \leq \frac{2}{n}(1-x^2)^{\frac{1}{2}}16\pi(n-1) + \sin\theta \leq (32\pi+1)\sin\theta .$$

From this we may conclude that

$$(1-x_{kn}^2)^{\frac{1}{4}}|\ell_{kn}(x)| \leq 11\;(1-x^2)^{\frac{1}{4}}, \; k = 2,3,\ldots, n-1. \tag{3.4}$$

Now, on using (2.1), (2.2) and (2.9)

$$\sum_{k=2}^{n} |x-x_{kn}|h_{kn}^2(x) = \sum_{k=2}^{n-1} |x-x_{kn}|\ell_{kn}^4(x)$$

$$= \sum_{k=2}^{n-1} \frac{(1-x^2)|P'_{n-1}(x)|}{n(n-1)|P_{n-1}(x_{kn})|}\;|\ell_{kn}(x)|\ell_{kn}^2(x)|$$

$$\leq \sum_{k=2}^{n-1} \frac{(1-x^2)^{1/4}\;\sqrt{2n}}{n(n-1)}\;\frac{11(1-x^2)^{1/4}}{(1-x_{kn}^2)^{1/4}|P_{n-1}(x_{kn})|}\;\ell_{kn}^2(x)$$

$$\leq \frac{(1-x^2)^{1/2}}{n}\sum_{k=2}^{n-1} \frac{11\sqrt{2n}}{(n-1)}\;4\sqrt{\pi}\;(n-1)^{1/2}\ell_{kn}^2(x)$$

$$\leq \frac{c(1-x^2)^{1/2}}{n}\sum_{k=2}^{n-1} \ell_{kn}^2(x) \leq \frac{c(1-x^2)^{1/2}}{n} .$$

This proves (3.2). This completes the proof of the lemma.

Next, we will prove that

$$\sum_{k=1}^{n-1} h_{kn}^2(x) \geq \frac{1}{8}, \quad -1 \leq x \leq 1. \tag{3.5}$$

Let $x_{j+1} \leq x \leq x_j$ for some $j$, $j = 1,2,\ldots, n-1$. Then

$$\begin{aligned}\sum_{k=1}^{n} h_{kn}^2(x) &\geq h_{jn}^2(x) + h_{j+1,n}^2(x) \geq \ell_{jn}^4(x) + \ell_{j+1,n}^4(x) \\ &\geq \frac{1}{2}[\ell_{jn}^2(x) + \ell_{j+1,n}^2(x)]^2 \\ &\geq \frac{1}{2}\left[\frac{(\ell_{jn}(x) + \ell_{j+1,n}(x))^2}{2}\right]^2 \\ &= \frac{(\ell_{jn}(x) + \ell_{j+1,n}(x))^4}{8}.\end{aligned} \tag{3.6}$$

On using the well known theorem of Erdős and Turan [4, p. 529] it follows that

$$\ell_{jn}(x) + \ell_{j+1,n}(x) \geq 1 \quad x_{j+1,n} \leq x \leq x_{jn}. \tag{3.7}$$

From (3.6) and (3.7) we obtain (3.5).

## 4. Proof of Theorem 1

From (1.14) we have

$$L_n[f,1] = f(1), \; L_n[f,-1] = f(-1).$$

Also from (1.13) and (1.14) it follows that

$$L_n[f,x] - f(x) = \frac{\sum_{k=1}^{n} [f(x_{kn}) - f(x)]h_{kn}^2(x)}{\sum_{k=1}^{n} h_{kn}^2(x)}.$$

Therefore, from (3.5) we obtain

$$|L_n[f,x] - f(x)| \leq 8 \sum_{k=1}^{n} |f(x_{kn}) - f(x)|h_{kn}^2(x). \tag{4.2}$$

For $|x| < 1$, we can express

$$|f(x_{kn}) - f(x)| \le \omega(f\ ,\ |x-x_{kn}|) \tag{4.3}$$

$$\le \left(1 + \frac{n|x-x_{kn}|}{\sqrt{1-x^2}}\right) \omega\left(f\ ,\ \frac{\sqrt{1-x^2}}{n}\right) .$$

From (4.2), (4.3) and Lemma 3.1 we obtain

$$|L_n[f,x] - f(x)| \le 8 \sum_{k=1}^{n} \left(1 + \frac{n|x-x_{kn}|}{\sqrt{1-x^2}}\right) h_{kn}^2(x)\ \omega\left(f\ ,\ \frac{\sqrt{1-x^2}}{n}\right)$$

$$\le 8\omega\left(f\ ,\ \frac{\sqrt{1-x^2}}{n}\right) [1 + 3 + c + 3] .$$

This proves our theorem.

REFERENCES

[1] J. Balazs and P. Turan, Notes on interpolation III, *Acta Math. Sci. Hung.* 9(1958), 195-214.

[2] R. Bojanic, A note on the precision of interpolation by Hermite-Fejer polynomials, Proceedings of the Conference on Constructive Theory of Functions, Akademiaikiado Budapest (1972), pp. 69-76.

[3] R. A. DeVore, Degree of approximation, Approximation theory II, Edited by G. G. Lorentz, C. K. Chui, L. L. Schumaker, Academic Press (1976), pp. 117-161.

[4] P. Erdös and P. Turan, On interpolation III, Annals of Math., Vol. 41, No. 3, (1940), pp. 510-553.

[5] L. Fejer, Uber interpolation, Nachrichten d.k. Gesellschaft zu Gottingen (1916), pp. 66-91.

[6] A. Meir, An interpolatory rational approximation, *Canad. Math. Bull, Vol. 21* (2), (1978).

[7] G. Szego, Orthogonal polynomials, Amer. Math. Soc. Coll. Pub., Vol. 23, (1967).

[8] S. A. Teljakovski, Two theorems on the approximation of functions by algebraic polynomials, A.M.Š. Translation Series 2, Vol. 77, pp. 163-178.

[9] A. K. Varma and J. Prasad, Contributions to the problem of L. Fejer on Hermite-Fejer interpolation, to appear in *J. of Approx. Theory*.

[10] P. Vertesi, Simultaneous approximation by interpolating polynomials, *Acta Math. Acad. Sci. Hung., Vol. 31* (3-4), (1978), pp. 287-294.

# DESIGN PROBLEMS FOR OPTIMAL SURFACE INTERPOLATION

Charles A. Micchelli

IBM
T.J. Watson Research Center
Yorktown Heights, New York

Grace Wahba[1]
Department of Statistics
University of Wisconsin
Madison, Wisconsin

We consider the problem of interpolating a surface given its values at a finite number of points. We place a special emphasis on the question of choosing the location of the points where the function will be sampled.

Using minimal norm interpolation in reproducing kernel Hilbert spaces, equivalently Bayesian interpolation, and N-widths, we provide lower bounds for interpolation error relative to certain error criteria. These lower bounds can be used when evaluating an existing design, or when attempting to obtain a good design by iterative procedures to decide whether further minimization is worthwhile. The bounds are given in terms of the eigenvalues of a relevant reproducing kernel and the asymptotic behavior of these eigenvalues for certain tensor product spaces in the unit d-dimensional cube is obtained.

We demonstrate that for $H_m$, the d-dimensional tensor product of Sobolev spaces $W_2^{(m)}[0,1]$ and $P_N g$, the minimal norm interpolant to g at N given data points, the uniform convergence of $\|g - P_N g\|_{H_m}$ over g in the unit ball in

[1]Research of this author was supported by the Office of Naval Research Grant No. 00014-77-C-0675.




ISBN 0-12-780650-4

$H_{2m}$ cannot proceed at a rate faster than $((\log N)^{d-1}/N)^{2m}$. Certain conjectures concerning designs converging at this rate are made.

## 1. INTRODUCTION

We are interested in the problem of recovering a surface $g(t), t \in T$, from observations of g at a discrete set $T_N = \{t_i\}_{i=1}^N$ of points in T (called the "design"). In particular, we are interested in choosing $T_N$ so that an estimate, say, $g_N$ of g from the data $\{g(t_i)\}_{i=1}^N$ is closest to g in some appropriate sense among all designs $T_N$.

This problem arises in numerous applications. To cite one group of examples, T may be a sphere (the surface of the earth) or a rectangle and g(t) the 500 millibar height or the temperature, or the concentration of some air pollutant at position t. The interpolation problem requires an estimate of g over the entire surface given its values on $T_N$ while the design problem concerns optimal or nearly optimal choices of $T_N$.

In this introduction we shall briefly survey several different ways of viewing the interpolation problem, i.e., reconstructing the function from its sample values, and then follow this discussion with a description of some known results for the design problem in one dimension.

In this discussion of interpolation we will distinguish between the Bayesian approach and the function-analytic or deterministic approach. We further distinguish the problem of estimating g(t), for all $t \in T$ from estimating g(t) at a single point in T as well as introduce the possibility that our observations are distorted with errors. However, this latter feature is not primary for our objectives here.

The Bayesian approach is as follows: We suppose $g(t), t \in T$, is a Gaussian stochastic process, or "random field" with zero mean and given strictly positive definite (prior) covariance $K(s,t) = Eg(s)g(t), s,t \in T$. Given the data $g(t_1),\ldots,g(t_N)$ the Bayesian estimator for g(t) is

$$g_N(t) = E\{g(t)\,|\,g(t_1),\ldots,g(t_N)\}$$

$$= (K_{t_1}(t),\ldots,K_{t_N}(t))K_N^{-1}\begin{pmatrix} g(t_1) \\ \cdot \\ \cdot \\ \cdot \\ g(t_N)\end{pmatrix}$$

where $K_{t_i}(t) = K(t_i,t)$ and $K_N$ is the $N \times N$ matrix with i, jth entry $K(t_i,t_j)$, and thus

$$\min_a E\Big(g(t) - \sum_{i=1}^{N} a_i g(t_i)\Big)^2$$

$$= E(g(t) - g_N(t))^2.$$

The functional analytic approach is closely related to the Bayesian approach. Instead of assuming that g is a stochastic process, suppose g is a fixed element of $H_K$, the reproducing kernel Hilbert space with space reproducing kernel K.

Then $g_N$ may be shown to be the minimal norm interpolant to g on $T_N$ in $H_K$, the Hilbert space with reproducing kernel K. Observing that $\langle g,K_{t_i}\rangle_K = g(t_i)$ where $\langle\cdot,\cdot\rangle_K$ is the inner product in $H_K$ it can be verified that if $P_N g$ is the minimal norm interpolator of g on $T_N$, i.e.,

$$\|P_N g\|_K = \min_{h(t_i)=g(t_i)} \|h\|_K, \quad (P_N g)(t_i) = g(t_i)$$

then $P_N g$ is the orthogonal projection of g onto span $\{K_{t_i}\}_{i=1}^N$ and that $P_N g = g_N$, see Kimeldorf and Wahba [6]. In particular,

$$\min_a E\Big(g(t) - \sum_{j=1}^{N} a_j g(t_j)\Big)^2 = \min_a \|K_t - \sum_{j=1}^{N} a_j K_{t_j}\|_K \tag{1.1}$$

$$= \min_a \max_{\langle f,f\rangle_K \le 1} \Big|f(t) - \sum_{j=1}^{N} a_j f(t_j)\Big|.$$

Minimal norm interpolation also has the striking property that it furnishes the best estimator for $g(t), g \in H_K$ among all estimators (linear or nonlinear in $g(t_1),\ldots,g(t_N)$) which uses the information $g(t_1),\ldots,g(t_N)$ with $\langle g,g\rangle_K \leqslant 1$, that is,

$$\min_{A} \max_{\langle f,f\rangle_K \leqslant 1} |f(t) - A(f(t_1),\ldots,f(t_N))|$$

where A is any map from $(f(t_1),\ldots,f(t_N))$ into the real line, is achieved for $A(g(t_1),\ldots,g(t_N)) = g_N$. This property and various extensions and related matters in other normed spaces is described in C.A. Micchelli and T.J. Rivlin [13].

In each instance above the data is viewed as known exactly. Frequently in applications only "noisy" data is available and this leads to the problem of data smoothing. We briefly discuss this problem in Section 5.

As a criterion for choosing $t_1,\ldots,t_N$ we minimize

$$E\int_T (g(t) - g_N(t))^2 dt \quad = J(T_N) \tag{1.2}$$

where the expectation is taken with respect to the prior covariance $K(s,t)$. It is not hard to show that

$$J = J(T_N) = \tag{1.3}$$
$$= \int_T \{K(t,t) - (K_{t_1}(t),\ldots,K_{t_N}(t))K_N^{-1}(K_{t_1}(t),\ldots,K_{t_N}(t))'\}dt.$$

In practice K may have to be estimated by use of a finer trial grid of points than will ultimately be used, or from physical principles governing the phenomena under study. The covariance of air pollution measurements for example surely depends on the local geography. If K is known, then frequently the minimization of J will have to be carried out numerically.

In this paper we will provide a lower bound for J in terms of the eigenvalues associated with the integral operator induced by K. Thus, trial solutions for the design $T_N$ minimizing J may be compared against the lower bound to decide whether the further minimization is worthwhile.

Theorem 1. Let the operator K defined by $(Kf)(t) = \int_T K(t,s)f(s)ds$ be a symmetric compact operator of $L_2(T)$ into itself and have eigenvalues $\lambda_1 \geq \lambda_2 \geq \ldots$ Then

$$\inf_{T_N} J(T_N) \geq \sum_{i=N+1}^{\infty} \lambda_i = \int_T K(t,t)dt - \sum_{i=1}^{N} \lambda_i .$$

It is not known whether or not his lower bound can be achieved.

A fair amount is known about optimum designs for $T = [0,1]$, see Sacks and Ylvisaker [15], Wahba [20], Hajek and Kimeldorf [3]. A sequential procedure for choosing an optimum design for $T = [0,1]$ is given in Athavale and Wahba [1]. The sequential procedure depends heavily on properties of optimal designs known from the earlier papers and does not at present generalize to $T = [0,1] \times [0,1]$ or the sphere. In fact it appears that nothing is known about best possible convergence rates in several dimensions for $\|g - P_N g\|_K^2$, see Ylvisaker [22].

Sacks and Ylvisaker [16] have shown that $\|g - P_N g\|_K^{-2}$ is the variance of the Gauss-Markov estimate of $\theta$ in the model

$$Y(t) = \theta g(t) + X(t),\ t \in T,\ EX(s)X(t) = K(s,t)$$

given $Y(t)$, $t \in T_N$.

If it is known that g is in some class C then it might be desirable to choose $T_N$ to minimize $\sup_{g \in C} \|g - P_N g\|_K^2$. Through the notion of N-widths, introduced by Kolmogorov [7], and asymptotic estimates of the eigenvalues of certain integral operators we will provide lower bounds for the supremum of the design error for g in a certain class C. The class we will consider here is the natural generalization of the function class for which optimal one-dimensional designs were obtained in [3, 15, 20]. Before stating this result we review briefly some results for optimal experimental design from [20] for $T = [0,1]$.

The basic assumption made in [20] about K is that it has the characteristic discontinuity of a Green's function for a 2m-th order self-adjoint differential operator,

$$\alpha(t) = \lim_{s \downarrow t} \frac{\partial^{2m-1}}{\partial s^{2m-1}} K(s,t) - \lim_{s \uparrow t} \frac{\partial^{2m-1}}{\partial s^{2m-1}} K(s,t). \tag{1.4}$$

Suppose that g has a representation

$$g(t) = \int_0^1 K(t,s)\rho(s)ds, \qquad \rho \in L_2[0,1],$$

for some $\rho \in L_2[0,1]$ and let $T_N = \{t_{iN}\}_{i=1}^N$ be determined by a strictly positive density f,

$$\int_0^{t_{iN}} f(s)ds = i/(N+1) \qquad i = 1,2,\ldots N; \quad N = 1,2,\ldots$$

Under various regularity conditions including $\alpha, \rho > 0$

$$\|g - P_{T_N} g\|_K^2 =$$

$$= \frac{C_m}{N^{2m}} \left\{ \int_0^1 \frac{\rho^2(s)\alpha(s)}{f^{(2m)}(s)} ds \right\} \left(1 + o(1)\right)$$

where $C_m$ is a constant depending on m. The density f is chosen to minimize the quantity in brackets, see [20]. Thus the rate of decay of $\|g - P_{T_N} g\|_K^2$ is asymptotically the same as the decay of the eigenvalues of K (see Naimark [14]).

We return now to a general set T and reproducing kernel K.

Theorem 2. Let K be a symmetric compact operator from $L_2(T)$ into itself with eigenvalues $\lambda_1 \geq \lambda_2 \geq \ldots$ Let
$C = \{g : g(t) = \int_T K(t,s)\rho(s)ds, \int_T \rho^2(s)ds \leq 1\}$.
Then

$$\sup_{g \in C} \|g - P_N g\|_K^2 \geq \lambda_{N+1} .$$

Next (Section 3) we investigate the eigensequences for certain useful reproducing kernels on $T = \otimes^d [0,1]$, the tensor product of [0,1] d times. We will prove

Theorem 3. Let $H_K = \otimes^d H_Q$, where $H_Q$ is an r.k.h.s. on [0,1] with Q satisfying (1.4). Then

$$\lambda_N = 0\left((\log N)^{d-1}/N\right)^{2m}. \tag{1.5}$$

Based on this result, we make some conjectures (Section 4) concerning good designs in $H_K$ using results from the multi-dimensional quadrature literature. In particular, we conjecture (1.5) is the optimal rate, which has only been proved for $d = 1$, as explained above. Finally in Section 5 we make some observations concerning noisy data.

## 2. LOWER BOUNDS FOR OPTIMAL DESIGNS

We begin with the proof of

Theorem 1. Let the (symmetric) operator K defined by $(Kf)(t) = \int_T K(t,s)f(s)ds$ be compact with eigenvalues $\lambda_1 \geq \lambda_2 \geq \ldots$ Then

$$\inf_{T_N} J(T_N) = \inf_{T_N} \int_T \|K_t - P_N K_t\|_K^2 \, dt \geq \sum_{\nu=N+1}^{\infty} \lambda_\nu .$$

Proof. The equality is immediate from (1.2). Since

$$\int_T \|K_t - P_N K_t\|_K^2 \, dt$$

$$= \int_T \|K_t\|_K^2 \, dt - \int_T \|P_N K_t\|_K^2 \, dt$$

$$= \sum_{\nu=1}^{\infty} \lambda_\nu - \int_T \|P_N K_t\|_K^2 \, dt,$$

it suffices to show that $\int_T \|P_N K_t\|_K^2 \, dt \leq \sum_{\nu=1}^{N} \lambda_\nu$.

Let $\phi_1, \ldots, \phi_N$ by any N orthonormal functions in $H_K$. Then the projection of $K_t$ onto span $\{\phi_i\}_{i=1}^N$ is

$$P_N K_t = \sum_{i=1}^{N} \phi_i(t) \phi_i ,$$

and

$$\int_T \|P_N K_t\|_K^2 = \sum_{i=1}^{N} \|\phi_i\|_{L_2}^2 .$$

Let $K^{\frac{1}{2}}$ be the symmetric square root of the operator K. Then by the properties of the reproducing kernel norm,

$$\|\phi_i\|_{L_2}^2 = \|K^{\frac{1}{2}} \phi_i\|_K^2 .$$

Now by the extremal properties of the eigenvalues of K,

$$\sup_{\phi \in H_K} \|K^{\frac{1}{2}} \phi\|_K^2 / \|\phi\|_K^2 = \lambda_1 ,$$

$$\sup_{\substack{\phi \in H_K \\ (\phi, \psi_1)_{L_2} = 0}} \|K^{\frac{1}{2}} \phi\|_K^2 / \|\phi\|_K^2 = \lambda_2 ,$$

where $\psi_1$ is the maximizing element for the first equality above, etc. Thus,

$$\sum_{i=1}^{N} \|\phi_i\|_{L_2}^2 \leqslant \sum_{\nu=1}^{N} \lambda_\nu .$$

This result is also a consequence of a classical result from the theory of integral equations (see [18], p. 149).

<u>Theorem 2</u>. Let $H_N$ be any N dimensional subspace in $H_K$ and $P_N$ be the orthogonal projection onto $H_N$. Then there exists a function g,

$$g(t) = \int_T K(t,s)\rho(s)ds,$$

such that

$$\|g - P_{H_N} g\|_K^2 \geqslant \lambda_{N+1} \int_T \rho^2(s)ds. \tag{2.1}$$

<u>Proof</u>. The proof of this theorem also follows directly from the extremal properties for the eigenvalues of $K^{\frac{1}{2}}$ and has an interpretation in the theory of N-widths, [17]. Specifically we have

$$\inf_{H_N} \sup_{g \in C} \|g - P_{H_N} g\|_K^2$$

$$= \inf_{H_N} \sup_{\|\rho\|_{L_2}=1} \|K\rho - P_{H_N} K\rho\|_K^2 =$$

$$= \inf_{H_N} \sup_{\|\rho\|_{L_2}=1} \|K^{\frac{1}{2}}\rho - P_{H_N} K^{\frac{1}{2}}\rho\|_{L_2}^2 .$$

The extremal properties of eigenvalues and eigenfunctions of symmetric operators imply that

$$\sup_{\|\rho\|_{L_2}=1} \|K^{\frac{1}{2}}\rho - P_{H_N} K^{\frac{1}{2}}\rho\|_{L_2}^2 ,$$

achieves its minimum for $H_N$ equal to the span of the first N eigenfunctions of $K^{\frac{1}{2}}$ and the value of the minimum is $\lambda_{N+1}$.

To prove the existence of an optimal design we must find a subspace of the form span $\{K_t : t \in T_N\}$ for some design $T_N$ which achieves the lower bound in (2.1) which would be expected to be close to the span of the first n eigenfunctions.

It should not be expected that for an arbitrary covariance kernel K an optimal design exists for each N. However, for certain classes of kernels existence of optimal designs has been shown, see Melkman [10]; Melkman and Micchelli [11].

## 3. GOOD DESIGNS IN TENSOR PRODUCT SPACES

To make use of Theorem 2 we will obtain the asymptotic rate of decay of the eigenvalues of the r.k. for $H_K$ of the form

$$H_K = \otimes^d H_Q \text{ (tensor product of d copies of } H_Q)$$

where $H_Q$ is an r.k.h.s of functions on [0,1] with eigenvalues that decay as a power, $\lambda_\nu = c\nu^{-2m}(1 + o(1))$, $\nu \to \infty$. For instance, if Q behaves as a Green's function for a 2m-th order linear differential operator this condition is satisfied. As a simple example of this possibility, let $H_Q = \{f : f, f', \ldots, f^{(m-1)}$ abs. cont. $f^{(m)} \in L_2[0,1]$, $f^{\nu}(0) = f^{\nu}(1)$, $\nu = 0,1,\ldots,m-1\}$ with inner product,

$$\langle f,g \rangle = \left(\int_0^1 f(u)du\right)\left(\int_0^1 g(u)du\right) + \int_0^1 f^{(m)}(u)g^{(m)}(u)du.$$

Then the r.k. Q is

$$Q(s,t) = 1 + \sum_{\nu \neq 0} \frac{e^{2\pi i\nu(s-t)}}{(2\pi\nu)^{2m}}$$

and the corresponding eigenvalues and eigenfunctions are

$$\left\{1, (2\pi\nu)^{-2m} : \nu = \pm 1, \ldots\right\}, \quad \left\{e^{2\pi i\nu s}; \quad \nu = 0, \pm 1,\right\}.$$

<u>Theorem 3.</u> Let $H_K = \otimes^d H_Q$ where the eigenvalues $\{\lambda_\nu\}$ of $H_Q$ satisfy $\lambda_\nu = \nu^{-2m}\big(1 + o(1)\big)$ then the eigenvalues $\{\xi_\nu\}$ of $H_K$ satisfy

$$\xi_N = \left(\frac{(\log N)^{d-1}}{N}\right)^{2m}\big(1 + o(1)\big) .$$

<u>Proof.</u> Since $K = \otimes^d Q$, the eigenvalues of K are the tensor products of the eigenvalues of Q, i.e., if $\xi_1 \geq \xi_2 \geq \ldots$ are the eigenvalues of K then

$$\{\xi_N\} = \left\{\lambda_{j_1}\lambda_{j_2} \cdots \lambda_{j_d} : (j_1,\ldots,j_d),\ j_1,\ldots,j_d = 1,2,\ldots\right\}$$

To estimate the decay of $\xi_N$ we observe that the number of lattice points $(j_1,\ldots,j_d)$ satisfying $\prod_{\ell=1}^{d} j_\ell \leq k$ is $k(\log k)^{d-1}\big(1 + o(1)\big)$.

Hence, since $\lambda_\nu = \nu^{-2m}\big(1 + o(1)\big)$, we have

$$\xi_{[k(\log k)^{d-1}]} = k^{-2m}\big(1 + o(1)\big)$$

($[X]$ = greatest integer $\leq X$). Choosing $k = [N(\log N)^{d-1}]$ gives the desired conclusion,

$$\xi_N = \left(\frac{(\log N)^{d-1}}{N}\right)^{2m}\big(1 + o(1)\big).$$

It is not known for $d > 1$ whether there exists a design for which

$$\|g - P_N g\|_K^2 \leq \text{const} \left(\frac{(\log N)^{d-1}}{N}\right)^{2m} \int_{T_N} \rho^2(u)\,du.$$

However designs with a convergence rate approaching the optimum rate have been given in Wahba [21] for d = 2.

Define

$$Z_n^j = \left\{\frac{k}{n^j} : k = 1,\ldots,n^j\right\}$$

and

$$T_{N,\ell} = \bigcup_{j=1}^{\ell+1} Z_n^j \otimes Z_n^{\ell+1-j} .$$

In [21] it is shown that $T_{N,\ell}$ has $N = (\ell + 1)n^{\ell+2} - \ell n^{\ell+1}$ distinct points and for $H_K = H_Q \otimes H_Q$,

$$\|g - P_{T_{N,\ell}} g\|^2 \leqslant \text{const}\ \frac{(\ell+2)^2}{n^{(\ell+1)2m}} \left(\int_T \rho^2(u)du\right)\left(1 + o(1)\right)$$

$$\leqslant \text{const}\ \frac{(\ell+2)^{2+\left(\frac{\ell+1}{\ell+2}\right)2m}}{N^{\left(\frac{\ell+1}{\ell+2}\right)2m}} \left(\int_T \rho^2(u)du\right)\left(1 + O(1)\right) \tag{3.1}$$

where $g = K\rho$ and $o(1) \to 0$ as $n \to \infty$.

Choosing $\ell = (\log N)^p\left(1 + o(1)\right)$ for any $p, 0 < p < 1$ we have

$$N = (\ell+1)n^{\ell+2}\left(1 + o(1)\right)$$

or $\log N = p \log\log N + (\log N)^p \log n\left(1 + o(1)\right)$. Hence

$$\log n = \left(1 + o(1)\right) \frac{\log N - p \log\log N}{(\log N)^p}$$

and $n \to \infty$ provided $0 < p < 1$ (for $p=1$ this conclusion fails). Setting $p = \frac{m}{m+1}$ into (3.1) gives

$$\|g - P_N g\|_K^2 = 0\left[\frac{(\log N)^{\left(1+\frac{\ell+1}{\ell+2}\right)/(m+1)}}{N^{\frac{\ell+1}{\ell+2}}}\right]^{2m},$$

a convergence rate which approaches the optimum rate of $(\log N/N)^{2m}$ implied by Theorems 2 and 3.

## 4. OPTIMAL QUADRATURE - A CONJECTURE

A quadrature formula for $\int_T g(t)dt$ can be obtained by setting $\int_T (P_N g)(t)dt = \sum_{i=1}^{N} c_i g(t_i)$. Then

$$|\int_T g(t)dt - \int_T (P_N g)(t)dt|$$

$$= |\langle \eta, g - P_N g \rangle_K|$$

$$= |\langle \eta - P_N \eta, g - P_N g \rangle_K|$$

$$\leq \|\eta - P_N \eta\|_K \|g - P_N g\|_K$$

$$\leq \|\eta - P_N \eta\|_K \|g\|_K$$

where $\eta$ is the representer of integration in $H_K$,

$$\eta(s) = \langle \eta, K_s \rangle_K = \int_T K(s,u)du .$$

An optimal quadrature problem may be formulated as: Find $t_1,\ldots,t_N$ to minimize $\|\eta - P_N \eta\|_K$.

There is a large literature on choosing sequences in the d-dimensional unit cube which makes the error for the special quadrature formula $\frac{1}{N}\sum_{i=1}^{N} g(t_i)$ asymptotically small. This work has focused on finding sequences $T_N = \{t_1,\ldots,t_N\}$ for which the discrepancy $D_N$ defined by

$$D_N = \sup_t |F_N(t) - F(t)|$$

is small. Here $F_N$ is the cumulative distribution function of the point set and F is the cumulative distribution of the uniform density, see Kuipers and Neiderreiter [8], Halton [4], Halton and Zaremba [5], Zaremba [23].

It is known that the Hammersley sequences defined below, have discrepancy

$$D_N = \left(\frac{\log^{d-1} N}{N}\right)\left(1 + o(1)\right)$$

see Halton [4].

These sequences are defined (in d-dimensions) by

$$\left\{\frac{n}{N}, \phi_2(n), \phi_3(n), \ldots, \phi_{p_d}(n)\right\}_{n=0}^{N-1}$$

where the subscripts in the $\phi$'s are successive primes and if $n = \sum_{j=0}^{M} n_j p^j$ where $M = [\log_p n]$, then $\phi_p(n) = \sum_{j=1}^{M+1} n_j p^{-j}$.

Bounds on $\varepsilon_N = |\int_T g(t)dt - \frac{1}{N}\sum_{i=1}^{N} g(t_i)|$ in terms of the discrepancy appear in the literature, see Kuipers and Neiderreiter [8, p.157], Zaremba [23] and references therein.

In [8] it is shown for certain sequences that $\varepsilon_N = O(D_N^q)$ where g has Fourier coefficients $c_{\nu_1,\nu_2,\ldots,\nu_d}$ satisfying

$$|c_{\nu_1,\nu_2,\ldots,\nu_d}| \leq \frac{M}{\left(\prod_{i=1}^{d} \bar{\nu}_i\right)^q}$$

where

$$\bar{\nu}_i = \begin{cases} |\nu_i| & \nu_i \neq 0 \\ 1 & \nu_i = 0 \end{cases}.$$

We conjecture that similar results obtain for Hammersley sequences and that for spaces $H_K$ satisfying the hypothesis of Theorem 2 the optimal convergence rate $\|g - P_N g\|_K^2 = \lambda_{N+1}(1 + o(1))$

$$= \left(\frac{(\log N)^{d-1}}{N}\right)^{2m}(1 + o(1))$$

will hold for $T_N$ the Hammersley sequence and g of the form $g = K\rho$.

## 5. NOISY DATA

In this section we include some remarks concerning estimation based on inaccurate data. If instead of observing $g(t)$, $t \in T_N$, we assume that the data is given by

$$y_i = g(t_i) + \varepsilon_i\ , \qquad E\varepsilon_i\varepsilon_j = \delta_{ij}\ ,$$

then the minimum norm estimator

$$\min_a E\left(g(t) - \sum_{i=1}^N a_i y_i\right)^2 = \min_a \|K_t - \sum_{j=1}^N a_j K_{t_j}\|_K^2 + \sigma^2 \sum_{j=1}^N a_j^2$$

leads in the functional-analytical approach to

$$= \min_a \max_{\langle f,f\rangle_K \leq 1} |f(t) - \sum_{i=1}^N a_i f(t_i)|^2 + \sigma^2 \sum_{i=1}^N a_j^2 \qquad (5.1)$$

Recently, this variational problem has been solved by Laurent [9] (see also Speckman [19]). It has been shown that the minimum norm estimator is the smoothing "spline" in $H_K$ with parameter $\sigma^{-2}$, that is, if $g_N$ minimizes

$$\min\left\{\|f\|_K^2 + \sigma^{-2} \sum_{i=1}^N \left(f(t_i) - g(t_i)\right)^2\right\} \qquad (5.2)$$

then $g_N(t) = \sum_{i=1}^N c_i(t)g(t_i)$ is the minimum norm estimator for g when we have noisy data. Note that the smoothing parameter $\sigma^2$ does not depend on the value $t \in T$ at which we choose to estimate $g(t)$. The following short proof of this result is instructive: We wish to determine

$$\min_a \|K_t - \sum_{i=1}^N a_i K_{t_i}\|_K^2 + \sigma^2 \sum_{j=1}^N a_j^2$$

To this end, we introduce the tensor product space $H_K \times R^N = \{(f,a) | f \in H_K, \ a \in R^N\}$ with the norm

$$\|(g,a)\|_\sigma^2 = \|g\|_K^2 + \sigma^2 \sum_{i=1}^N a_i^2$$

Then the above problem in $H_K \times R^N$ becomes

$$\min_a \|h - \sum_{j=1}^N a_j h_j\|_\sigma^2$$

$$h = (K_t, 0), \quad h_j = (K_{t_j}, -e_j), \quad (e_j)_k = \delta_{jk}.$$

But from the theory for estimating exactly given data, as in (1.1), the minimum $a = (a_1, \ldots, a_N)$ may be obtained from the best interpolant,

$$\min_{\langle (f,a),(K_{t_i},-e_i)\rangle = g(t_i)} \|(f,a)\|_\sigma^2 =$$

$$\min_f \left\{\|f\|_K^2 + \sigma^{-2} \sum_{i=1}^N \left(g(t_i) - f(t_i)\right)^2\right\}$$

in agreement with (5.2).

It has not yet been determined if the optimality of smoothing "splines" persists when an estimator for the full function $g(t)$, $t \in T$, when the error criterion (1.2) is used. However, let us replace (5.1) by

$$\min_{a} \max_{\substack{\langle f,f\rangle \le 1 \\ \sum_{i=1}^{N} r_i^2 \le 1}} \left| f(t) - \sum_{i=1}^{N} a_i \left(f(t_i) + \varepsilon r_i\right) \right|^2$$

that is, we minimize the worst least square error when we know the noise in the data is in the region $y_i = f(t_i) + \varepsilon r_i$, $\sum_{i=1}^{N} r_i^2 \le 1$. It has been shown that in this setting the smoothing spline is also optimal. However, unlike (5.1), the smoothing parameter depends on $\varepsilon$ as well as $t \in T$. Moreover, this theory holds in great generality, including, in particular, estimating the full function $g(t)$, $t \in T$ (see Melkman and Micchelli [13] for the details). For methods of choosing the smoothing parameter using a cross-validation procedure based on the data see Craven and Wahba [2].

The design problem of Theorem 1 has an analogue for noisy data which may be described as follows. Let $g(t)$, $t \in T$, be a stochastic process as before and let

$$g_N(t) = E\{g(t)|y_1,\ldots,y_N\} = \left(K_{t_1}(t),\ldots,K_{t_N}(t)\right)\left(K_N + \sigma^2 I\right)^{-1} (y_1,\ldots,y_N)'.$$

Then $J_N$ becomes

$$J_N = E\int_T \left(g(t) - g_N(t)\right)^2 dt$$

$$= \int_T \{K(t,t) - \left(K_{t_1}(t),\ldots,K_{t_N}(t)\right)\left(K_N + \sigma^2 I\right)^{-1}\left(K_{t_1}(t),\ldots,K_{t_N}(t)\right)'\}dt.$$

which may be compared to equation (1.3).

## REFERENCES


1. Athavale, M., and Wahba, G., Determination of an optimal mesh for a collocation-projection method for solving two-point boundary value problems. Department of Statistics, University of Wisconsin, Madison T.R. No. 530, to appear *J. Approx. Theory*.
2. Craven, P., and Wahba, G., Smoothing noisy data with spline functions: Estimating the correct degree of smoothing by the method of generalized cross-validation, *Numer. Math. 31* (1979), 377-403.
3. Hajek, J. and Kimeldorf, G., Regression designs in autoregressive stochastic processes. *Ann. Statist. 2* (1974), 520-527.
4. Halton, J.H., On the efficiency of certain quasi-random sequences of points in evaluating multi-dimensional integrals, *Numer. Math. 2,* (1960), 84-90.
5. Halton, J.H. and Zaremba, S.K., The extreme and $L^2$ discrepancies of some plane sets. *Monatshefte für Mathematik 73* (1969), 316-328.
6. Kimeldorf, G.S. and Wahba, G., A correspondence between Bayesian estimation on stochastic processes and smoothing by splines. *Ann. Math. Statist., 41* (1970), 2.
7. Kolmogorov, A., über die best annäherung von funktioner einer gegebenen funktionenklasse, *Ann. of Math., 37* (1936), 107-110.
8. Kuipers, L. and Neiderreitter, H., "Uniform distribution of Sequences," John Wiley & Sons, New York, 1974.
9. Laurent, P.J., Colloquium talk, Univ. of Wisconsin, Department of Statistics, April 1978.
10. Melkman, A.A., n-widths and optimal interpolation of time- and band-limited functions, in "Optimal Estimation in Approximation Theory," C.A. Micchelli and T.J. Rivlin, eds., Plenum Press, New York, 1977, 55-68.

11. Melkman, A.A. and C.A. Micchelli, Spline spaces are optimal for $L^2$ n-widths, *Illinois Journal of Mathematics, 22* (1978), 541-564.

12. Melkman, A.A. and C.A. Micchelli, Optimal estimation of linear operators in Hilbert spaces from inaccurate data, to appear *SIAM Journal of Numer. Anal.*

13. Micchelli, C.A. and T.J. Rivlin, A survey of optimal recovery, in "Optimal Estimation in Approximation Theory," C.A. Micchelli and T.J. Rivlin, Eds., Plenum Press, New York, 1977, 1-54.

14. Naimark, M.A., Linear Differential Operators, Frederick Ungar Publishing Co., New York, 1967.

15. Sacks, J. and Ylvisaker, D., Designs for regression problems with correlated errors, *Ann. Math. Statist. 49* (1969), 2057-2074.

16. Sacks, J. and Ylvisaker, D., Statistical designs and integral approximation, Proc. of the 12th Biennial Seminar of the Canadian Mathematical Congress, 1970.

17. Shapiro, Harold,"Topics in Approximation", Lecture Notes in Mathematics, 187, Springer, New York, 1971, p. 187.

18. Smithies, F., "Integral Equations," Cambridge Tracts in Mathematics and Mathematical Physics, No.49, Cambridge University Press, 1965.

19. Speckman, P., On minimax estimation of linear operators in Hilbert spaces from noisy data, preprint, 1980.

20. Wahba, G., On the regression design problem of Sacks and Ylvisaker, *Ann. Math. Statist. 42* (1971), 1035-1053.

21. Wahba, G., Interpolating surfaces: High order convergence rates and their associated designs, with application to X-ray image reconstruction. Department of Statistics, University of Wisconsin - Madison T.R. No. 523, to appear in *SIAM J. Numer.*

22. Ylvisaker, D., Designs on random fields, in "A Survey of Statistical Design and Linear Models," J. Srivastava, ed., North-Holland, 1975, 593-607.

23. Zaremba, S.K., The Mathematical basis of Monte Carlo and quasi-Monte Carlo methods. *SIAM Review 3* (1968), 303-314.

# OPEN PROBLEMS

The following problems were posed by several groups working in industrial and defense related research.

1. Is there any research done about vector valued splines (splines having their values in $R^n$). For example, it seems logical that three coordinates of a falling body as computed by some tracking system should be simultaneously smoothed.

2. Are there any efficient algorithms for a recursive spline smoothing with a smoothing power comparable to that of the overall smoothing?

3. What are the least bounds for the errors in the estimation of the first two derivatives by spline smoothing ( in terms of the noise, sampling frequency etc.)?

4. Are there any methods for smoothing of functions with a finite number of discontinuities in the derivatives?

5. Are there any methods for recovering the second derivative at the end points of the data (such as splines without boundary constraints on the second derivative)?

6. We would like to smooth noisy tracking radar position data (time, range (±5m), azimuth (±300 μrad), elevation (±300 μrad)) to obtain accurate position and velocity (and perhaps

acceleration) estimates. Tracked targets may be of a low acceleration nature, or a highly accelerating nature. Data is recorded at a rate of 10 points per second on magnetic tape and is to be smoothed on a large computer either interactively or automatically.

The following alternative approaches were considered:

A. Find cubic spline with knot at every data point which satisfies:

$$\sum_{i=1}^{N} (X_i(t) - S_i(t))^2 = N\delta^2$$

$$\int (S''(t))^2 dt = \text{minimum}$$

where: $X_i(t)$ measured data

$S_i(t)$ smoothed value

Vary estimate of $\delta$ to give "best" function S, displaying the results on graphical display. Acceleration values at end points may be varied as well as $\delta$.

B. Calculate an estimate for $\delta$ from a polynomial smoothing approximation and calculate the above spline.

C. Calculate a variable knot, least means square spline approximation displaying the result and varying the number of knots to achieve the "best" result.

D. Kalman filter approach.

The specific questions we would like to raise, are:

a. Which of the above 4 methods (A, B, C, D) should give the best position and velocity estimate?

b. What spline smoothing method can you recommend for totally automatic non-interactive smoothing of large amounts of data (30000 data points)?

Several of these problems were partially resolved by C. de Boor, P. J. Laurent and G. Wahba.

E. W. Cheney, University of Texas, Austin, (U.S.A.)

I. Simultaneous Chebyshev Approximation

A set V in a Banach space X is said to be proximinal if the infinimum of $\| x-v \|$ is attained as v ranges over V (for each $x \in X$). In other words, each x in X has at least one best approximation in V. Let S and T be compact Hausdorff spaces, and let V be a proximinal set in C(T). Is V necessarily proximinal when considered as a subset of $C(T \times S)$? Thus we ask whether for each $z \in C(T \times S)$ the infimum

$$\inf_{v \in V} \sup_{t \in T} \sup_{s \in S} |z(t,s) - v(t)|$$

is attained.

II. On Proximinality in Banach Spaces

Let $V_1$ and $V_2$ be proximinal subspaces in a Banach space. (For the definition of proximinality, refer to a previous problem.) Assume that $V_1 + V_2$ is closed. Does it follow that $V_1 + V_2$ is proximinal?

## III.

Let T be a compact metric space, X a Banach space, and Y a subspace of X. Let C(T,X) be the Banach space of all continuous functions f from T into X, with norm $\|f\| = \sup_t \|f(t)\|$.

If C(T,Y) is proximinal in C(T,X), does it follow that there exists a continuous proximity map $P : X \to Y$? (A proximity map has the property $\|x-Px\| = \text{dist}\ (x,y)$ for all $x \in X$).

## IV.

Adopt the notation of Problem III. If X is finite-dimensional, does it follow that C(T,Y) is proximinal in C(T,X) for all subspaces Y in X?

I.J. Schoenberg, University of Wisconsin, Madison, WI (U.S.A.)

(An open problem sent by mail.)

A conjecture on the Landau-Kolmogorov problem concerning uniform rotations on a circle.

It is a classical result that if we have a real-valued function f(t) which is n times differentiable, then, with $M_k = \sup|f^{(k)}(t)|$ on the real axis, we have the inequality

$$M_k \leqq C_{n,k} M_0^{1-(k/n)} M_n^{k/n} \quad \text{for} \quad 1 \leqq k \leqq n-1. \tag{1}$$

It is easy to see that (1) remains valid even if f(t) is a complex-valued function of the real variable t. Let me call this class (K).

Let us now consider the subclass of circular motions

$$(C) : |f(t)| = 1 \quad \text{for all real } t. \tag{2}$$

Conjecture 1. *For the subclass* (C) *the constant* $C_{n,k}$ *in* (1) *may be replaced by* 1, *so that*

$$M_k \leqq (M_n)^{k/n} \quad \text{if } f \in (C). \tag{3}$$

That (3) can not be further improved is shown by the uniform rotation

$$f(t) = e^{it} , \ (-\infty < t < \infty), \tag{4}$$

when (3) becomes an equality. In other words, the uniform rotation (4) plays the role of the Euler splines for the general case of complex-valued functions. My own contributions to this problem are almost trivial, as I wish to settle the following two cases:

Case 1: $k = 1,\ n = 2,$ (5)

and

Case 2: $k = 1,\ n = 3.$ (6)

Of course (2) is equivalent to the representation

$$f(t) = e^{ig(t)}, \text{ where } g(t) \text{ is real valued.} \tag{7}$$

But then

$$\dot{f}(t) = i\,\dot{g}\,e^{ig}$$
$$\ddot{f}(t) = (i\ddot{g} - \dot{g}^2)e^{ig}$$
$$\dddot{f}(t) = \{i(\dddot{g} - \dot{g}^3) - 3\dot{g}\,\ddot{g}\}\,e^{ig}.$$

This gives

$$\begin{aligned} M_0 &= 1,\ M_1 = \sup|\dot{g}| \\ M_2 &= \sup(\ddot{g}^2 + \dot{g}^4)^{1/2} \\ M_3 &= \sup\{(\dddot{g} - \dot{g}^3)^2 + 9\,\dot{g}^2\ddot{g}^2\}^{1/2}. \end{aligned} \tag{8}$$

Setting

$$\dot{g}(t) = h(t), \tag{9}$$

our conjecture (3) is changed into differential inequalities for the arbitrary smooth function $h(t)$. For the special cases (5) and (6) these are as follows: For

*Case 1*: *To show that*

$$\sup|h| \leqq \sup\{h^4 + \dot{h}^2\}^{1/4}, \tag{10}$$

*Case 2*: *To show that*

$$\sup h^6 \leqq \sup\{(\ddot{h} - h^3)^2 + 9\,h^2\,\dot{h}^2\}. \tag{11}$$

Since (10) is evidently true, this *case 1 is settled*.

*A proof of the inequality (11).*

This is the first case which is not quite trivial. We rewrite (11) as

$$\sup h^6 \leqq \sup\{h^6 + \ddot{h}^2 - 2h^3\,\ddot{h} + 9\,h^2\,\dot{h}^2\}. \tag{12}$$

Setting

$$E(t) = \ddot{h}^2 - 2h^3 \ddot{h} + 9 h^2\dot{h}^2, \tag{13}$$

we are to show that

$$\sup h^6 \leqq \sup\{h^6 + E(t)\}. \tag{14}$$

<u>Lemma 1</u>. <u>We assume that</u>

$$h(t) \geqq 0 \text{ in the interval } P = \{a \leqq t \leqq b\}. \tag{15}$$

<u>Then</u>

$$\sup_{t \in P} h^6 \leqq \sup_{t \in P} \{h^6 + E(t)\}. \tag{16}$$

<u>Proof of Lemma 1</u>.

We may assume that $h(t)$ is of the class $C^2$, hence that it is twice continuously differentiable, since we could reduce the case of $\ddot{h}(t)$ having isolated discontinuities to this case by "rounding the corners" of the graph of $h(t)$. Suppose that $P$ is partitioned into consecutive intervals $I_i$ and $J_j$ such that

$h(t)$ is a <u>concave</u> function (in the wide sense) in all $I_i$,

$h(t)$ is a <u>convex</u> function (in the wide sense) in all $J_j$.

This means that

$$\ddot{h}(t) \leqq 0 \text{ if } t \in I_i \quad \text{and} \quad \ddot{h}(t) \geqq 0 \text{ if } t \in J_j. \tag{17}$$

<u>If $t \in J_j$</u>, where $h$ is convex, the relations

$$D\, h^6 = 6h^5\dot{h}\ , \ D^2\, h^6 = 30\, h^4\, \dot{h}^2 + 6\, h^5\, \ddot{h}$$

show that <u>also</u> $h^6$ <u>is convex in all the intervals</u> $J_j$. In forming the supremum of $h^6$ in the interval $P$, we may therefore ignore all of the intervals $J_j$, and look only at the $I_i$ because

$$\sup_P h^6 = \sup_{\cup I_i} h^6. \tag{18}$$

If $t \in I_i$, then $\ddot{h}(t) \leqq 0$, and therefore

$$E(t) = \ddot{h}^2 - 2\, h^3\, \ddot{h} + 9\, h^2\, \dot{h}^2 \geqq 0,$$

and so

$$h^6 \leqq h^6 + E(t) \quad \text{if} \quad t \in \cup\, I_i. \tag{19}$$

This shows that in all the intervals $I_i$ the graph of the right side of (19) is on top of the graph of the left side. Clearly (18) and (19) establish (16).

Lemma 2. We assume that

$$h(t) \leqq 0 \quad \text{in the interval} \quad N = \{c \leqq t \leqq d\}. \tag{20}$$

Then

$$\sup_N h^6 \leqq \sup_N \{h^6 + E(t)\}. \tag{21}$$

Proof of Lemma 2. Let

$$s(t) = -h(t).$$

Then $s(t) \geqq 0$ in the interval N and

$$s^3 = -h^3, \quad \ddot{s} = -\ddot{h} \quad \text{and so} \quad s^3\, \ddot{s} = h^3\, \ddot{h}.$$

To $s(t)$ we can therefore apply Lemma 1 so that

$$\sup s^6 \leqq \sup \{s^6 + \ddot{s}^2 - 2\, s^3\, \ddot{s} + 9\, s^2\, \dot{s}^2\} \quad \text{in } N.$$

Therefore

$$\sup h^6 \leqq \sup\{h^6 + \ddot{h}^2 - 2\, h^3\, \ddot{h} + 9\, h^2\, \dot{h}^2\} \quad \text{in } N.$$

Proof of (14).

Decompose the real axis into consecutive intervals in which $h(t)$ is $\geqq 0$ or $\leqq 0$. In each of these intervals (14) holds by Lemmas 1 and 2. Therefore it holds on the real axis.

A general proof of Conjecture 1 would give a new extremum property of the humble uniform rotation known since the dawn of Astronomy.

# PARTICIPANTS*

M. Alon *Israel*
V. Amdursky *Israel*
D. Amir *Israel*
E. Artzy *Israel*
E. Barta *Israel*
N. Ben–Yoseph *Israel*
J. Bogin *Israel*
C. de Boor *United States*
D. Braess *West Germany*
P. Bruner *Israel*
L. Brutman *Israel*
Y. Censor *Israel*
E. W. Cheney *United States*
Chr. Coatmelec *France*
L. Diamant *Israel*
N. Dyn *Israel*
B. Epstein *Israel*
P. Erdös *Hungary*
P. D. Feigin *Israel*
S. D. Fisher *United States*
A. Fogel *Israel*
S. Gal *Israel*
E. Geresh *Israel*
D. Getz *Israel*
R. Godel *Israel*
I. Gohberg *Israel*
B. Granovsky *Israel*
M. Herbst *Israel*
A. Inbar *Israel*
A. Jakimovsky *Israel*
J. W. Jerome *United States*
R. Landsman *Israel*
E. Lapidot *Israel*
P. J. Laurent *France*
A. Le Mehaute *France*
D. Levin *Israel*
W. A. Light *United Kingdom*
H. Loeb *United States*
T. Lyche *Norway*
J. C. Mason *United Kingdom*
A. Melkman *Israel*
J. Meinguet *Belgium*
B. Neta *United States*
D. J. Newman *United States*
P. Ney *Israel*
E. Passow *United States*
R. Pieterkowski *Israel*
A. M. Pinkus *Israel*
E. Rakotch *Israel*
D. Ram *Israel*
L. Risman *Israel*
T. J. Rivlin *United States*
M. Rozenfeld *Israel*
E. Schatz *Israel*
W. Schempp *West Germany*
S. Sela *Israel*
R. Senderovizh *Israel*
I. Shapira *Israel*
Y. Shimoni *Israel*
O. Shisha *United States*
A. Sidi *Israel*
I. Stoltz *Israel*
A. Studnicky *Israel*
H. G. ter Morsche *Netherlands*

*There were numerous sessions and formal registration was conducted only at some of them.

G. N. Trytten *United States*
P. K. Varma *United States*
G. Wahba *United States*
M. Wald *Israel*
Z. Yerushalmy *Israel*
L. Zalcman *United States*
Z. Ziegler *Israel*
A. Ziv *Israel*
I. Zmora *Israel*
D. Zwick *United States*